D1618654

Ute Schepers
RNA Interference in Practice

Further Titles of Interest

E. Westhof, A. Bindreif, A. Schön, R. K. Hartmann (Eds.)

Handbook of RNA Biochemistry

2004
ISBN 3-527-30826-1

D. Kambhampati (Ed.)

Protein Microarray Technology

2003
ISBN 3-527-30597-1

R. Westermeier, T. Naven

Proteomics in Practice

2002
ISBN 3-527-30354-5

D. T. Gjerde, C. P. Hanna, D. Hornby

DNA Chromatography

2002
ISBN 3-527-30244-1

R. Westermeier

Electrophoresis in Practice

3rd edition, 2001
ISBN 3-527-30300-6

Ute Schepers

RNA Interference in Practice

WILEY-
VCH

WILEY-VCH Verlag GmbH & Co. KGaA

Author

Dr. Ute Schepers
University of Bonn
Kekulé Institute for Organic Chemistry
and Biochemistry
Gerhard-Domagk-Str. 1
53121 Bonn
Germany
schepers@uni-bonn.de

Cover Illustration:
Gene silencing by double stranded RNA.
This cover presents current group members
from left to right: Mustapha Diallo,
Sven Hoffman, Frank Hahn, Sabine Hanke,
Michaela Smuda, Katja Schmitz,
Michael Schleeger, Susi Anheuser,
Henning Breyhan, Albina Cryns,
Christoph Arenz, and myself.

■ This book was carefully produced. Nevertheless, authors and publisher do not warrant the information contained therein to be free of errors. Readers are advised to keep in mind that statements, data, illustrations, procedural details or other items may inadvertently be inaccurate.

Library of Congress Card No.: applied for

British Library Cataloguing-in-Publication Data: A catalogue record for this book is available from the British Library.

Bibliographic information published by Die Deutsche Bibliothek
Die Deutsche Bibliothek lists this publication in the Deutsche Nationalbibliografie; detailed bibliographic data is available in the Internet at <http://dnb.ddb.de>.

Printed in the Federal Republic of Germany
Printed on acid-free paper

Composition   ProSatz Unger, Weinheim
Printing   Strauss GmbH, Mörlenbach
Bookbinding   Litges & Dopf, Buchbinderei GmbH, Heppenheim

ISBN   3-527-31020-7

Preface

When I first gave lectures on RNA interference (RNAi), I was asked by many scientists, "What is the meaning of 'interference'", and "How does RNA fit into this phenomenon, which has long been recognized by physicists?" Even if the two phenomena appear to have nothing in common, the physical definition of interference can be easily converted to fit RNAi:

"Interference is the process in which waves (RNAs) of the same frequency (sequence) combine to reinforce or cancel (delete) each other".

With their discovery that double-stranded RNA (dsRNA) can efficiently delete homologous mRNA in *Caenorhabditis elegans*, Craig Mello and Andrew Fire commenced a new era on gene silencing and created a major "hype" in the scientific community. Since then, a flood of reports has been published describing vast amounts of data and many techniques for performing RNAi in many organisms and, as a therapeutical tool, making it almost impossible to follow up the basic and striking news. *RNA Interference in Practice* is targeted at all scientists – including students, novices, and regular users of RNAi – who wish either to apply or to expand the use of RNAi in their laboratories.

RNA Interference in Practice is the most recent issue of the successful *"... in Practice"* series published by Wiley-VCH. As with previous issues such as *Proteomics in Practice and Electrophoresis in Practice*, this book contains a comprehensive theoretical introduction to guide the user through the practical protocols employed. These protocols are supported by many notes that lead to improvements of the procedures. The success of the other books of the *"... in Practice"* series has shown that laboratory manuals comprising both theory and practice serve as useful guides for daily laboratory investigations. Although the present book will doubtless never be completed due to the rapid development of RNAi research, it will nonetheless provide scientists with a summary of current literature up to 2004, the basic techniques of RNAi, and the common drawbacks for the successful application of RNAi in worms, flies, and mammals.

What are the prerequisites for a successful RNAi experiment? Success depends not only on knowledge of the mechanisms themselves, but also on the technical requirements and limitations of the practical applications. One challenge for every scientist is to evaluate critically the methods described in the primary literature or in the manufacturer's manuals. Therefore, this book – in one issue – comprises critical

RNA Interference in Practice: Principles, Basics, and Methods for Gene Silencing in C. elegans, Drosophila, and Mammals. Ute Schepers
Copyright © 2005 WILEY-VCH Verlag GmbH & Co. KGaA, Weinheim
ISBN: 3-527-31020-7

steps, from the design of siRNAs or dsRNAs to the delivery into the respective organism, and the design of important controls. It further describes cloning strategies of hairpin constructs, analytical tools, and many new perspectives. Each chapter contains appendices with useful web sites, book and literature references, together with company addresses. A comprehensive glossary is also provided for the reader. Most of the protocols have been carried out in our laboratory, and are being constantly updated. *RNA Interference in Practice* differs from other books on the subject of RNAi in as much it is has been created especially for use at the bench, thus facilitating daily laboratory studies. My aim is to encourage readers of the book to stroll through all of its chapters, as there is much to learn from RNAi in different organisms, and this in turn might inspire the creation of novel experimental set-ups.

Finally, I would like to thank all of those people who provided me with constant encouragement during the writing of this book. I would like to thank Frank Weinreich of Wiley-VCH for his great patience and continuous motivation. Likewise, the production of this book would not have been possible without the constant support of my mentor Konrad Sandhoff and my co-workers, who developed and tested most of the protocols and designed the wonderful illustration for the book's cover, and especially Katja Schmitz, who was always there when I needed a helping hand or fruitful criticism. Last - but not least - I cannot close without thanking my husband S. Braese and my family for criticising and pampering me during the exciting experience of writing a book.

I would also like to thank: Susi Anheuser, Christoph Arenz, Henning Breyhan, Albina Cryns, Mustapha Diallo, Frank Hahn, Sabine Hanke, Sven Hoffman, Konrad Sandhoff, Katja Schmitz, Michael Schleeger, Michaela Smuda, and all the companies that supported me with copyrights.

August 2004   *Ute Schepers*

Contents

RNA Interference in Practice: Principles, Basics, and Methods for Gene Silencing in C. elegans, Drosophila, and Mammals. Ute Schepers
Copyright © 2005 WILEY-VCH Verlag GmbH & Co. KGaA, Weinheim
ISBN: 3-527-31020-7

1
Introduction: RNA interference, the "Breakthrough of the Year 2002"

When in 2001, with the sequencing of the human genome, the sequencing projects of many organisms reached a summit, there was no doubt among the scientific community that nothing in the near future would be as spectacular. It was called one of the biggest milestones of the 21st century, and the most important achievement in biology. However, this excitement did not last long. Already in the same year, the discovery of RNA interference (RNAi) in mammals created a similar hype, which is now experiencing explosive growth. When *Science* nominated RNAi as the "Breakthrough of the year 2002" (Couzin 2002), it was already clear that RNAi will revolutionize biomedical research during the next few years.

Although RNAi is being used mainly to unravel the functions of genes by switching them "off" at the post-transcriptional level, it offers a novel approach for disease therapy, by shutting off unwanted genetic activity in a targeted manner. It can be applied to targets ranging from rogue genes in cancer to genes of viruses, such as hepatitis B or C virus or HIV.

With the knowledge of the genome sequence of many species, RNAi can contribute to a more detailed understanding of complicated physiological processes, and also to the development of many more new drugs since it connects genomics, proteomics, and functional genomics (functionomics).

Today, RNAi is known as a common denominator for several post-transcriptional gene silencing (PTGS) processes observed in a variety of eukaryotic organisms (Hannon 2002). It is induced by double-stranded RNA (dsRNA).

It comprises phenomena such as co-suppression (Napoli et al. 1990), quelling (Cogoni and Macino 1997), and transgene-induced silencing (Baulcombe 1999), even if those processes are not completely identical. Some scientists prefer the name RNA silencing rather than RNAi, which is solely specific for invertebrates such as worms and flies or vertebrates. This book will further refer to the phenomenon as RNAi, as the majority of the published reports are using this as a general term to describe dsRNA-induced RNA silencing.

The dsRNA-induced gene-silencing effects were first discovered in plants (Napoli et al. 1990) and *Neurospora crassa* (Cogoni and Macino 1997), where they serve as an antiviral defense system. The viruses encoding for the silencing transgenes were known to produce dsRNA during their replication. However, the decisive discovery in RNA silencing was made when Andrew Fire and Craig Mello tried to explain the

RNA Interference in Practice: Principles, Basics, and Methods for Gene Silencing in C. elegans, Drosophila, and Mammals. Ute Schepers
Copyright © 2005 WILEY-VCH Verlag GmbH & Co. KGaA, Weinheim
ISBN: 3-527-31020-7

unusually high silencing activity of sense control RNA found in a previously reported antisense experiment in the worm *Caenorhabditis elegans* (Guo and Kemphues 1995). Simultaneous injection of sense and antisense RNA exhibited a tenfold stronger effect than antisense RNA alone, which led to the conclusion that dsRNA triggers an efficient silencing mechanism in which exogenous dsRNA significantly reduces the overall level of target-mRNA. (Fire et al. 1998). This newly discovered phenomenon was termed RNA interference (RNAi) (for a review, see Arenz and Schepers 2003).

1.1
RNAi as a Tool for Functional Genomics

A number of fundamental features were soon caught up by various research groups, who started to develop RNAi as a tool to study gene function and to interfere with pathogenic gene expression in diseases (Schmitz and Schepers 2004)

RNAi is highly selective upon degrading an mRNA target if the exogenously added dsRNA shares sequences of perfect homology with the target. Whereas the transcription of the gene is normal, the translation of the protein is prevented by selective degradation of its encoded mRNA. Further, it turned out that sequences with homology to introns or promoter regions as contained in the DNA sequence showed no effect at all, indicating that the silencing was taking place at the post-transcriptional level (Fire et al. 1998; Montgomery and Fire 1998; Montgomery et al. 1998).

With the full sequence of the human genome and many well-studied model organisms available, it is now possible to choose dsRNAs that selectively degrade the mRNA of a gene of interest, leading to a corresponding loss-of-function phenotype without affecting other or related genes. As a response to substoichiometric amounts of dsRNA, levels of homologous mRNA will be drastically decreased within 2–3 h. In some species, the RNAi phenotype can cross cell boundaries and is inherited to the progeny of the organism (Zamore et al. 2000). The latter observations are referred to as systemic RNAi, and are described in more detail for the nematode *C. elegans* in Chapter 2. Moreover, cultured cells transfected with dsRNA can maintain a loss-of-function phenotype for up to nine cell divisions (Tabara et al. 1998).

This disproportion between input dsRNA and its long lived-effects seen in *C. elegans* and plants (Grishok et al. 2000; Wassenegger and Pelissier 1998) suggests that the mechanism of RNAi is catalytic and does not function by titrating endogenous mRNA, as was proposed for antisense RNA.

Today, RNA silencing including RNAi is assumed to be an ancient self-defense mechanism of eukaryotic cells to combat infection by RNA viruses (Ruiz et al. 1998; Voinnet 2001) and transposons (mobile parasitic stretches of DNA that can be inserted into the host's genome) (Ketting et al. 1999; Tabara et al. 1999). The trigger for this cellular defense mechanism is dsRNA, which occurs during replication of those elements but never from tightly regulated endogenous genes. Intermediate dsRNA will be recognized and degraded by a multipart protein machinery. Furthermore, RNAi is presumed to carry out numerous additional functions in depending on the organism. There is evidence that it eliminates defective mRNAs by degradation (Plasterk 2002), as there is overlapping activity of *C. elegans* genes for RNAi and Nonsense-mediated mRNA decay (NMD) (Domeier et al. 2000). RNAi is further assumed to tightly regulate protein levels in response to various environmental stimuli, although the extent to which this mechanism is employed by specific cell types remains to be discovered (McManus et al. 2002).

Later, the real RNAi technology arose from the observation that exogenously applied naked dsRNAs induce specific RNA silencing in plants and *C. elegans*, when the nucleotide sequence of the dsRNA is homologous to the respective mRNA.

1.2
Mechanism of RNAi

Since its discovery, much progress has been made towards the identification and characterization of the genes implicated in the RNAi events in *C. elegans* (Qiao et al. 1995; Smardon et al. 2000), *Arabidopsis* (Mourrain et al. 2000), *N. crassa* (Cogoni and Macino 1997, 1999), *Drosophila*, and mammals. Most of the important mechanistic steps and molecular components were discovered in *C. elegans*, *D. melanogaster*, and in plants. Far from being understood, RNAi has emerged as a more complex mechanism than expected, as it involves several different proteins and small RNAs. Even if it shares common features with established dsRNA-induced RNA silencing phenomena such as "co-suppression" in plants and "quelling" in fungi, it is not known if RNAi uses identical mechanisms.

In fact, genetic studies in RNA silencing-deficient mutants of *Arabidopsis* (Mourrain et al. 2000), *N. crassa* (Cogoni and Macino 1997, 1999), and *C. elegans* (Qiao et al. 1995; Smardon et al. 2000) revealed several genes involved in quelling, co-suppression and RNAi-including members of the helicase family, RNaseIII-related nucleases, members of the Argonaute family, and RNA-dependent RNA polymerases (RdRp). So far it is known that, despite all differences and similarities, the process of RNA silencing consists of an initiator step, in which long dsRNA is cleaved into short dsRNA fragments, and an effector step in which these fragments are incorporated into a protein complex, unwound and used as a guiding sequence to recognize homologous mRNA that is subsequently cleaved (Schmitz and Schepers 2004)

1.3
Dicer – the Initiator to "Dice" the dsRNA?

A common characteristic of all RNA silencing pathways initiated by dsRNA is the cleavage of long dsRNA by a double strand-specific RNase called "Dicer" (Bernstein et al. 2001). Dicer cleaves dsRNA into so-called small interfering RNA duplexes (siRNAs) encompassing a length of 21 to 25 nt (Hamilton and Baulcombe 1999; Zamore et al. 2000). Such small dsRNAs, which are complementary to both strands of the silenced gene, have been initially observed by Baulcombe and co-workers in plants undergoing transgene- or virus-induced post-transcriptional gene silencing or co-suppression. These first experiments in plants revealed that the small dsRNAs – later termed siRNAs – are the active components of the RNA silencing pathway (Hamilton and Baulcombe 1999), leading later on to their discovery in many other species such as *Drosophila* embryos (Yang et al. 2000) and *C. elegans* (Parrish et al. 2000) that were injected with dsRNA, as well as in *Drosophila* Schneider 2 (S2) cells that were transfected with long dsRNA (Hammond et al. 2000). Surprisingly, endo-genously expressed siRNAs have not been observed in mammals, indicating that there are slightly modified mechanisms for different species.

The mechanism by which these siRNAs mediate the cleavage and degradation of RNA has been thoroughly investigated by several groups. Various studies have shown that this process is restricted to the cytosol (Hutvagner and Zamore 2002a; Kawasaki et al. 2003; Zeng and Cullen 2002) facilitating the experimental set-up. Based on these results, processing of long dsRNAs to 21–23-nt RNAs was repeated in vitro, using RNase III enzyme from *Drosophila* extract (cytosol).

Precise studies of these so-called "short interfering RNAs" (siRNAs) revealed characteristic 3′-overhangs of two nucleotides on both strands (Hamilton and Baulcombe 1999; Parrish et al. 2000), and unphosphorylated hydroxyl groups (Elbashir et al. 2001b) that play a crucial role in the recognition by the other RNAi components. The specific features of siRNA resemble the characteristic cleavage pattern of nu-cleases of the RNase III family that specifically cuts dsRNAs (Bernstein et al. 2001; Billy et al. 2001; Robertson et al. 1968) and leaves them with staggered cuts on each side of the RNA (Zamore 2001). The RNase III family is divided into three classes, depending on their domain organization. While members of class I from bacteria and yeasts contain only one conserved RNase III domain and an adjacent dsRNA-binding domain (dsRBD), class II enzymes have tandem RNase III domains and one dsRBD, as well as an extended amino-terminal domain of unknown function (Filippov et al. 2000; Fortin et al. 2002; Lee et al. 2003).

Beside the already characterized classes of RNase III enzymes such as the regular canonical RNase III (Filippov et al. 2000) and Drosha – a member of the class II en-zymes localized to the nucleus (Wu et al. 2000) – homology screens of genomic data from *Drosophila* revealed many new candidate genes carrying RNase III-like do-mains. Among those candidates, Hannon and colleagues (Bernstein et al. 2001) identified a nuclease with 2249 amino acids predicted from *Drosophila* sequence data containing two RNase III domains (Mian 1997; Rotondo and Frendewey 1996), a dsRNA-binding motif (DSRM) (Aravind and Koonin 2001), an amino-terminal DexH/DEAH RNA helicase/ATPase domain, and a motif called "PAZ domain" (Cerutti et al. 2000) – all properties that characterize class III of the large noncanon-ical ribonucleases (RNase) III family. Due to the capability of producing fragments from long dsRNA that comprise a uniform size, the newly discovered enzyme was called Dicer (Bernstein et al. 2001; Ketting et al. 2001b). So far, it is loosely associated with ribosomes in the cytoplasm (endoplasmic reticulum–cytosol interface) (Billy et al. 2001; Provost et al. 2002).

Usually, bacterial RNase III-type enzymes cleave dsRNA by building a dimeric structure comprising two active centers that embrace a cleft in the protein structure that can accommodate a dsRNA substrate. The presence of divalent cations, including Mg^{2+}, has significant impact on crystal packing, intermolecular interactions, thermal stability, and the formation of two RNA-cuffing sites within each active center for catalysis (Blaszczyk et al. 2001; Zamore et al. 2000). Modeling and comparison of a RNase III structure led to the proposal of a working model of Dicer. The first assumption was that the enzyme presumably aggregates as an anti-parallel homodimer, in which only two of four catalytic actives sites are involved in dsRNA cleavage leading to ~22-mers, whereas the activity of all four sites would lead to the production of 11-mers (Blaszczyk et al. 2001). The central pair of active sites should be then replaced by a noncanonical motif making it inactive, whereas the 5'- and 3'-site remains active (Blaszczyk et al. 2001). Another working model proposes a monomeric action of Dicer in a semireciprocal fashion, cleaving the dsRNA during translocation of the enzyme down its substrate (Bernstein et al. 2001). Thus, the helicase/ATPase domain of Dicer is supposed to either induce structural rearrangement of the dsRNA template or to drive movement of the enzyme along the dsRNA in an ATP-dependent manner (Bernstein et al. 2001; Hutvagner et al. 2001; Ketting et al. 2001b; Myers et al. 2003; Provost et al. 2002; Zhang et al. 2002). The unwinding of such an RNA-duplex by a helicase homologue would require at least a temporary energy-consuming step as a prerequisite. Studies performed with *Drosophila* and *C. elegans* Dicer indicated that generation of siRNAs from dsRNA is ATP-dependent (Bernstein et al. 2001; Ketting et al. 2001a; Nykänen et al. 2001; Zamore et al. 2000). It has been shown that Dicer from enriched fractions of *Drosophila* extracts could be inactivated by ATP withdrawal (Zamore et al. 2000). However, experiments with purified recombinant human Dicer recently showed that Dicer is preferentially cleaving dsRNAs at their termini into 22 nt-long siRNAs, which is clearly an ATP-independent process (Zhang et al. 2002). Testing the human recombinant Dicer in the presence of human cell extracts without ATP revealed the same nuclease activity as in the presence of ATP, but compared to the *Drosophila* Dicer the catalytic efficiency is much lower. Since common RNasesIII do not show an ATP requirement, it might be specific for the RNAi mechanism in *Drosophila*. The results from mammalian Dicer suggest that direct dsRNA cleavage by Dicer may not involve ATP, but do not exclude the necessity of an ATP-dependent catalytic activity in the RNAi pathway (Provost et al. 2002). Further results suggest that, if ATP is necessary for the Dicer cleavage reaction, it might be involved in the siRNA release – a process which is also Mg^{2+}-regulated (Zhang et al. 2002).

Evolutionarily conserved homologues of Dicer exist in *C. elegans* (Grishok et al. 2001; Hutvagner et al. 2001), *Arabidopsis thaliana* (Jacobsen et al. 1999), mammals, and *Schizosaccharomyces pombe* (Volpe et al. 2002), where they might share similar biochemical functions. Recently, the cDNAs of murine and human Dicer were identified. The mouse cDNA spans 6.13 kilobases (kb), and encodes for a polypeptide of 1906 amino acids (Figure 5), which shares 92% sequence homology with the human orthologue (Nicholson and Nicholson 2002).

In *Arabidopsis*, two species of siRNAs have been detected, of which the shorter 21-mers appears to be responsible for mRNA degradation, while the longer 24- to 25-nt species are held responsible for the systemic spreading of the effect (Hamilton

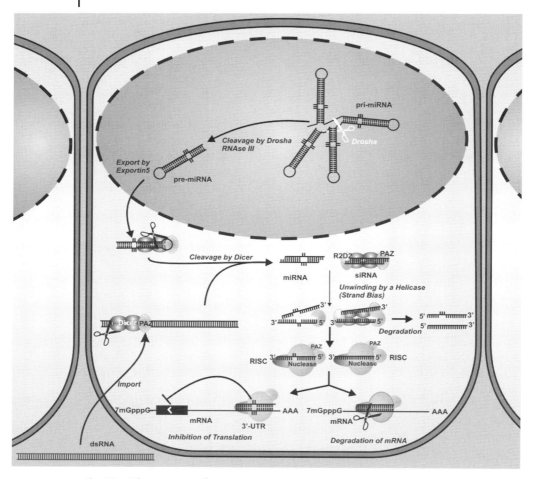

Fig. 1.1 Schematic view of siRNA- and miRNA-induced gene silencing.

et al. 2002). Mutants deficient of the CARPEL/FACTORY gene are deficient of siRNA production, indicating that the plant homologue of Dicer is encoded by this locus. Studies in wheat germ extracts led to the assumption that the two species of siRNAs might originate from the action of two distinct Dicer orthologues, one favoring the production of 21-mers from exogenous dsRNA, the other being responsible for the production of 24–25-mers from dsRNA derived from transgenic mRNA (Tang et al. 2003)

Although only one Dicer enzyme is found in *C. elegans* and humans, two Dicer homologues, DCR-1 and DCR-2, have been identified in *Drosophila* (Bernstein et al. 2001). In fact, recent findings clearly demonstrate that Dicer is involved in more processes than cleavage of dsRNA after a viral attack.

1.4
miRNAs versus siRNAs: Two Classes of Small RNAs Using the RNAi Pathway?

During mechanistic studies of RNAi in *C. elegans*, another species of small RNAs was discovered which resembled the cleavage pattern of an RNaseIII. lin-4 (lineage-abnormal-4) and let-7 (lethal-7) RNAs are expressed as 22-nt RNAs, having been processed from a ~70-nt precursor hairpin RNA. Additionally, Dicer-1-deficient (Dcr-1) *C. elegans* mutants show deficiencies in development, fertility and in RNAi (Grishok et al. 2001; Ketting et al. 2001a; Knight 2001). The phenotypes resembled the ones observed with let-7- and lin-4-deficient worms that exhibit heterochronicity (Reinhart et al. 2000) and affect larval transition (Lee et al. 1993). More remarkably, the Dcr-1 phenotype could be rescued by the application of short RNA transcripts encoded by the let-7 and lin-4 loci (Hutvagner et al. 2001). It was assumed that inside the nucleus a longer precursor is encoding the ~70-nt hairpin RNAs that form by folding back to a stem-loop structure bearing one or two mismatches in the double-stranded region (Lee and Ambros 2001). Besides its role in initiating RNAi, Dicer also cleaves these 70-nt precursor RNA stem-loop structures known as small temporal RNAs (stRNAs) or others known as microRNAs (miRNAs) derived from larger stem-loop precursors into single-stranded 21- to 23-nt RNAs during germline development of *C. elegans* (Grishok et al. 2001; Hutvagner et al. 2001; Ketting et al. 2001b; Reinhart and Bartel 2002). They do not trigger RNA degradation, but rather bind with partially complementary binding sites at the 3'-UTR of the mRNA to inhibit the translation of specific genes (Olsen and Ambros 1999; Seggerson et al. 2002; Slack et al. 2000).

Hundreds of small RNAs of miRNAs have been discovered recently in animals and plants (Lagos-Quintana et al. 2001, 2002, 2003; Lau et al. 2001; Lee and Ambros 2001; Lee et al. 1993; Mourelatos et al. 2002; Park et al. 2002; Reinhart et al. 2000, 2002). Although some of their functions are being unraveled (Brennecke et al. 2003; Kawasaki and Taira 2003a, b; Llave et al. 2002; Tang et al. 2003), their mechanism of biogenesis remains poorly understood. The generation of miRNAs occurs via sequential processing and maturation of long primary transcripts (pri-miRNAs). A pre-processing happens in the nucleus, where the pri-miRNAs are cleaved into stem-loop precursors of ~70 nt (pre-miRNAs), which are eventually exported into the cytosol by Exportin 5 (Lund et al. 2003). As reported recently, Exportin 5 specifically binds correctly processed pre-miRNAs, while interacting only weakly with extended pri-miRNAs (Lund et al. 2003). Dicer is now mediating the final cleavage of the pre-miRNAs into mature miRNAs (Grishok et al. 2001; Hutvagner et al. 2001; Ketting et al. 2001b; Knight and Bass 2001; Lee et al. 2002). It was previously speculated that Dicer is also actively involved in the processing of pri-miRNAs in the nucleus, since in-silico analysis of the various Dicer orthologues identified several nuclear localization signals (NLS) within each sequence (Nicholson and Nicholson 2002). Although experimental evidence indicates that RNAi operates in the cytoplasm, the predicted NLS suggest possible additional functions for Dicer in the nucleus.

In-vitro digestion of pri-miRNAs using Dicer as a nuclease revealed an incorrect miRNA processing (Lund et al. 2003), leading to the speculation that the nuclear processing enzyme is different from Dicer. Recently, human Drosha – another

RNase III type nuclease, which localizes predominantly to the nucleus (Wu et al. 2000) – was found to be the core nuclease that executes the initiation step of miRNA processing in the nucleus (Lee et al. 2003). Immunopurified Drosha cleaves pri-miRNA to release pre-miRNA in vitro. Furthermore, RNA interference of Drosha resulted in the strong accumulation of pri-miRNA and the reduction of pre-miRNA and mature miRNA in vivo, showing a collaboration of Drosha and Dicer in miRNA processing (Lee et al. 2003).

Like siRNAs, miRNAs show the Dicer-specific staggered cut and bear 5'-monophosphate and 3'-hydroxyl groups (Elbashir et al. 2001 a, b; Hutvagner et al. 2001). However, miRNAs seem to recognize their targets by imperfect base pairing, with the exception of those occurring in plants, where it has been shown that miRNAs with high complementarity direct RNAi by guiding an endonuclease to cleave efficiently mRNA for correct plant development. The imperfect base pairing occurring in animals makes it very difficult to locate their targets and to predict the miRNA function. *Drosophila* lacks the miRNA precursor completely, but it can process the transgenic precursor to mature miRNA, supporting the idea of Dicer being the responsible factor.

However, siRNAs and miRNAs were found to be functionally interchangeable. If synthetic siRNAs bear a sufficiently low degree of complementary bases, target translation will be inhibited without degradation (Ambros et al. 2003 b), whereas miRNAs will lead to mRNA degradation if a target with perfect complementarity is provided (Doench et al. 2003; Hutvagner and Zamore 2002 a; Zeng and Cullen 2003).

In human cell extracts, the miRNA *let*-7 naturally enters the RNAi pathway, suggesting that only the degree of complementarity between an miRNA and its RNA target determines its function (Hutvagner and Zamore 2002 a).

Very early on, it was assumed that the distinction of the two mechanisms could be conveyed by the presence of wobble-base pairs resulting from mismatches in the homology region of miRNAs and their targets (Ha et al. 1996).

1.5
RISC – the Effector to "Slice" the mRNA?

In vivo, Dicer is part of a protein complex. Today, even if the molecular mechanism of Dicer-mediated dsRNA cleavage is partially unraveled, it is still not fully clear how the initiator step is connected to the effector step of the RNAi process, since Dicer is not directly involved in the target cleavage process (Martinez et al. 2002). During the past two years, several protein factors have been identified that seem to play a role as interaction partners or even RNAi signal transporters.

During the early mechanistic studies on RNAi it was assumed that the newly generated siRNAs form a ribonucleotide protein complex (RNP) with some unknown proteins. This promotes unwinding of the RNA duplex, presumably in an ATP-dependent manner, and leads to the final activation of the RNA-induced silencing complex called RISC. Eventually, this complex presents the antisense strand of the siRNA to the target mRNA and guides mRNA degradation (Zamore et al. 2000).

Several recent studies have shown that Dicer and several components of the RISC could be co-purified, suggesting an association between the initiation and effector complex, although Dicer is not required for the final target cleavage. The connection between the two reaction steps is the transfer of siRNA, which are not assumed to move freely throughout the cytoplasm.

To date, the RISC complex is barely characterized, but it appears that RISC from *Drosophila* is a ~500 kDa complex bound to ribosomes in cell-free extracts (Nykänen et al. 2001). Closer studies of the protein complex revealed that RISC contains a DEAD-box helicase and an elusive nuclease. These constituents seem to be conserved in *Drosophila*, *C. elegans*, and mammals, although the overlap is not complete (Carmell et al. 2002).

The helicase domain is probably required to unwind the siRNA, as the tight binding of the complementary strands would prevent any specific target recognition. This is achieved in an ATP-dependent step that leads to the remodeling of the complex into its active form referred to as RISC* (Hammond et al. 2000; Nykänen et al. 2001). The antisense strand then serves as a template for the recognition of homologous mRNA (Martinez et al. 2002; Tijsterman et al. 2002) which, upon binding to RISC*, is cleaved in the center of the recognition sequence 10 nt from the 5'-end of the siRNA antisense strand (Hutvagner and Zamore 2002b) by the nuclease activity of the complex. The two fragments are subject to degradation by unspecific exonucleases. The template siRNA is not affected by this reaction, so that the RISC can undergo numerous cycles of mRNA cleavage that comprise the high efficiency of RNAi. Recently, a nuclease was purified in association with the RISC complex. This was an evolutionarily conserved 103-kDa protein comprising five repeats of a nuclease domain usually found in *Staphylococcus* bacteria. While four of the five RNase domains remain active, the fifth is fused to a Tudor domain which is involved in the binding of modified amino acids, which gives the nuclease its name, Tudor-staphylococcal nuclease (Tudor-SN) (Caudy et al. 2003). The nuclease has been shown to be conserved in plants (Ponting 1997), *C. elegans*, *Drosophila* (Callebaut and Mornon 1997; Caudy et al. 2003; Ponting 1997), and mammals (Callebaut and Mornon 1997), but is rather responsible for the unspecific degradation of the mRNA remainder than for the siRNA-targeted specific mRNA cleavage (Caudy et al. 2003).

Further compounds of RISC are siRNAs and proteins, one of which was identified as Argonaute-2 (Hammond et al. 2001). Like Dicer, Argonaute-2 contains a PAZ domain and appears to be essential for the nuclease activity of RISC (Hammond et al. 2001). Moreover, using affinity-tagged siRNAs, Tuschl and colleagues showed that single-stranded siRNA resides in the RISC together with mammalian homologues of Argonaute proteins Ago-2, eIF2C1 and/or eIF2C2 (Martinez et al. 2002), which contain two characteristic domains, PAZ and PIWI. The PAZ domain plays an essential role in RNAi, since a mutation in the PAZ domain of the *C. elegans* RDE-1 gene correlates with an RNAi-deficient phenotype (Cerutti et al. 2000). It is highly conserved and is found only in Argonaute proteins and Dicer. Structural analysis revealed highly conserved structural residues, suggesting that PAZ domains in all members of the Argonaute and Dicer families adopt a similar fold with a nucleic-acid binding function (Lingel et al. 2003). Even though the binding affinity for nucleic acids is usually low, PAZ domains exhibit enhanced affinity siRNA binding, most likely interacting with the extended 3' ends or the 5'-phosphorylated ends of siRNAs for their specific incorporation into the RNAi pathway (Song et al. 2003; Yan et al. 2003). Recently, several reports proposed the atomic structure of the PAZ domain to contain a six-stranded β-barrel with an additional appendage, to bind both single- and double-stranded RNA in a sequence-independent manner (Lingel et al. 2003; Song et al. 2003; Yan et al. 2003). This revealed a nucleic acid-binding fold that is stabilized by conserved hydrophobic residues. NMR studies on the PAZ-siRNA complex suggest two modes of possible binding mechanisms: The lack of sequence specificity suggests either multiple PAZ domain molecules binding to a single RNA molecule, forming a complex analogous to "protein beads on an RNA string" (Yan et al. 2003), or a single PAZ domain is engaged in different modes of interactions with a single RNA molecule such as "sliding" through the RNA sequence, resulting in the co-existence of different complex species. RISC, which can be separated from the Dicer fraction by centrifugation of *Drosophila* S2 lysates at 100 000 g, is not able to cleave dsRNA. There was a speculation that Dicer and the RISC complex physically interact between the two PAZ domains of Argonaute-2 and Dicer, facilitating incorporation of siRNAs out of the Dicer complex into RISC (Hammond et al. 2001). In parallel to the solution of the PAZ structure, another protein – R2D2 – was found which is probably the key player in the Dicer-RISC interaction. Wang and coworkers termed this the 36-kDa small protein with tandem dsRNA binding domains (R2) and a *Drosophila* Dicer-2 binding domain (D2) R2D2 (Liu et al. 2003). Like its putative *C. elegans* homologue RDE-4 (Grishok et al. 2000; Tabara et al. 2002), it forms a stable complex with the nascent siRNAs, and has been shown to be essential for transfer of the siRNA from the initiator complex Dicer to the molecular components responsible for the effector step (Liu et al. 2003). While Dicer alone is sufficient to cleave dsRNA, it needs R2D2 to bind not only the nascent siRNAs but also synthetic siRNAs.

This model is supported by previous studies, in which it was shown that if the dsRNA was processed from the 5' → 3' direction of the sense strand, it would generate RISC that can mediate degradation of the sense but not antisense target mRNA, and vice versa (Elbashir et al. 2001b). Since synthetic siRNAs do not need a cleavage process by Dicer rather than a binding by Dicer/R2D2 complex, they can be transferred to RISC in an nonoriented fashion, leading to the degradation of either sense or antisense target mRNA. This suggests that newly synthesized symmetric siRNA generated from a longer dsRNA is not released from the complex, but rather is retained by DCR-2/R2D2 in a fixed orientation, which is determined by the direction of dsRNA processing. Then, only the antisense strand can become the guiding RNA for RISC (Liu et al. 2003).

If this mechanism is homologous to *C. elegans*, where RDE-4 also interacts with RDE-1, an AGO2 homologue and a RISC component, it can be proposed that R2D2 play a similar role in bridging the initiation and effector steps of the *Drosophila* (Liu et al. 2003; Tabara et al. 2002).

In contrast to the results in *Drosophila*, the human RISC is found in the 100 000 g fraction of HeLa cells (Hutvagner and Zamore 2002b), revealing a slightly different localization of the complex and a variation of the mechanism. Dicer, as well as the human RISC, are both localized in the cytosol. Determination of the RNAi mechanism in invertebrates does not necessarily imply that it is the same in humans.

Studies with chemically synthesized short dsRNAs showed that only siRNAs with lengths between 20 and 23 base pairs are able to integrate into the RISC and to guide this complex to its substrate mRNA by conventional base-pairing (Hammond et al. 2000). Recognition of mRNAs by RISC eventually triggers their destruction. Common models of RNAi propose that only the antisense strand of siRNAs is part of the RISC complex, thus provoking the question of whether ssRNA of appropriate size can mediate RNAi.

An important result which has emerged from recent studies is that single-stranded antisense RNA ranging from 19 to 29 nt can also enter the RNAi pathway, albeit less efficiently, than the double-stranded siRNA (Martinez et al. 2002; Schwarz et al. 2002). Zamore and colleagues showed that with siRNA there is a profound strand bias in the mRNA targeting and cleavage. Even if the separate antisense and sense strand of a distinct siRNA reveal a similar intrinsic efficacy in targeting the mRNA, they show different activities when hybridized to a duplex siRNA. The stability of the 5′-end determines which strand enters into RISC, whereas the other strand is degraded (Schwarz et al. 2003). 5′-ends starting with an A-U base pairing are preferred over those beginning with G-C, the hypothesis being that a less stable 5′-end will be preferentially accepted by RISC. Even an energy difference corresponding to a single hydrogen bond can largely favor the incorporation of one strand over the other (Schwartz et al. 2003). Statistical analysis of the internal energies of a vast number of naturally occurring siRNAs and synthetic siRNAs has recently revealed a decreased stability at the 5′-ends of the functional duplexes and a slightly decreased stability between base pairs 9–14 counting from the 5′-terminus (Khvorova et al. 2003). The 5′-instability is assumed to facilitate duplex unwinding by the DEAD box helicase. Mutational analysis of the siRNA strands revealed that the decreased stability between base pairs 9–14 might also facilitate the dissociation–association reaction observed for the DEAD box helicase (Nykänen et al. 2001). It is also likely to play a role in target cleavage that takes place between the 9th and 10th base pair from the 5′-end of the guiding ssRNA strand, or in the release of the mRNA fragments. From siRNAs isolated from cytosolic extracts it was concluded that the natural selection of siRNAs is based on thermodynamic properties rather than mere function. Those studies raised the question whether the asymmetry found in the miRNA strand selection of RISC is closely related to the asymmetric incorporation of siRNA strands. Very early on, it was assumed that the distinction of the two mechanisms could be conveyed by the presence of wobble-base pairs resulting from mismatches in the homology region of miRNAs and their targets (Ha et al. 1996). This goes along with studies showing that duplex unwinding plays a crucial role in the processing of siRNAs and miRNAs and their incorporation into RISC (Bernstein et al. 2001; Nicholson and Nicholson 2002). Mature miRNAs are usually unstable at their 5′-end and present a lower stability near their center.

For miRNAs, it is the miRNA strand of a short-lived, siRNA duplex-like intermediate that assembles into a RISC complex, causing miRNAs to accumulate in vivo as single-stranded RNAs. Alternatively, both strands of siRNA could be integrated into RISC and form a triple helix with the target mRNA.

From this, it was concluded that the effector complexes containing siRNAs and miRNAs are related, but function by different mechanisms. Exogenously supplied siRNAs and shRNAs with single mismatches fail to repress the translation of their target gene (Elbashir et al. 2001 a; Paddison et al. 2002). However, siRNAs and miRNAs were found to be functionally interchangeable. If synthetic siRNAs bear a sufficiently low degree of complementarity, target translation will be inhibited without degradation (Ambros et al. 2003 a, b), whereas miRNAs will lead to mRNA degradation if a target with perfect complementarity is provided (Doench et al. 2003; Hutvagner and Zamore 2002 a; Zeng and Cullen 2003).

1.6
Are RNA-Dependent RNA Polymerases (RdRps) Responsible for the Catalytic Nature of RNAi?

Considering the high efficiency of RNA degradation, as was first observed in *C. elegans* (Fire 1994), the question arises as to whether this is due to a form of catalysis or to an amplification mechanism.

The main question of why RNAi is so much more powerful than the antisense approach led to investigations of RNA-dependent RNA polymerases (RdRps). In 2000, Dalmay reported that the suppression of transgenes in *Arabidopsis* is disrupted in sgs2/sde2 mutants. This locus encodes a putative RNA-dependent RNA polymerase (RdRp) (Dalmay et al. 2000; Mourrain et al. 2000). Since then, various RdRp proteins have been identified in a number of organisms, such as EGO-1, RRF-1 and RRF-2 in *C. elegans* (Sijen et al. 2001; Simmer et al. 2002; Smardon et al. 2000) and QDE-1 in *Neurospora* (Cogoni and Macino 1999).

RdRps are enzymes that are characteristically involved in RNA-virus replication by synthesizing complementary RNA molecules using RNA as a template.

In cells displaying RNAi, RdRp is assumed to convert the single-stranded target mRNA to dsRNA using the antisense strands of primary siRNAs as primers (Lipardi et al. 2001; Sijen et al. 2001). After Dicer-mediated cleavage of dsRNA, the resulting primary siRNAs are proposed to bind to their complementary target mRNA and to be extended by nucleotide addition in a target-dependent manner to form dsRNA. The resulting dsRNA can then be cleaved by Dicer to form siRNAs that lead to degradation of the mRNA. Since RdRp should be capable of transforming all targeted mRNA to dsRNA, the nuclease activity of Dicer would be sufficient to completely cleave trigger-dsRNA and also the target-mRNA.

In the course of this process – termed random degradative PCR (Lipardi et al. 2001) – the regions upstream of the primary dsRNA sequence are also amplified, leading to a set of secondary siRNAs that mediate cleavage of sequences that do not show homology to the primary dsRNA sequence. Notably, such a mechanism would not necessarily include the RNA-induced silencing complex (RISC) as an additional nuclease. The model is supported by a report of an antisense RNA ranging from 19 to 40 nt effectively triggering germline RNAi in *C. elegans* in the presence of Dicer (Tijsterman et al. 2002). In several organisms such transitive RNAi has been observed, in which siRNAs of a sequence beyond the targeted region of homology are detected (Sijen et al. 2001).

So far, the role of various RdRps remains to be clarified. Mutation studies in *C. elegans* showed that the RdRp rff-1 is essential for RNAi (Sijen et al. 2001), suggesting that primary siRNAs are neither quantitatively nor qualitatively sufficient for RNAi, and that RdRp plays an additional role for RNAi distinct from synthesis of secondary siRNAs.

Further studies in plants revealed that transitive RNAi was found to proceed in both $5' \rightarrow 3'$ or $3' \rightarrow 5'$ directions, pointing out that aberrant mRNAs from altered chromatin structures serve as substrates for RdRps. Experiments with wheat germ extracts have shown that ssRNA originating from transgenes is amplified by an RdRP, albeit no corresponding siRNAs are present (Tang et al. 2003). However, previous reports from experiments in *Drosophila* indicated an mRNA cleavage only within the homology sequence that it shares with the siRNA (Nykänen et al. 2001; Schwarz et al. 2002). In flies and mammals, no cellular RdRp for the generation of secondary siRNAs has been found (Stein et al. 2003). Thus, the high efficiency of RNAi can only be accounted for by the catalytic nature of RISC.

In these organisms, RISC is assumed to turn over many times, thereby presenting evidence for its catalytic nature. Again, it should be mentioned that different mechanisms have apparently evolved in different species for amplification of the silencing effect.

1.7
Is RNAi Involved in the Regulation of Gene Expression?

A few years after the discovery of co-suppression, it was found that in plant dsRNA also leads to methylation of genomic DNA (Wassenegger et al. 1994). No effect on transcription was observed if stretches of the coding sequence were affected, but it was terminated upon the methylation of the promoter sequence (Mette et al. 2000). This so-called transcriptional gene silencing (TGS) is not only stable but also heritable (Jones et al. 2001; Pal-Bhadra et al. 2002). The findings that mutations in methyltransferases (MET1) and chromatin remodeling complexes (DDM1) in *Arabidopsis* have an influence on the efficiency and stability of post-transcriptional gene silencing (PTGS) (Jones et al. 2001) suggested a link between dsRNA-induced gene silencing and gene regulation in plants. Studies in *C. elegans* revealed that this connection also exists in animals. Mutations in mut-7 and rde-2 release the repression of transgenes that are otherwise silenced on the level of transcription due to the reorganization of chromatin by polycomb proteins (Tabara et al. 1999). Adding to this finding, it has recently been reported that proteins of the same family are required for RNAi under some conditions (Dudley et al. 2002; Hannon 2002).

Now, more recent studies – especially in *Schizosaccharomyces pombe* (Volpe et al. 2002, 2003) and *Tetrahymena* (Mochizuki et al. 2002; Taverna et al. 2002) – revealed functional small RNAs complementary to centromeric repeats directly interacting with the chromatin remodeling (Reinhart and Bartel 2002) and chromosome dynamics (Hall et al. 2003).

In these organisms, these small RNAs seem to be part of mechanism that is responsible for the methylation of histone H3 at lysine 9 (H3K9) on genes corresponding to the small RNA. Thus far, H3K9 methylation triggers the formation of heterochromatin, leading to the repression of gene expression at this site. The generation of small RNAs and the H3K9 methylation require Dicer and putative RdRps. This observation suggests an implication of the RNAi pathway in the regulation of gene expression. Presumably, RdRp is recruited to chromatin by priming itself with small RNAs or siRNAs: If RdRp is primed while the repeats are being transcribed, coupling of RNA-dependent and DNA-dependent transcription would tether RdRP to the chromatin (Martienssen 2003). To date, some working models have been proposed, one of which is the idea that RISC-bound small RNAs are guiding a H3K9 methyltransferase to the respective DNA via their associated RdRps.

Nevertheless, a direct interaction between the small RNAs and the putative binding domains on the H3K9-methyltransferase was not observed (Hall et al. 2003).

Additionally, RNAi itself is not needed for the maintenance and inheritance of heterochromatin domains (Hall et al. 2002), suggesting a different mechanism of RISC in the nucleus. A further link between chromatin remodeling and RNAi was also found in *Drosophila* where mutations of PIWI, that is related to one of the RISC components, reduces the degree of silencing on both the transcriptional and posttranscriptional level (Pal-Bhadra et al. 2002).

All of these findings indicate that genomic DNA is affected by the cell's response to dsRNA. It is assumed that a nuclear variant of RISC exists that bears a chromatin remodeling complex instead of a nuclease activity.

The latest findings indicate that tandem repeats play a role in heterochromatin silencing (Martienssen 2003). It has been shown that tandemly repeated genes form heterochromatin and are prone to epigenetic silencing in many organisms (Dorer and Henikoff 1994). If those tandem repeats are transcribed and subsequently amplified by RdRps, dsRNA is generated that is cleaved into siRNAs by Dicer (Matzke et al. 2001; Mourrain et al. 2000). These siRNAs then not only lead to the degradation of the transcripts but also serve as primers to the RdRp so that a pool of siRNAs covering the full sequence is maintained and the stretch of heterochromatin remains silenced.

1.8
RNAi in Mammals

The interesting features of RNAi in *C. elegans* and *Drosophila* led to many research projects, which focused on adapting this technique to mammalian and human cell lines.

It appears that mammals have developed different pathways to combat parasitic and viral dsRNA, however. In mammalian cells, dsRNA causes an interferon response, which leads to the activation of RNase H degrading all mRNA transcripts in

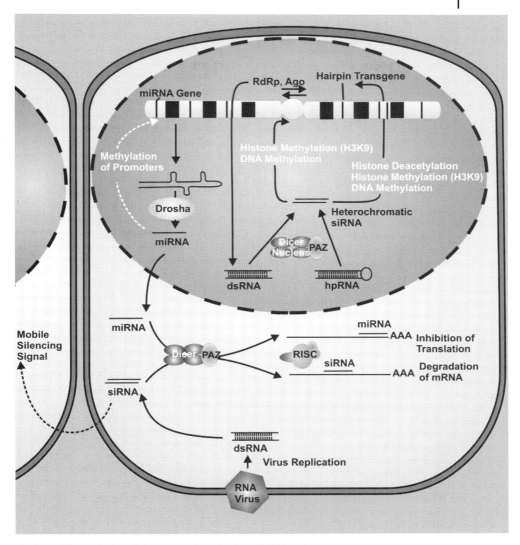

Fig. 1.2 Schematic view of the proposed mode of small RNA action in the nucleus (modified from Finnegan and Matzke 2003).

an unspecific fashion. At the same time, γ-interferon triggers the activation of protein kinase R (PKR), which phosphorylates and thus inactivates transcription factor EIF2α. This leads to a global shutdown of protein biosynthesis and, as a result, to apoptosis (Clemens 1997; Clemens and Elia 1997). For this reason it was believed that RNAi could not be induced in mammalian cells.

Despite the arguments that RNAi would not function in mammalian cells, several independent groups proved the existence of mammalian RNAi pathways by the introduction of dsRNA or vectors producing dsRNA into cell lines lacking the interferon machinery; examples were mice oocytes or mice embryonic cancer cell lines (Billy et al. 2001; Wianny and Zernicka-Goetz 2000). However, in most somatic mammalian cells this approach provokes a strong cytotoxic response. Unlike plants and nematodes, RNAi in mammalian cell lines underlies some serious limitations, and in most mammalian cells the approach of transient introduction of large dsRNAs is not feasible. The decisive breakthrough in acquiring the new RNAi technique in the field of mammalian functional genomics was the studies by Tuschl and co-workers (Elbashir et al. 2001 a). This group found that transiently applied siRNAs of 21–23 nt are able to trigger the RNA interference machinery in cultured mammalian cells, without initiating the programmed cell death response. Although these siRNAs are probably too short to trigger the interferon response, they are able to direct sequence-specific cleavage of homologous mRNAs in mammalian cells (Hutvagner 2000). Clearly, the siRNAs produced from long dsRNA by the enzyme Dicer are too short for the activation of PKR. Further studies by Tuschl and co-workers showed that dsRNAs shorter than 21 bp and longer than 25 bp are inefficient in initiating RNAi (Elbashir et al. 2001 c) as well as siRNAs with blunt ends. Only short dsRNAs with a 2-nt 3'-overhang, which resembles the naturally active products of Dicer, are efficient mediators of RNAi. With this technology even somatic primary neurons have been successfully treated to produce knock-down RNAi phenotypes (Krichevsky and Kosik 2002).

Until now, efforts to synthesize modified siRNAs more potent in inducing RNAi have failed. The replacement of siRNA 3' two ribouridine overhangs at the 3'-end by two deoxy-thymidine overhangs resulted in a decreased induction of RNAi. Furthermore, complete replacement of either sense or antisense strand of siRNAs by DNA resulted in the reduction or complete loss of RNAi activity (Hohjoh 2002).

Several groups have investigated the minimal chemical requirements for siRNAs to function in RNAi by altering the 3'-end of the antisense strand with either 2',3'-dideoxy cytidine, amino modifier (Schwarz et al. 2002), puromycin, or biotin (Chiu and Rana 2002; Martinez et al. 2002). This does not inhibit siRNA action either in vivo or in vitro in *Drosophila* and human systems. However, data from Zamore's group revealed an absolute requirement for a 5'-phosphate residue for siRNAs to direct target-RNA cleavage in *Drosophila* embryo lysates, which is thought to be necessary for the so-called authentication in the assembly of the RISC by building noncovalent interactions with other components of the RNAi (Schwarz et al. 2002). Nonetheless, the 5'-phosphate requirement might also reflect a requirement for the phosphate group in covalent interactions, such as the ligation of multiple siRNAs to generate cRNA (Nishikura 2001).

The striking results from Philipp Zamore and co-workers of the siRNA strand bias will also help with the design of synthetic siRNAs with a high degree of silencing efficiency. In this context, the design of synthetic siRNAs that more closely resemble these double-stranded miRNA intermediates reveals highly functional siRNAs, even when targeting mRNA sequences apparently refractory to cleavage by siRNAs selected by conventional siRNA design rules (Schwarz et al. 2003).

1.9
Practical Approaches

RNAi procedures are much more rapid and straightforward than traditional genomics approaches, such as the generation of knock-out animal models or the study of inherited diseases (Arenz and Schepers 2003).

Beyond its biological relevance, PTGS is emerging as a powerful tool to study the function of individual proteins or sets of proteins. User-friendly technologies for introducing siRNA into cells, in culture or in vivo, to achieve a selective reduction of single or multiple proteins of interest are rapidly evolving.

This chapter has focused mainly on the mechanism and cellular requirements of RNAi. Further details – especially on the practical aspects derived from many recent publications – will be provided elsewhere in this book in order to avoid duplication. This comprises topics from systemic gene silencing and high-throughput applications in *C. elegans* to endogenous expression of short hairpin RNAs (shRNAs) in mammals. This book is intended as a laboratory manual and will provide useful protocols and working notes, not only for those who are novices in the RNAi field but also for those experts and users who might discover some new tricks.

References

AMBROS V, BARTEL B, BARTEL DP, BURGE CB, CARRINGTON JC, CHEN X, DREYFUSS G, EDDY SR, GRIFFITHS-JONES S, MARSHALL M, MATZKE M, RUVKUN G, TUSCHL T (**2003**a) A uniform system for microRNA annotation. *Rn.a* 9: 277–279.

AMBROS V, LEE RC, LAVANWAY A, WILLIAMS PT, JEWELL D (**2003**b) MicroRNAs and other tiny endogenous RNAs in *C. elegans*. *Curr Biol* 13: 807–818.

ARAVIND L, KOONIN EV (**2001**) A natural classification of ribonucleases. *Methods Enzymol* 341: 3–28.

ARENZ C, SCHEPERS U (**2003**) RNA interference: An ancient mechanism or a state of the art therapeutic application. *Naturwissenschaften* 90: 345–359.

BAULCOMBE D (**1999**) Viruses and gene silencing in plants. *Arch Virol* 15: 189–201.

BERNSTEIN E, CAUDY AA, HAMMOND SM, HANNON GJ (**2001**) Role for a bidentate ribonuclease in the initiation step of RNA interference. *Nature* 409: 363–366.

BILLY E, BRONDANI V, ZHANG H, MULLER U, FILIPOWICZ W (**2001**) Specific interference with gene expression induced by long, double-stranded RNA in mouse embryonal teratocarcinoma cell lines. *Proc Natl Acad Sci USA* 98: 14428–14433.

BLASZCZYK J, TROPEA JE, BUBUNENKO M, ROUTZAHN KM, WAUGH DS, COURT DL, JI XH (**2001**) Crystallographic and modeling studies of RNase III suggest a mechanism for double-stranded RNA cleavage. *Structure* 9: 1225–1236.

BRENNECKE J, HIPFNER DR, STARK A, RUSSELL RB, COHEN SM (**2003**) bantam encodes a developmentally regulated microRNA that controls cell proliferation and regulates the proapoptotic gene hid in *Drosophila*. *Cell* 113: 25–36.

CALLEBAUT I, MORNON JP (**1997**) The human EBNA-2 coactivator p100: multidomain organization and relationship to the staphylococcal nuclease fold and to the tudor protein involved in *Drosophila melanogaster* development. *Biochem J* 321 (Pt 1): 125–132.

CARMELL MA, XUAN Z, ZHANG MQ, HANNON GJ (**2002**) The Argonaute family: tentacles that reach into RNAi, developmental control, stem cell maintenance, and tumorigenesis. *Genes Dev* 16: 2733–242.

CAUDY AA, KETTING RF, HAMMOND SM, DENLI AM, BATHOORN AM, TOPS BB, SILVA JM, MYERS MM, HANNON GJ, PLASTERK RH (**2003**)

A micrococcal nuclease homologue in RNAi effector complexes. *Nature* 425: 411–414.

CERUTTI L, MIAN N, BATEMAN A (2000) Domains in gene silencing and cell differentiation proteins: the novel PAZ domain and redefinition of the Piwi domain. *Trends Biochem Sci* 25: 481–482.

CHIU YL, RANA TM (2002) RNAi in human cells: Basic structural and functional features of small interfering RNA. *Mol Cell* 10: 549–561.

CLEMENS MJ (1997) PKR – a protein kinase regulated by double-stranded RNA. *Int J Biochem Cell Biol* 29: 945–949.

CLEMENS MJ, ELIA A (1997) The double-stranded RNA-dependent protein kinase PKR – structure and function [Review]. *J Interferon Cytokine Res* 17: 503–524.

COGONI C, MACINO G (1997) Isolation of quelling defective (QDE) mutants impaired in posttranscriptional transgene-induced gene silencing in *Neurospora crassa*. *Proc Natl Acad Sci USA* 94: 10233–10238.

COGONI C, MACINO G (1999) Gene silencing in *Neurospora crassa* requires a protein homologous to RNA-dependent RNA polymerase. *Nature* 399: 166–169.

COUZIN J (2002) Breakthrough of the year. Small RNAs make big splash. *Science* 298: 2296–2297.

DALMAY T, HAMILTON A, RUDD S, ANGELL S, BAULCOMBE DC (2000) An RNA-Dependent RNA polymerase gene in *Arabidopsis* is required for posttranscriptional gene silencing mediated by a transgene but not by a virus. *Cell* 101: 543–553.

DOENCH JG, PETERSEN CP, SHARP PA (2003) siRNAs can function as miRNAs. *Genes Dev* 17: 438–442.

DOMEIER ME, MORSE DP, KNIGHT SW, PORTEREIKO M, BASS BL, MANGO SE (2000) A link between RNA interference and nonsense-mediated decay in *Caenorhabditis elegans*. *Science* 289: 1928–1930.

DORER DR, HENIKOFF S (1994) Expansions of transgene repeats cause heterochromatin formation and gene silencing in *Drosophila*. *Cell* 77: 993–1002.

DUDLEY NR, LABBE JC, GOLDSTEIN B (2002) Using RNA interference to identify genes required for RNA interference. *Proc Natl Acad Sci USA* 99: 4191–4196.

ELBASHIR SM, HARBORTH J, LENDECKEL W, YALCIN A, WEBER K, TUSCHL T (2001a)

Duplexes of 21-nucleotide RNAs mediate RNA interference in cultured mammalian cells. *Nature* 411: 494–498.

ELBASHIR SM, LENDECKEL W, TUSCHL T (2001b) RNA interference is mediated by 21- and 22-nucleotide RNAs. *Genes Dev* 15: 188–200.

ELBASHIR SM, MARTINEZ J, PATKANIOWSKA A, LENDECKEL W, TUSCHL T (2001c) Functional anatomy of siRNAs for mediating efficient RNAi in *Drosophila melanogaster* embryo lysate. *EMBO J* 20: 6877–6888.

FILIPPOV V, SOLOVYEV V, FILIPPOVA M, GILL SS (2000) A novel type of RNase III family proteins in eukaryotes. *Gene* 245: 213–221.

FINNEGAN EJ, MATZKE MA (2003) The small RNA world. *J Cell Sci* 116: 4689–4693.

FIRE A (1994) A four-dimensional digital image archiving system for cell lineage tracing and retrospective embryology. *Computer Appl Biosci* 10: 443–447.

FIRE A, XU SQ, MONTGOMERY MK, KOSTAS SA, DRIVER SE, MELLO CC (1998) Potent and specific genetic interference by double-stranded RNA in *Caenorhabditis elegans*. *Nature* 391: 806–811.

FORTIN KR, NICHOLSON RH, NICHOLSON AW (2002) Mouse ribonuclease III. cDNA structure, expression analysis, and chromosomal location. *BMC Genomics* 3: 26.

GRISHOK A, PASQUINELLI AE, CONTE D, LI N, PARRISH S, HA I, BAILLIE DL, FIRE A, RUVKUN G, MELLO CC (2001) Genes and mechanisms related to RNA interference regulate expression of the small temporal RNAs that control *C. elegans* developmental timing. *Cell* 106: 23–34.

GRISHOK A, TABARA H, MELLO CC (2000) Genetic requirements for inheritance of RNAi in *C. elegans*. *Science* 287: 2494–2497.

GUO S, KEMPHUES KJ (1995) Par-1, a gene required for establishing polarity in *C. elegans* embryos, encodes a putative Ser/Thr kinase that is asymmetrically distributed. *Cell* 81: 611–620.

HA I, WIGHTMAN B, RUVKUN G (1996) A bulged lin-4/lin-14 RNA duplex is sufficient for Caenorhabditis elegans lin-14 temporal gradient formation. *Genes Dev* 10: 3041–3050.

HALL IM, NOMA K, GREWAL SI (2003) RNA interference machinery regulates chromosome dynamics during mitosis and meiosis in fission yeast. *Proc Natl Acad Sci USA* 100: 193–198.

HALL IM, SHANKARANARAYANA GD, NOMA K, AYOUB N, COHEN A, GREWAL SI (2002) Establishment and maintenance of a heterochromatin domain. *Science* 297: 2232–2237.

HAMILTON A, VOINNET O, CHAPPELL L, BAULCOMBE D (2002) Two classes of short interfering RNA in RNA silencing. *EMBO J* 21: 4671–4679.

HAMILTON AJ, BAULCOMBE DC (1999) A species of small antisense RNA in posttranscriptional gene silencing in plants. *Science* 286: 950–952.

HAMMOND SM, BERNSTEIN E, BEACH D, HANNON GJ (2000) An RNA-directed nuclease mediates post-transcriptional gene silencing in *Drosophila* cells. *Nature* 404: 293–296.

HAMMOND SM, BOETTCHER S, CAUDY AA, KOBAYASHI R, HANNON GJ (2001) Argonaute2, a link between genetic and biochemical analyses of RNAi. *Science* 293: 1146–1150.

HANNON GJ (2002) RNA interference. *Nature* 418: 244–251.

HOHJOH H (2002) RNA interference (RNA(i)) induction with various types of synthetic oligonucleotide duplexes in cultured human cells. *FEBS Lett* 521: 195–199.

HUTVAGNER G, MCLACHLAN J, PASQUINELLI AE, BALINT E, TUSCHL T, ZAMORE PD (2001) A cellular function for the RNA-interference enzyme Dicer in the maturation of the let-7 small temporal RNA. *Science* 293: 834–838.

HUTVAGNER G, ZAMORE PD (2002a) A MicroRNA in a multiple-turnover RNAi enzyme complex. *Science* 1: 1.

HUTVAGNER G, ZAMORE PD (2002b) RNAi: nature abhors a double-strand. *Curr Opin Genet Dev* 12: 225–232.

HUTVAGNER GM, L.; NAP, J. P. (2000) Detailed characterization of the posttranscriptional gene-silencing-related small RNA in a GUS gene-silenced tobacco. *RNA – A publication of the RNA Society* 6: 1445–1454.

JACOBSEN SE, RUNNING MP, MEYEROWITZ EM (1999) Disruption of an RNA helicase/RNase III gene in *Arabidopsis* causes unregulated cell division in floral meristems. *Development* 126: 5231–5243.

JONES L, RATCLIFF F, BAULCOMBE DC (2001) RNA-directed transcriptional gene silencing in plants can be inherited independently of the RNA trigger and requires Met1 for maintenance. *Curr Biol* 11: 747–757.

KAWASAKI H, SUYAMA E, IYO M, TAIRA K (2003) siRNAs generated by recombinant human Dicer induce specific and significant but target site-independent gene silencing in human cells. *Nucleic Acids Res* 31: 981–987.

KAWASAKI H, TAIRA K (2003a) Functional analysis of microRNAs during the retinoic acid-induced neuronal differentiation of human NT2 cells. *Nucleic Acids Res Suppl*: 243–244.

KAWASAKI H, TAIRA K (2003b) Hes1 is a target of microRNA-23 during retinoic-acid-induced neuronal differentiation of NT2 cells. *Nature* 423: 838–842.

KETTING RF, FISCHER SE, BERNSTEIN E, SIJEN T, HANNON GJ, PLASTERK RH (2001a) Dicer functions in RNA interference and in synthesis of small RNA involved in developmental timing in *C. elegans*. *Genes Dev* 15: 2654–2659.

KETTING RF, HAVERKAMP TH, VAN LUENEN HG, PLASTERK RH (1999) Mut-7 of *C. elegans*, required for transposon silencing and RNA interference, is a homolog of Werner syndrome helicase and RNaseD. *Cell* 99: 133–141.

KHVOROVA A, REYNOLDS A, JAYASENA SD (2003) Functional siRNAs and miRNAs exhibit strand bias. *Cell* 115: 209–216.

KNIGHT SW, BASS BL (2001) A role for the RNase III enzyme DCR-1 in RNA interference and germ line development in *Caenorhabditis elegans*. *Science* 293: 2269–2271.

KRICHEVSKY AM, KOSIK KS (2002) RNAi functions in cultured mammalian neurons. *Proc Natl Acad Sci USA* 99: 11926–11929.

LAGOS-QUINTANA M, RAUHUT R, LENDECKEL W, TUSCHL T (2001) Identification of novel genes coding for small expressed RNAs. *Science* 294: 853–858.

LAGOS-QUINTANA M, RAUHUT R, MEYER J, BORKHARDT A, TUSCHL T (2003) New microRNAs from mouse and human. *Rn.a* 9: 175–179.

LAGOS-QUINTANA M, RAUHUT R, YALCIN A, MEYER J, LENDECKEL W, TUSCHL T (2002) Identification of tissue-specific microRNAs from mouse. *Curr Biol* 12: 735–739.

LAU NC, LIM LP, WEINSTEIN EG, BARTEL DP (2001) An abundant class of tiny RNAs with probable regulatory roles in *Caenorhabditis elegans*. *Science* 294: 858–862.

LEE RC, AMBROS V (2001) An extensive class of small RNAs in *Caenorhabditis elegans*. *Science* 294: 862–864.

LEE RC, FEINBAUM RL, AMBROS V (1993) The *C. elegans* heterochronic gene lin-4 encodes small RNAs with antisense complementarity to lin-14. *Cell* 75: 843–854.

LEE Y, AHN C, HAN J, CHOI H, KIM J, YIM J, LEE J, PROVOST P, RADMARK O, KIM S, KIM VN (**2003**) The nuclear RNase III Drosha initiates microRNA processing. *Nature* 425: 415–419.

LEE Y, JEON K, LEE JT, KIM S, KIM VN (**2002**) MicroRNA maturation: stepwise processing and subcellular localization. *EMBO J* 21: 4663–4670.

LINGEL A, SIMON B, IZAURRALDE E, SATTLER M (**2003**) Structure and nucleic-acid binding of the *Drosophila* Argonaute 2 PAZ domain. *Nature* 426: 465–469.

LIPARDI C, WEI Q, PATERSON BM (**2001**) RNAi as random degradative PCR: siRNA primers convert mRNA into dsRNAs that are degraded to generate new siRNAs. *Cell* 107: 297–307.

LIU Q, RAND TA, KALIDAS S, DU F, KIM HE, SMITH DP, WANG X (**2003**) R2D2, a bridge between the initiation and effector steps of the *Drosophila* RNAi pathway. *Science* 301: 1921–1925.

LLAVE C, XIE ZX, KASSCHAU KD, CARRINGTON JC (**2002**) Cleavage of Scarecrow-like mRNA targets directed by a class of *Arabidopsis* miRNA. *Science* 297: 2053–2056.

LUND E, GUTTINGER S, CALADO A, DAHLBERG JE, KUTAY U (**2004**) Nuclear Export of Micro-RNA Precursors. *Science* 303: 95–98.

MARTIENSSEN RA (**2003**) Maintenance of heterochromatin by RNA interference of tandem repeats. *Nature Genet* 35: 213–214.

MARTINEZ J, PATKANIOWSKA A, URLAUB H, LUHRMANN R, TUSCHL T (**2002**) Single-stranded antisense siRNAs guide target RNA cleavage in RNAi. *Cell* 110: 563–574.

MATZKE M, MATZKE AJ, KOOTER JM (**2001**) RNA: guiding gene silencing. *Science* 293: 1080–1083.

MCMANUS MT, PETERSEN CP, HAINES BB, CHEN J, SHARP PA (**2002**) Gene silencing using micro-RNA designed hairpins. *Rn.a* 8: 842–850.

METTE MF, AUFSATZ W, VAN DER WINDEN J, MATZKE MA, MATZKE AJM (**2000**) Transcriptional silencing and promoter methylation triggered by double-stranded RNA. *EMBO J* 19: 5194–5201.

MIAN IS (**1997**) Comparative sequence analysis of ribonucleases Hii, Iii, Ii Ph and D. *Nucleic Acids Res* 25: 3187–3195.

MOCHIZUKI K, FINE NA, FUJISAWA T, GOROVSKY MA (**2002**) Analysis of a piwi-related gene implicates small RNAs in genome rearrangement in *Tetrahymena*. *Cell* 110: 689–699.

MONTGOMERY MK, FIRE A (**1998**) Double-stranded RNA as a mediator in sequence specific genetic silencing and co-suppression. *Trends Genet* 14: 255–258.

MONTGOMERY MK, XU SQ, FIRE A (**1998**) RNA as a target of double-stranded RNA-mediated genetic interference in *Caenorhabditis elegans*. *Proc Natl Acad Sci USA* 95: 15502–15507.

MOURELATOS Z, DOSTIE J, PAUSHKIN S, SHARMA A, CHARROUX B, ABEL L, RAPPSILBER J, MANN M, DREYFUSS G (**2002**) miRNPs: a novel class of ribonucleoproteins containing numerous microRNAs. *Genes Dev* 16: 720–728.

MOURRAIN P, BECLIN C, ELMAYAN T, FEUERBACH F, GODON C, MOREL JB, JOUETTE D, LACOMBE AM, NIKIC S, PICAULT N, REMOUE K, SANIAL M, VO TA, VAUCHERET H (**2000**) *Arabidopsis* SGS2 and SGS3 genes are required for posttranscriptional gene silencing and natural virus resistance. *Cell* 101: 533–542.

MYERS JW, JONES JT, MEYER T, FERRELL JE, JR. (**2003**) Recombinant Dicer efficiently converts large dsRNAs into siRNAs suitable for gene silencing. *Nature Biotechnol* 21: 324–328.

NAPOLI C, LEMIEUX C, JORGENSEN R (**1990**) Introduction of a chimeric chalcone synthase gene into petunia results in reversible co-suppression of homologous genes in trans. *Plant Cell* 2: 279–289.

NICHOLSON RH, NICHOLSON AW (**2002**) Molecular characterization of a mouse cDNA encoding Dicer, a ribonuclease III ortholog involved in RNA interference. *Mammalian Genome* 13: 67–73.

NISHIKURA K (**2001**) A short primer on RNAi: RNA-directed RNA polymerase acts as a key catalyst. *Cell* 107: 415–418.

NYKÄNEN A, HALEY B, ZAMORE PD (**2001**) ATP requirements and small interfering RNA structure in the RNA interference pathway. *Cell* 107: 309–321.

OLSEN PH, AMBROS V (**1999**) The lin-4 regulatory RNA controls developmental timing in *Caenorhabditis elegans* by blocking LIN-14 protein synthesis after the initiation of translation. *Dev Biol* 216: 671–680.

PADDISON PJ, CAUDY AA, BERNSTEIN E, HANNON GJ, CONKLIN DS (**2002**) Short hairpin RNAs (shRNAs) induce sequence-specific silencing in mammalian cells. *Genes Dev* 16: 948–958.

PAL-BHADRA M, BHADRA U, BIRCHLER JA (2002) RNAi related mechanisms affect both transcriptional and posttranscriptional transgene silencing in *Drosophila*. *Mol Cell* 9: 315–327.

PARK W, LI J, SONG R, MESSING J, CHEN X (2002) CARPEL FACTORY, a Dicer homolog, and HEN1, a novel protein, act in microRNA metabolism in *Arabidopsis thaliana*. *Curr Biol* 12: 1484–1495.

PARRISH S, FLEENOR J, XU S, MELLO C, FIRE A (2000) Functional anatomy of a dsRNA trigger: differential requirement for the two trigger strands in RNA interference. *Mol Cell* 6: 1077–1087.

PLASTERK RHA (2002) RNA silencing: The genome's immune system. *Science* 296: 1263–1265.

PONTING CP (1997) Evidence for PDZ domains in bacteria, yeast, and plants. *Protein Sci* 6: 464–468.

PROVOST P, DISHART D, DOUCET J, FRENDEWEY D, SAMUELSSON B, RADMARK O (2002) Ribonuclease activity and RNA binding of recombinant human Dicer. *EMBO J* 21: 5864–5874.

QIAO L, LISSEMORE JL, SHU P, SMARDON A, GELBER MB, MAINE EM (1995) Enhancers of glp-1, a gene required for cell-signaling in *Caenorhabditis elegans*, define a set of genes required for germline development. *Genetics* 141: 551–569.

REINHART BJ, BARTEL DP (2002) Small RNAs correspond to centromere heterochromatic repeats. *Science* 297: 1831.

REINHART BJ, SLACK FJ, BASSON M, PASQUINELLI AE, BETTINGER JC, ROUGVIE AE, HORVITZ HR, RUVKUN G (2000) The 21-nucleotide let-7 RNA regulates developmental timing in *Caenorhabditis elegans*. *Nature* 403: 901–906.

REINHART BJ, WEINSTEIN EG, RHOADES MW, BARTEL B, BARTEL DP (2002) MicroRNAs in plants. *Genes Dev* 16: 1616–1626.

ROBERTSON HD, WEBSTER RE, ZINDER ND (1968) Purification and properties of ribonuclease III from *Escherichia coli*. *J Biol Chem* 243: 82–91.

ROTONDO G, FRENDEWEY D (1996) Purification and characterization of the Pac1 ribonuclease of *Schizosaccharomyces pombe*. *Nucleic Acids Res* 24: 2377–2386.

RUIZ MT, VOINNET O, BAULCOMBE DC (1998) Initiation and maintenance of virus-induced gene silencing. *Plant Cell* 10: 937–946.

SCHMITZ K, SCHEPERS U (2004) Silencio: RNA interference, the tool of the new millennium. *Biol Chem* (in press)

SCHWARTZ A, RAHMOUNI AR, BOUDVILLAIN M (2003) The functional anatomy of an intrinsic transcription terminator. *EMBO J* 22: 3385–3394.

SCHWARZ DS, HUTVAGNER G, DU T, XU Z, ARONIN N, ZAMORE PD (2003) Asymmetry in the assembly of the RNAi enzyme complex. *Cell* 115: 199–208.

SCHWARZ DS, HUTVAGNER G, HALEY B, ZAMORE PD (2002) Evidence that siRNAs function as guides, not primers, in the *Drosophila* and human RNAi pathways. *Mol Cell* 10: 537–548.

SEGGERSON K, TANG LJ, MOSS EG (2002) Two genetic circuits repress the *Caenorhabditis elegans* heterochronic gene lin-28 after translation initiation. *Dev Biol* 243: 215–225.

SIJEN T, FLEENOR J, SIMMER F, THIJSSEN KL, PARRISH S, TIMMONS L, PLASTERK RH, FIRE A (2001) On the role of RNA amplification in dsRNA-triggered gene silencing. *Cell* 107: 465–476.

SIMMER F, TIJSTERMAN M, PARRISH S, KOUSHIKA SP, NONET ML, FIRE A, AHRINGER J, PLASTERK RH (2002) Loss of the putative RNA-directed RNA polymerase RRF-3 makes *C. elegans* hypersensitive to RNAi. *Curr Biol* 12: 1317–1319.

SLACK FJ, BASSON M, LIU Z, AMBROS V, HORVITZ HR, RUVKUN G (2000) The lin-41 RBCC gene acts in the *C. elegans* heterochronic pathway between the let-7 regulatory RNA and the LIN-29 transcription factor. *Mol Cell* 5: 659–669.

SMARDON A, SPOERKE JM, STACEY SC, KLEIN ME, MACKIN N, MAINE EM (2000) EGO-1 is related to RNA-directed RNA polymerase and functions in germ-line development and RNA interference in *C. elegans*. *Curr Biol* 10: 169–178.

SONG JJ, LIU J, TOLIA NH, SCHNEIDERMAN J, SMITH SK, MARTIENSSEN RA, HANNON GJ, JOSHUA-TOR L (2003) The crystal structure of the Argonaute2 PAZ domain reveals an RNA binding motif in RNAi effector complexes. *Nature Struct Biol* 10: 1026–1032.

STEIN P, SVOBODA P, ANGER M, SCHULTZ RM (2003) RNAi: Mammalian oocytes do it without RNA-dependent RNA polymerase. *Rn.a – A Publication of the Rn.a Society* 9: 187–192.

TABARA H, GRISHOK A, MELLO CC (1998) RNAi in *C. elegans*: soaking in the genome sequence. *Science* 282: 430–431.

TABARA H, SARKISSIAN M, KELLY WG, FLEENOR J, GRISHOK A, TIMMONS L, FIRE A, MELLO CC (**1999**) The rde-1 gene, RNA interference, and transposon silencing in *C. elegans. Cell* 99: 123–132.

TABARA H, YIGIT E, SIOMI H, MELLO CC (**2002**) The dsRNA binding protein RDE-4 interacts with RDE-1, DCR-1, and a DExH- box helicase to direct RNAi in *C. elegans. Cell* 109: 861–871.

TANG GL, REINHART BJ, BARTEL DP, ZAMORE PD (**2003**) A biochemical framework for RNA silencing in plants. *Genes Dev* 17: 49–63.

TAVERNA SD, COYNE RS, ALLIS CD (**2002**) Methylation of histone h3 at lysine 9 targets programmed DNA elimination in *Tetrahymena. Cell* 110: 701–711.

TIJSTERMAN M, KETTING RF, OKIHARA KL, SIJEN T, PLASTERK RHA (**2002**) RNA helicase MUT-14-dependent gene silencing triggered in *C. elegans* by short antisense RNAs. *Science* 295: 694–697.

VOINNET O (**2001**) RNA silencing as a plant immune system against viruses. *Trends Genet* 17: 449–459.

VOLPE T, SCHRAMKE V, HAMILTON GL, WHITE SA, TENG G, MARTIENSSEN RA, ALLSHIRE RC (**2003**) RNA interference is required for normal centromere function in fission yeast. *Chromosome Res* 11: 137–146.

VOLPE TA, KIDNER C, HALL IM, TENG G, GREWAL SIS, MARTIENSSEN RA (**2002**) Regulation of heterochromatic silencing and histone H3 lysine-9 methylation by RNAi. *Science* 297: 1833–1837.

WASSENEGGER M, HEIMES S, RIEDEL L, SANGER HL (**1994**) RNA-directed de novo methylation of genomic sequences in plants. *Cell* 76: 567–576.

WASSENEGGER M, PELISSIER T (**1998**) A model for RNA-mediated gene silencing in higher plants. *Plant Mol Biol* 37: 349–362.

WIANNY F, ZERNICKA-GOETZ M (**2000**) Specific interference with gene function by double-stranded RNA in early mouse development. *Nature Cell Biol* 2: 70–75.

WU H, XU H, MIRAGLIA LJ, CROOKE ST (**2000**) Human RNase III is a 160-kDa protein involved in preribosomal RNA processing. *J Biol Chem* 275: 36957–36965.

YAN KS, YAN S, FAROOQ A, HAN A, ZENG L, ZHOU MM (**2003**) Structure and conserved RNA binding of the PAZ domain. *Nature* 426: 468–474

YANG D, LU H, ERICKSON JW (**2000**) Evidence that processed small dsRNAs may mediate sequence-specific mRNA degradation during RNAi in *Drosophila* embryos. *Curr Biol* 10: 1191–1200.

ZAMORE PD (**2001**) Thirty-three years later, a glimpse at the ribonuclease III active site. *Mol Cell* 8: 1158–1160.

ZAMORE PD, TUSCHL T, SHARP PA, BARTEL DP (**2000**) RNAi: Double-stranded RNA directs the ATP-dependent cleavage of mRNA at 21 to 23 nucleotide intervals. *Cell* 101: 25–33.

ZENG Y, CULLEN BR (**2002**) RNA interference in human cells is restricted to the cytoplasm. *Rn.a* 8: 855–860.

ZENG Y, CULLEN BR (**2003**) Sequence requirements for micro RNA processing and function in human cells. *Rn.a – A Publication of the Rn.a Society* 9: 112–123.

ZHANG H, KOLB FA, BRONDANI V, BILLY E, FILIPOWICZ W (**2002**) Human Dicer preferentially cleaves dsRNAs at their termini without a requirement for ATP. *EMBO J* 21: 5875–5885.

2
RNAi in *Caenorhabditis elegans*

2.1
Introduction

More than a decade of genetic studies using *Drosophila melanogaster* has provided invaluable insights into how animals – humans included – regulate gene function in a variety of processes such as developmental biology, signal transduction, and those related to diseases. However, attention has also turned to another simple animal model, the nematode worm *Caenorhabditis elegans*.

During the past few years, the completion of two major projects has provided new insights into *C. elegans* development. One is the completion of cell lineages, from zygote to adult, with description of the complete anatomy at the level of electron microscope resolution, that provides a complete connection diagram of cells and organs. The second is the complete description of the genome (100.3 Mbp) in 1998 (Consortium 1998). Although *C. elegans* was the first multicellular organism for which the genome has been sequenced completely, some small gaps remain to be closed. Furthermore, decoding the program embedded in the genetic sequence remains a challenge (Stein et al. 2003), and the biology of *C. elegans* is far from being completely solved. One surprising result of this approach was that ~65% of the human disease genes have a counterpart in the worm.

Currently, the genome predicts 19 427 genes that are mostly well annotated in a database named WormBase (http://www.wormbase.org). This WormBase comprises very detailed descriptions of many gene functions, knock-out phenotypes, the animal's morphology, development, and physiology. Thus, almost the complete phenotype and genotype of an animal is known, but unfortunately the correlation between both is still lacking. In *C. elegans*, much progress has been made by using classical or forward genetics: mutagenesis experiments have identified genes and their products involved in a specific feature. Many of these genetic functions have already been defined at the molecular level, and the genome sequence will certainly support the identification of many more. Likewise, phenotyping the mutant morphology and behavior is greatly facilitated by the complete cell lineage map. Organogenesis and even complex behaviors can be studied and dysfunctions can be attributed to defects in individual cells. In addition, the animals can be analyzed by molecular, genetic, and biochemical methods, thereby allowing the identification of

RNA Interference in Practice: Principles, Basics, and Methods for Gene Silencing in C. elegans, Drosophila, and Mammals. Ute Schepers
Copyright © 2005 WILEY-VCH Verlag GmbH & Co. KGaA, Weinheim
ISBN: 3-527-31020-7

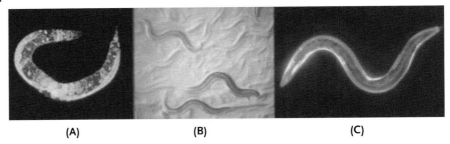

(A)   **(B)**   **(C)**

Fig. 2.1 Images of *C. elegans*. (A) DAPI staining of the cells,
(B) *C. elegans* on NG/agar/bacteria plates, fluorescence staining
of the worm. (Image courtesy of the Baumeister laboratory
http://neurogenetik.klinikum.uni-muenchen.de/).

complex protein interactions. Thus, the dissection of entire regulatory pathways is now a possible task.

Meanwhile, the opposite approach – reverse genetics – has becomes increasingly important since, with the introduction of RNA interference (RNAi), a method has been developed that allows easy and straightforward generation of knock-out phenotypes.

Screens to identify gene knock-outs can be automated, and large-scale set-ups have been devised with the recently developed RNA interference (RNAi) approaches that allow the temporal gene inactivation of many genes in many parallel experiments.

RNAi was originally discovered in *C. elegans* by Fire and colleagues in 1998 (Fire et al. 1998). For the following years, the technology was well established in the invertebrate field but was not well known by other scientists. Following on from an antisense experiment of Guo and Kemphues (Guo and Kemphues 1995) in *C. elegans*, who surprisingly observed that sense RNA is as effective as antisense RNA, it was found that simultaneous injection of both, sense and antisense RNA, in *C. elegans* reveals at least a 10-fold higher potency as a silencing trigger than the separate injection of the sense or antisense strand. Most likely, the Guo and Kemphues observation can be explained by a contamination of the sense strand with traces of antisense RNA. Hence, the active species was double-stranded RNA (dsRNA) (Fire et al. 1998). Injecting or feeding *C. elegans* with dsRNA resulted in a specific and long-lasting interference with gene expression. Even offspring of the transiently treated worms still showed silenced phenotypes (Fire et al. 1998). Moreover, cultured cells transfected with dsRNA can keep a loss-of-function phenotype for up to nine cell divisions (Tabara et al. 1998). Because of its nature, in *C. elegans* this type of RNA silencing based on the response especially to the injection of large dsRNA was termed RNA interference.

Since RNAi was first discovered in *C. elegans*, intensive studies to understand its mechanism have been carried out in this organism (Qiao et al. 1995; Smardon et al. 2000). To date, most of the important molecular components that contribute to the phenomenon of RNAi have been discovered in *C. elegans* or adapted from plant research on RNA silencing (see Chapter 1). One of the first reported – and still not fully known – aspects of RNAi in *C. elegans* is that it is systemic. The injection of gene-specific dsRNA into one tissue leads to the post-transcriptional silencing of the same gene in other tissues, and also in the worm's progenies, assuming a direct transport of the signal via some types of receptor (Fire et al. 1998; Tabara et al. 1998; Tavernarakis et al. 2000). At present, systemic RNA silencing has not been demonstrated in any other animal but plants, where it is well established (Fagard and Vaucheret 2000; Palauqui et al. 1997). Since *C. elegans* does not have an active circulatory system similar to the vascular system in vertebrates and mammals, the distribution of dsRNA to the cells takes place via the coelomic fluid. The mobile behavior of the RNA silencing signal could reflect a combination of different transport mechanisms, including cellular uptake of dsRNA from the coelomic fluid, exit of dsRNA from cells, direct intercellular trafficking of dsRNA between coupled cells, and separation of the dsRNA pool upon cell division.

These plausible explanations are contradictory to the observation that specific RNAs are generally not noted outside cells in animal systems. Because RNAi and related mechanisms are thought to be a response to challenges from viral and transposon parasites (Plasterk and Ketting 2000), it was reasoned that the systemic character of the response might depend rather on physiological conditions. An increasing number of studies are giving rise to the assumption that genes are involved in encoding transmembrane proteins that function as active transporters. Based on experiments that injected dsRNA is not uniformly effective in disrupting gene expression in the nervous system, transgenic *C. elegans* strains were constructed that allow simultaneous monitoring of localized and systemic RNAi (Timmons et al. 2003; Winston et al. 2002). The assay relies on the intracellular transgene-driven expression of dsRNA within a subset of cells, followed by monitoring of the interference in neighboring cells that lack the transgene. The assay uses green fluorescent protein (GFP) as an RNAi target that allows the effects on GFP fluorescence to be monitored. In order to distinguish between unidirectional uptake of dsRNA and bidirectional movement of dsRNA across cell boundaries, Timmons and Fire generated *C. elegans* strains harboring two different transgenes: one transgene produced GFP in all cells, and a second transgene produced a double-stranded gfp hairpin RNA in a subset of cells. However, no systemic silencing was observed unless those transgenic strains were additionally fed with unrelated dsRNA, demonstrating that dsRNA derived from the environment can partially trigger RNA-silencing mechanisms with a limited response in the organism. It is also possible that the immediate physiological inducer of systemic silencing may not be dsRNA, but rather some other molecule produced during the experiments.

There are several possible explanations for this systemic response: systemic RNAi may be part of a general mechanism for sensing and responding to environmental pathogens contaminating the growth media. There may also be a dsRNA or a virus present in the media that provide both with an abundant source of dsRNA. Further, some environmental components might condition *C. elegans* to handle intracellular dsRNA. Alternatively, some environments might allow systemic silencing after processes such as perforation of cells.

However, systemization of RNA-silencing signals has been observed previously using similar transgene configurations in *C. elegans*. The *myo-2* promoter was used to drive expression of a gfp hairpin, and a weak loss of fluorescence was observed outside the normal expression range of this promoter (pharyngeal muscle) (Winston et al. 2002). The systemic silencing effect observed in the body wall muscle was very weak, but it could be used to identify systemic RNAi-defective (*sid*) mutants, by screening for animals resistant to systemic RNAi of GFP in the body wall muscles of the worm. Those mutants are still able to exhibit RNAi when injected with GFP dsRNA, while they are unable to respond to dsRNA delivered by feeding or soaking. RNA silencing in this system was dependent upon a functional *sid-1* gene, which encodes a protein with 11 putative transmembrane-spanning regions and localizes a GFP protein fusion to the cell periphery of most non-neuronal cells. It was assumed that this may act as a channel for dsRNA, siRNAs, or some undiscovered RNAi signal, or it may be necessary for endocytosis of the systemic RNAi signal, by functioning as a receptor.

Meanwhile, the details of other mutants have been published that express similar phenotypes. These are mainly defective in systemic RNA silencing when fed with dsRNA (*fed*-mutants). However, the phenotypes of the *fed* mutants differ in many details from that of *sid-1* mutants (Winston et al. 2002). The normal activities performed by these genes might include inhibiting the generation of a mobile silencing signal, or inhibiting the cellular exit or uptake of a mobile silencing signal.

Like *sid-1* mutants, the *fed* mutants are defective in responding to ingested dsRNAs, but these mutants respond to a much higher degree to injected dsRNA than *sid-1*, leading to silencing in the progeny of *fed* animals. It is assumed that the phenotypes exhibited by fed mutants reflect a deficiency of possible RNAi transporters in gut cells (Timmons et al. 2003).

The failure to detect any systemic silencing effect in nervous system seems to be consistent with the observation, that *sid-1* is not expressed in the majority of neuronal cells. There are hints that the lack of systemic RNAi in *Drosophila* is due to the absence of a *sid-1* homologue in *Drosophila* (Roignant et al. 2003), whereas homologous sequences could be detected in human and mouse proteins – suggesting the possibility that RNAi is systemic in mammals. However, expression of *sid-1* sensitizes *Drosophila* cells to soaking dsRNA (Feinberg and Hunter 2003).

One should keep in mind those observations, when embarking on *C. elegans* RNAi studies. A careful data evaluation must be performed considering the current research on systemic silencing, which surely will be investigated extensively in the next few years.

2.2
Application of RNAi in C. elegans

The first RNAi experiments in *C. elegans* were carried out by injecting dsRNA into the gut or other sites of the animal (Fire et al. 1998; Grishok et al. 2000). Since then, other methods have been developed to generate RNAi knock-out models of *C. elegans* by using RNAi. These comprise soaking of the worms in a dsRNA solution (Tabara et al. 1998) or feeding the worms on genetically modified *E. coli* transcribing dsRNA transcripts (Timmons and Fire 1998) and in-vivo transcription of dsRNA from transgene promoters (Tabara et al. 1999; Tavernarakis et al. 2000). In all approaches the RNAi phenotype is spread over almost the entire organism and is transmitted to the offspring (Dernburg et al. 2000; Sijen et al. 2001 b), indicating that the RNAi signal is highly mobile and can be taken up by different tissues

As described in Section 2.1, the mobile behavior of the RNA-silencing signal could reflect a combination of different transport mechanisms, including cellular uptake of dsRNA from the coelomic fluid, exit of dsRNA from cells, or direct intercellular trafficking of dsRNA between coupled cells.

On injection, needle-mediated tissue disruption undoubtedly facilitates access to the coelomic fluid. On delivery by feeding and soaking, dsRNA may be distributed to cells from the gut in the same manner as nutrients.

2.3
Target Sequence Evaluation

Before starting with the dsRNA synthesis, one should perform an extensive homology search with other mRNAs. To prohibit interfering with other mRNAs, it is necessary to exclude sequence stretches of complete homology, which comprise more then 15 bp. If there is a need to silence a whole family of genes, one must revert the process by searching for complete homologies. The chosen DNA template sequence must be analyzed in two ways: (1) the sense strand has priority in the homology search; (2) the antisense strand should also be screened, as there are indications that this strand can induce off-target effects (Schwarz et al. 2003). Programs for homology searches can be found at the NCBI webpage (*www.ncbi.nih.gov*), such as the Blastn program. A more reliable database for searching homologous domains is the Smith-Waterman-based algorithm (Sections 4.2.3 and 4.2.3.4).

2.4
dsRNA Synthesis

dsRNA synthesis can be performed using either of two methods. One method comprises the cloning of the chosen sequence into a plasmid's multiple cloning site that will be flanked with either T7, Sp6, or T3 RNA polymerase recognition sites (promoters). There are many commonly used plasmid vectors that are now commercially available (e. g., pBluescript SK, Stratagene). The generation of individual plasmid vectors that do not comprise those properties requires PCR amplification of the respective sequence using oligonucleotide primers with an additional T7, T3, or SP6 promoter sequence (Figure 2.2). Further descriptions of methods in this chapter will focus on the use of the T7 promoter as an example for all three systems. The cloning can be achieved in two ways. A single DNA template with opposing T7 promoters can be used (Figure 2.2A).

Alternatively, one can construct separate DNA templates, with one containing the target sequence in a sense orientation relative to the T7 promoter while the other is oriented in the opposite direction (Figure 2.2B). The plasmids can be stored and used as a template for numerous dsRNA productions. The second method, which makes the cloning step obsolete, is based on the amplification of the sequence using

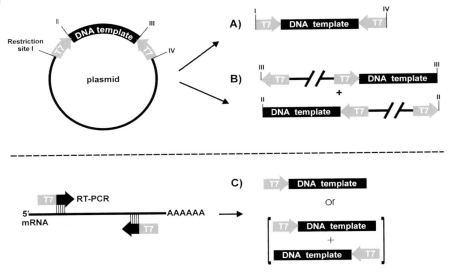

Fig. 2.2 Generation of the DNA template for dsRNA production.
(A) Restriction digest of a plasmid at the restriction sites I and IV reveals a
template flanked by two T7 promoters. (B) Restriction digest at either site III
or II reveals a template with only one T7 promoter for the separate production of the sense and antisense strand. (C) Template production by PCR.

oligonucleotide primers with an additional T7, T3, or Sp6 recognition sequence at
the 5'-end of the primer (Figure 2.2C). This allows the generation of several different
dsRNAs in a very short time. DNA templates can be amplified from total mRNA
using RT-PCR. However, the RT-PCR products are ready for direct use in dsRNA
production, but cannot be reused for further reactions.

2.4.2
Generation of the DNA Template

dsRNA production requires T7 RNA polymerase promoters at the 5'-ends of both
DNA target sequence strands.

2.4.2.1 Plasmid Templates
Using plasmid templates to generate dsRNA requires linearization with restriction
endonuclease at appropriate sites at the end of either the sense or the antisense orientation of the target sequence, as depicted in Figure 2.2A and B. This will produce RNA transcripts of defined length. Before transcription, the DNA template
should be purified either by agarose gel electrophoresis and gel extraction or by
phenol:chloroform extraction (Sambrook and Russell 2001). It is useful to start
with at least 30% more DNA than required for the transcription reaction to allow
for DNA loss during purification. It is important to avoid the use of restriction enzymes that produce 3'-overhangs. If these enzymes must be used, the ends of the

linearized template can be filled up to form blunt ends prior to transcription using DNA Polymerase I Large (Klenow) Fragment or T4 DNA Polymerase (www. promega.com).

Transcription of templates prepared by digestion with restriction enzymes that leave 3' protruding ends can result in the production of significant amounts of long, template-sized RNA transcripts, which hybridize to vector DNA. Sequences copied from the noncoding template strand can be among the extraneous transcripts (Schenborn and Mierendorf 1985).

Single DNA Templates with Opposing T7 Promoters

Cloning of the plasmid just includes PCR-amplification of the target sequence with the T7-oligonucleotide primers, as depicted in Figure 2.5. The introduction of restriction sites at the very 5'-end of the primer facilitates cloning into the plasmid vector of choice. Likewise, a direct insertion of the PCR product into a nonspecific vector can be achieved using T/A cloning (for example, the TOPO™ cloning system from Invitrogen) (Figure 2.3A)

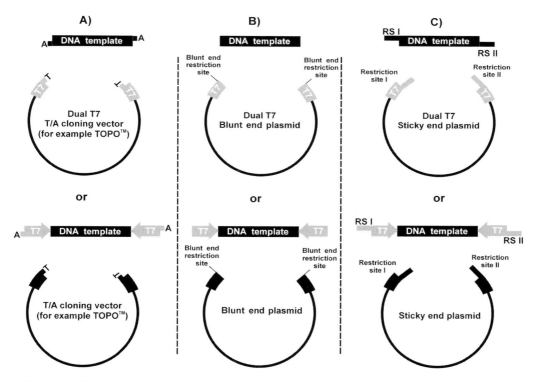

Fig. 2.3 Different cloning procedures for plasmids containing target DNA flanked by opposing T7 promoters. (A) Direct T/A cloning. (B) Blunt-end cloning. (C) Directed cloning using different restriction sites. The upper panel shows vectors already containing T7 promoters. The lower panel shows the cloning of PCR products containing the T7 promoters.

For most RNAi applications in non-mammalian systems, dsRNAs of 400 bp or larger are used (Hammond et al. 2000; Zamore et al. 2000). Data suggest that longer dsRNA molecules are more effective on a molar basis for silencing, but higher concentrations of smaller dsRNA molecules may have similar silencing effects. Indeed, newly acquired data even suggest that smaller dsRNAs can be as effective and efficient at inducing RNAi in non-mammalian systems (Betz 2003)

Separate Sense and Antisense Templates

Separate templates may either be transcribed in one reaction, or in two separate reactions. For the maximum yield of dsRNA separate in-vitro transcription reactions for each of the directions are recommended, followed by an annealing step to form the duplex. This method has shown approximately threefold higher dsRNA yields compared to using a single dual-opposed promoter template, and works better for targets that tend to form secondary structures. Templates with a high GC content are generally best expressed as single-promoter templates in separate reactions, and then pooled prior to annealing (Figure 2.2B and C).

2.4.2.2 DNA Templates Derived by PCR/RT-PCR

The second method to generate a DNA template for dsRNA production is based on PCR or RT-PCR using a T7 RNA polymerase promoter added to the 5'-end of either or both of the amplification primers (Figures 2.4 and 2.5)

The minimal T7 RNA polymerase promoter sequence requirement is shown in Figure 2.5. The +1 base here depicted in bold as a G in the T7 sequence will also be the first G in the RNA sequence. DNA templates either derived from a plasmid or from PCR are always double-stranded. However, only a double-stranded minimal promoter is sufficient for efficient binding of the T7 polymerase, whereas the rest of the sequence can be single-stranded (Figure 2.5)

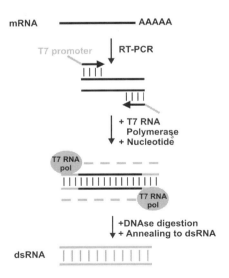

Fig. 2.4 Flow chart of the T7-based in-vitro transcription reaction by RT-PCR using two opposing T7 promoters.

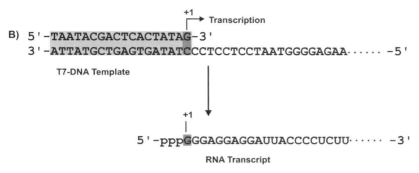

A) 5'-TAATACGACTCACTATAG-3'

+1
Transcription

B) 5'-TAATACGACTCACTATAG-3'
3'-ATTATGCTGAGTGATATCCCTCCTCCTAATGGGGAGAA······-5'

T7-DNA Template

+1

5'-pppGGGAGGAGGAUUACCCCUCUU······-3'

RNA Transcript

Fig. 2.5 (A) Sequence of the minimal T7 promoter, that must be introduced into the PCR primers. (B) Transcription of a T7 promoter-flanked DNA sequence. The last base of the minimal T7 promoter sequence is also encoding the first RNA nucleotide of the transcript, which will contain a triple phosphate tail at the 5'-end.

The start sequence should be optimized to reveal the highest amount of dsRNA. The first six bases are crucial for yield and success of the transcription reaction. In addition to full-length RNAs, the reaction also yields large amounts of abortive initiation products. Variants in the +1 to +6 region of the promoter are transcribed with reduced efficiency. Transcription reaction conditions have to be optimized to reach milligram amounts (Figure 2.6) (Milligan 1987).

The optimal sequence for a transcription start is GGG followed by GG or GC. Avoiding A and T in the first three base pairs would greatly enhance the yield. GGG should be followed by 17 to 22 gene-specific nucleotides. Moreover, extra bases upstream of the minimal T7 RNA polymerase promoter sequence may increase yield by allowing more efficient polymerase binding and initiation (Figure 2.7).

Typically, PCR product yields are higher when a single primer contains a T7 promoter than when both primers have a T7 promoter. However, this requires four PCR primers and two PCR amplifications to generate the necessary two DNA templates (Figure 2.2C). For reactions containing two primers that both have T7 promoter sequences, a primer concentration of 100–500 nM is recommended for the PCR amplification. Higher concentrations may result in significant primer–dimer formation. Amplification strategies using primers containing T7 promoter sequences may include an initial 5–10 cycles at an annealing temperature approximately 5 °C above the melting temperature of the gene-specific sequences, followed by 20–35 cycles of annealing approximately 5 °C above the melting temperature of the entire primer, including the T7 promoter (according to the Promega Ribomax™ protocol).

PCR products should be examined by agarose gel electrophoresis before transcription, in order to verify that a single PCR product of the expected size is generated.

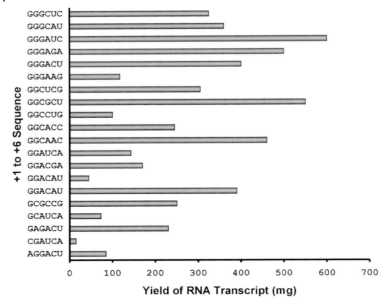

Fig. 2.6 Yields of T7-based RNA transcription of oligonucleotides between 12 and 18 nt. The first six bases are crucial for the transcription of the full-length transcripts. Each reaction contained 50 nM template, and incubation was performed at 37 °C for 4 h (Milligan 1987).

A) 5'-**TAATACGACTCACTATAG**-3' Minimal T7 sequence

B) 5´-GGATCC**TAATACGACTCACTATAG**-3' Elongated T7 sequence

Fig. 2.7 Extra bases added to the minimal T7 promoter (A) often increase the yield by allowing more efficient T7 polymerase binding and initiation (B).

2.4.2
In-vitro dsRNA Transcription

To date, many companies offer RNA production kits (Ambion, Roche, Stratagene, Promega, etc.), which reveal RNA or dsRNA with similar qualities. Here, we describe the procedure using the T7 RiboMAX™ System or the new variant T7 RiboMAX™ Express RNAi System (Promega). Both systems are widely used, and have been shown to synthesize milligram amounts of RNA. The RNA is used to generate dsRNA by simple annealing, nuclease digestion, and precipitation steps.

The T7 RiboMAX™ Express RNAi System is used to generate dsRNAs in the size range of 180 to 1000 bp. The yield of dsRNA is dependent on the first six bases or the GC content of the DNA-template, but can reach 1–2 mg dsRNA/ml. Extending the incubation time during the initial transcription reaction from 30 min to 12 h and incubation at 42 °C instead of 37 °C, usually increases the yield, particularly for GC-rich templates. Meanwhile, it is very important for a successful reaction to use highly purified DNA template, which can either be obtained by phenol:chloroform extraction (Sambrook and Russell 2001) or agarose gel extraction (QIAquick; Qiagen, Germany). After transcription, the resulting RNA strands are annealed to form dsRNA or siRNA, and the remaining single-stranded RNAs and the DNA template are removed by nuclease digestion. The dsRNA or siRNA is then purified by isopropanol precipitation and can be introduced into the organism of choice for RNAi applications (for a more detailed description, see the Promega manual for The T7 Ribomax™).

PROTOCOL 1

1. Prepare 1–8 µl of a DNA solution with a final DNA amount of 1 µg for a dual-opposed promoter PCR product or either 1 µg per separate reaction or 1 µg each in a combined reaction (2 µg total) for separate single-promoter templates.

2. Add the following reaction components into a DNAse- and especially RNase-free microcentrifuge tube and fill up to a final volume of 20 µl (for T7 RiboMAX™ Express RNAi System) or 100 µl (for T7 RiboMAX™ System). Mix by gently flicking the tube (see Table 2.1).

Tab. 2.1

RiboMAX™Express T7		*RiboMAX™ T7/Sp6/T3*	
T7 Reaction components	Reaction	T7 Reaction components	Reaction
RiboMax™Express T7 Buffer (2x)	10 µl	RiboMAX™ T7 Transcription Buffer (5x)	20 µl
Linear DNA template (1–2 µg)	1=8 µl	rNTPs (25 mM)	30 µl
Enzyme Mix T7 Express	2 µl	Linear DNA template (1–10 µg)	1–40 µl
Nuclease-free water	0–7 µl	Enzyme Mix T7 Express	10 µl
		Nuclease-free water	0–39 µl
Total volume	20 µl	Total volume	100 µl

RiboMAX™ T7 can be replaced by RiboMAX™Sp6 and RiboMAX™T3

▶ **Note:** It is important that no RNase is present in the DNA. If contamination of RNase is suspected, treat the DNA with proteinase K (100 µg/ml) and SDS (0.5 %) in 50 mM Tris-HCl (pH 7.5) and 5 mM $CaCl_2$ for 30 min at 37 °C.

3. Incubate at 37 °C for 30 min (for T7 RiboMAX™ Express RNAi System) or overnight (for T7 RiboMAX™ System).

▶ **Note:** In contrast to Promega Notes on T7 RiboMAX, a dramatic increase in yield for almost all templates can be observed when incubating for longer than 6 h.

▶ **Note:** Incubation at 42 °C may improve the yield of dsRNA for transcripts containing secondary structure, which is often due to the GC-rich templates. The use of separate single-promoter templates in separate transcription reactions has also been observed to increase yield of targets.

4. The reaction can be monitored measuring the viscosity. As the yield increases, the reaction mixture turns into a gelatinous translucent pellet.

▶ **Note:** If the sample is too viscous for the annealing step, add a few µl of water or annealing buffer (Table 2.2).

Tab. 2.2

Annealing buffer I (1x)		Annealing buffer II (10x)	
Potassium acetate	100 mM	Tris-HCl, pH 8.0	100 mM
HEPES-KOH, pH 7.4	30 mM	EDTA, pH 8.0	10 mM
Magnesium acetate	2 mM	NaCl	1 mM

5. After RNA synthesis, anneal both RNA strands, mix equal volumes of complementary RNA reactions, and incubate at 70 °C for 10 min.

6. Spin the samples very briefly! Then cool to room temperature very slowly.

▶ **Note:** It is essential to cool the reaction mixture very slowly. This can be performed either by heating the samples in a thermoblock (waterbath), or switching off the thermoblock (waterbath heater) and waiting until the temperature of the metal block (water) reaches ambient temperature.

7. Dilute the supplied RNase Solution 1:200 by adding 1 µl RNase Solution to 199 µl nuclease-free water. Add 1 µl freshly diluted RNase Solution and 1 µl RQ1 RNase-Free DNase (for T7 Ribomax™ add 1 µl RQ1 for 1 µg DNA) per 20 µl (100 µl) reaction volume, and incubate for 30 min at 37 °C. This will remove any remaining single-stranded RNA and the template DNA, leaving double-stranded RNA.

8. Add 0.1 volume of 3 M sodium acetate (pH 5.2) and 1 volume of isopropanol or 2.5 volumes of 100 % ethanol. Mix and place on ice for 1 h. The reaction will appear cloudy at this stage. Spin at top speed in a microcentrifuge for 10 min at 4 °C.

9. Carefully aspirate the supernatant, and wash the pellet with 0.5 ml of ice-cold 70 % ethanol, removing all ethanol following the wash. Air-dry the pellet for 15 min at room temperature, and resuspend the RNA sample in 100 µl DEPC- or nuclease-free water.

10. Aliquot the dsRNA and store at −20 °C or −70 °C.

11. Alternatively, further purify dsRNA following precipitation using a G25 micro spin column following the manufacturer's instructions (Amersham Biosciences, Cat.# 27-5325-01). This will remove any remaining rNTPs and allow accurate quantitation by absorbance at 260 nm.

▶ **Note:** Do not process more than an initial 40 µl reaction volume per spin column. A loss of yield can be expected following G25 purification (approximately 66% recovery).

▶ **Note:** Do not use those columns with water. The purification will also result in a desalting of the solution and water will decrease the dsRNA annealing efficiency.

12. Prepare a 1:100 to 1:300 dilution of the dsRNA and measure the concentration at 260 nm (OD (260) = 1 is equivalent to 40 µg RNA per ml).

13. Dilute 1 µl of dsRNA in 50–100 µl of 1x TAE buffer or nuclease-free water (DEPC water) and use 50–500 ng per lane and the respective amount of 10x DNA agarose loading dye (Table 2.4).

Tab. 2.3

Gel loading buffer (10x)	
Ficoll	25%
EDTA, pH 8.0	1 mM
Bromphenol blue	0.25%

14. Analyze the quality of the dsRNA on a 1–2% TAE-agarose gel using a DNA size marker (for example 1 KB plus; Invitrogen) (see Figure 2.8).

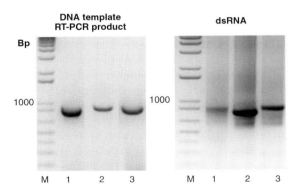

Fig. 2.8 Agarose gel analysis of dsRNA. The left panel shows the DNA template derived by RT-PCR; the right panel shows the corresponding dsRNA molecules generated using the T7 RiboMAX^{TM} System. dsRNA is separated on a 1.8% agarose/1x TAE gel. Lane designations: M, 1 kb+ DNA Ladder (Invitrogen); lanes 1–3, DNA (dsRNA).

▶ **Note:** Double-stranded RNA usually migrates more slowly than double-stranded DNA. Use 1–5 μl of diluted dsRNA per lane (dilute at least 1:50 with nuclease-free water) or use 50–500 ng per lane.

15. Staining of the gel can be performed using 0.5 mg/ml ethidium bromide in 1x TAE buffer (Sambrook and Russell 2001).

▶ **Note:** Ethidium bromide (0.1 μg/ml) can be added to the gel, but in some cases it can interfere with the migration of the dsRNA and the resolution of the bands.

2.5
Delivery of dsRNA

The introduction of dsRNA into the worm is based on the three methods mentioned above. With the exception of microinjection, all methods are easy to apply and can even be carried out in laboratories that are not specialized in *C. elegans* research. The worms can be cultured on *E. coli* OP50/NG-agar plates at temperatures between 15–25 °C, as described previously by IA Hope (*C. elegans*: A practical approach. Oxford University Press, 1999) or other *C. elegans* laboratory manuals such as WB Wood (1988) *The Nematode* Caenorhabditis Elegans (Cold Spring Harbor Laboratory Press) or D Riddle et al. (2003) *C. elegans* II (Cold Spring Harbor Laboratory Press). This chapter will only list some useful protocols and hints on strains, delivery issues, anatomy of the worm, etc., but the methods of how to work with worms should be taken from the literature described above.

2.5.1
General Information on the *C. elegans* Anatomy

It is highly recommended that the anatomy of the worm be studied before embarking on RNAi experiments such as the microinjection of dsRNA. This chapter provides some crude descriptions of the organs usually important for RNAi applications, but for more detailed information on the anatomy, morphology, physiology, and development of *C. elegans*, see the Worm Atlas database (http://www.wormatlas.org/index.htm), other websites mentioned at the end of this chapter, or textbooks on *C. elegans* (Hope IA (1999) *C. elegans*: A practical approach; Oxford University Press; WB Wood (1988) The Nematode *Caenorhabditis elegans*; Cold Spring Harbor Monograph Series; D Riddle et al. (2003) *C. elegans* II (Cold Spring Harbor Laboratory Press)).

The nematode *C. elegans* has a simple unsegmented, cylindrical body plan with a full set of differentiated tissues (neural, endoderm, ectoderm, and muscle), which is typical of many nematode species. At the anterior end of this simple elongated tube-like body, the pharynx opens, while the extreme tail tip ends in a tapering whip-like structure. Along the length of the body, the animal has a uniform diameter and typically adopts a posture with only one or two shallow body bends along the dorsal or

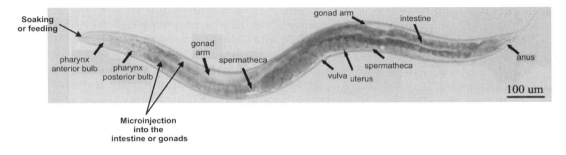

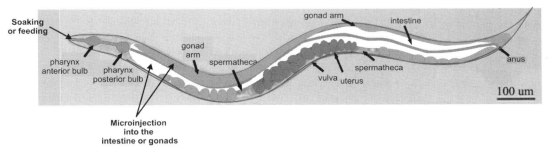

Fig. 2.9 Anatomy of the worm. The upper panel shows a Nomarski image of the *C. elegans*. The lower panel shows a scheme of a selection of organs superimposed to the image to explain the anatomy of the worm. (Image source: Y. Shibata and Takagi.)

ventral aspects. The male nematode undergoes a reorganization of the body shape during the last larval molt to develop specialized mating structures at the tail (http://www.wormatlas.org/index.htm) (Figure 2.9).

C. elegans possesses a two bulbed-pharynx that comprises almost 50% of the body length at birth, but grows more slowly than other organs as the animal ages. As the animal grows during larval development, the most substantial changes occur in the germline. In the hermaphrodite, the gonad begins to develop during the later larval stages into a two-lobed structure containing both sperm and oocytes, and this organ eventually fills a major portion of the body cavity, each lobe becoming reflexed into a U-shape, and joining the opposite lobe at the uterus and vulva.

The male *C. elegans* displays the same simple cylindrical body plan, but as the sexual organs develop the worm grows a single unreflexed gonad arm that opens in the tail at a cloacal structure that is part of the anal opening.

Another characteristic organ is the intestine. This is very interesting for RNAi scientists as the internalization of dsRNA or DNA inverted repeats follows the uptake via the intestinal tract either by microinjection, feeding, or soaking. The intestine is made of 20 large epithelial cells, which form a tube and are mostly situated as bilaterally symmetric pairs around the tubular lumen. Each of these cell pairs forms an intestinal ring. Intestinal cells contain large nuclei with large nucleoli and numerous autofluorescent granules in their cytoplasm; these must be considered when carrying out fluorescence microscopy studies. Although the intestine initially fills the entire body cavity behind the pharynx, it eventually becomes redirected to permit outgrowth of the gonad within the same cavity as the animal grows older. The intestine is not rigidly attached to the body wall, but is firmly anchored to the pharyngeal and rectal valves at either end (Hope et al. 1996; Molin et al. 1999). The intestine is not directly innervated, and has only one associated muscle at its posterior extreme.

Another interesting feature is the coelomic fluid, which is a fluidic phase in the animal. The fluid contains coelomocytes, which are free-floating spherical cells lying in the pseudocoelomic cavity of larvae and adult *C. elegans* that can endocytose many compounds, possibly for immune surveillance. There are six coelomocytes in adult hermaphrodites, often lying pairwise together, and they display prominent cytoplasmic inclusions and vacuoles (http://www.nematode.net/index.php).

2.5.2
C. elegans Strains for Silencing

A great variety of *C. elegans* strains are available at the *Caenorhabditis* Genetics Center (CGC), which is funded by the NIH National Center for Research Resources.

The center collects, maintains, and distributes stocks of *C. elegans*, maintaining a *C. elegans* Bibliography, and publishing and distributing the *Worm Breeder's Gazette*. Further, the center coordinates genetic nomenclature and maintains the *C. elegans* genetic map. Data can be obtained on the *C. elegans* web server (http://elegans. swmed.edu/), which was initiated by Avery and coworkers. To order strains from the CGC visit the website (http://biosci.umn.edu/CGC/CGChomepage.htm.) and choose the strains or genotypes of interest. Strains are sent growing (or starved) on agar plates via airmail, and delivery generally takes about two weeks (Figure 2.10). Regular RNAi experiments are carried out in the wild-type strain Bristol N2, characterized by Sidney Brenner (Brenner 1974).

Just recently, a very interesting mutant strain was characterized to be more sensitive for RNAi. The strain rrf-3(pk1426) (CGC # NL2099) has a mutation in *rrf-3*, a putative RNA-directed RNA polymerase (RdRP) (Sijen et al. 2001a; Simmer et al. 2002) (Figure 2.11). There are four RdRP-like genes in *C. elegans*. While mutations in other RdRP-like genes decrease the sensitivity of RNAi in *C. elegans*, (Sijen et al. 2001a; Smardon et al. 2000), the *rrf-3* strain shows an opposite response to dsRNA (Sijen et al. 2001a).

RNAi phenotypes in *rrf-3* animals are often stronger, and show a complete silencing of the genes compared to the wild-type strain. It even expresses neuronally RNAi phenotypes that are otherwise not detected.

CGC Strain N2

Number:
  N2
Species:
  Caenorhabditis elegans
Genotype:
  C. elegans wild type, DR subclone of CB original (Tc1 pattern I).
Descirption:
  C. elegans var Bristol. Generation time is about 3 days. Brood size is about 350. Also CGC
  reference 257. Isolated from mushroom compost near Bristol, England by L. N. Staniland.
  Cultured by W. L. Nicholas, identified to genus by Gunther Osche and species by Victor
  Nigon; subsequently cultured by C. E. Doutherty. Given to Sydney Brenner ca. 1966. Sub-
  cultured by Don Riddle in 1973. Caenorhabditis elegans wild isolate.
Mutagen:
Outcrossed:
Reference:
  CGC #31 Brenner S The genetics of Caenorhabditis elegans. Genetics 77: 71-94 1974
Made_by:
Received:
  // from Hodgkin J, Oxford University, Oxford, England

  Leon Avery (Leon@eatworms.swmed.edu)
  Last modified: Wed Sep 24 07:15:03 2003

Fig. 2.10 Datasheet for the CCG strain Bristol N2.

CGC Strain NL2099

Number:
  NL2099
Species:
  Caenorhabditis elegans
Genotype:
  rrf-3(pk1426) II.
Description:
  Homozygous rrf-3 deletion allele. Increased sensitivity to RNAi when compared to WT ani-
  mals.
Mutagen:
  UV/TMP
Outcrossed:
  2
Reference:
  CGC #4981 Sijen T; Fleenor J; Simmer F; Thijssen KL; Parrish S; Timmons L; Plasterk
  RHA; Fire A On the role of RNA amplification in dsRNA-triggered gene silencing. Cell
  107: 465-476 2001
Made_by:
  F. Simmer
Received:
  02/25/02 from Plasterk R, NKI, Amsterdam, The Netherlands

  Leon Avery (Leon@eatworms.swmed.edu)
  Last modified: Wed Sep 24 07:15:03 2003

Fig. 2.11 Datasheet for the CGC strain rff-3 (rrf-3(pk1426) (CGC # NL2099).

2.5.3
Culturing the Worms

The worms are usually cultured on NGM agar at 15 °C and propagated on *E. coli* strain OP50 by using established procedures as described by Brenner and others (Brenner 1974; Hope 1999; Riddle 2003; Wood 1988).

In order to minimize the contamination of stocks, ampicillin (100 µg/ml), tetracycline (10 µg/ml), and kanamycin (10 µg/ml) can be added to freshly thawed stocks that were recovered onto normal growth media seeded with wild-type OP50 bacteria. After recovery from freezing, animals can be moved to fresh plates, to increase the population. The animals are collected and lyzed in a solution of 10% bleach/1 N NaOH. The embryos that survived this treatment are washed with water and plated. L1–L3 larvae are then soaked in solutions of antibiotics in M9 media overnight and replated.

2.5.4
Microinjection Protocol

Usually, the injection of dsRNA will be performed in the intestinal tract or the gonads of adult hermaphrodites to obtain a better distribution of the RNAi signal. However, dsRNA can be injected into virtually any cell of the worm. *C. elegans* hermaphrodites possess two gonad arms. The *C. elegans* hermaphrodite gonad consists of two U-shaped arms. For most experiments dsRNA is injected into only one gonad arm, which has been demonstrated to be sufficient to target endogenous RNAs synthesized in both gonad arms and indeed throughout most cells and tissues of the animal. Germline transformation has been achieved by microinjection of DNA directly into oocyte nuclei (Fire, 1986), which are located near the bend of the gonad, or by microinjection of DNA into the cytoplasm of the syncytial gonad (Stinchcomb et al., 1985; Mello et al., 1991). Once injected into the oocyte nuclei, there are three forms of heritable DNA transformation: (1) extrachromosomal transformation; (2) nonhomologous integration; and (3) homologous integration, though spontaneous homologous insertions of injected DNA are extremely rare. In RNAi experiments most of the dsRNA producing transgenes, the inverted DNA repeats, were maintained as extrachromosomal arrays. Unfortunately, these can be lost at some frequency during meiotic and mitotic cell divisions (Stinchcomb et al. 1985).

The strategy of targeted gene interruption by homologous recombination with transgenic DNA has been attempted, but has never worked satisfactorily. The common way to introduce DNA into the germline of a large number of worms is microinjection into the gonads (Mello et al. 1992; Seydoux et al. 1996). It is possible that the frequency with which microinjected DNA integrates into the genome – and subsequently the number of progeny exposed to the transgenic DNA – is limited by the number of injections. Nevertheless, a few successful cases have been reported, using either plasmid DNA or oligonucleotides (Broverman et al. 1993).

Since microinjection is a more elaborate method that requires the laboratory to have microinjection equipment and the expertise of *C. elegans* injection, it is not an easy technique for the *C. elegans* novice worker. For a very detailed protocol, including the description of the required equipment, use the following sources:

Fire lab homepage (protocol section) (http://www.ciwebemb.edu). A very detailed description is also given in *Methods in Cell Biology* (Mello and Fire 1995). It contains the most detailed description of the method and its limitation. A further source is the Ambros lab homepage (protocol section) (http://cobweb.dartmouth.edu/āmbros/worms/index.html), as well as the *C. elegans* server (http://elegans.swmed.edu). The Ambros lab webpage also offers a great collection of other *C. elegans* protocols. A very short, but minimally detailed, protocol is provided below.

PROTOCOL 2

1. Prepare dsRNA by T7 in-vitro transcription.

2. Purify the dsRNA very carefully using phenol:chloroform extraction (Sambrook and Russell 2001), as impurities can spoil the results.

3. Dissolve dsRNA in injection buffer or water (Table 2.4) at a suitable concentration.

Tab. 2.4

Injection buffer (10x)	
Polyethylene glycol 8000	2%
Potassium phosphate	20 mM
Potassium citrate	3 mM

Adjust to pH 7.5 with KOH at 37 °C just prior to injection

4. Follow the instructions for needle preparation and preparation of the worms described by Mello and Fire (1995).

5. Inject dsRNA in concentrations from 100 to 1000 ng/µl into the gonad or the intestine (or other locations).

6. Let the animals recover, and then recover the progeny of the worms as described by Mello and Fire (1995).

2.5.5
Soaking Protocol

Another method of introducing dsRNA into the worms is the soaking method. There are different methods for soaking the animal: (1) L4 stage animals are soaked in several different control solutions (Stinchcomb et al. 1985) and a dilution of the dsRNA in water (Tabara et al. 1998); or (2) L4 stage animals are soaked in liposome-embedded dsRNA to enhance the uptake efficiency of for example inverted repeat plasmids.

PROTOCOL 3

A. Soaking plain dsRNA

1. Soaking of dsRNA can be performed in the following solutions:

Tab. 2.5

M9 medium (1x)		Injection buffer (10x)	
KH_2PO_4	3 g		
K_2HPO_4	6 g	Polyethylen glycol 8000	2%
NaCl	5 g	Potassium phosphate	20 mM
$MgSO_4$ (1M)	1 ml	Potassium citrate	3 mM
Add Water	to 1000 ml		

Filter the solution through a 0.2 μm bottle top filter to sterilize the solution. Do not autoclave.   Adjust the pH to pH 7.5 with KOH at 37 °C shortly befor use

2. In a microcentrifuge tube, prepare 10 μl of a 0.5–1 μg/μl dsDNA (inverted repeat DNA plasmid) solution in M9 media.

3. Add worms to the solution and incubate overnight at 15–20 °C.

4. Allow animals a 4-h recovery period in plain M9 medium before examination.

5. Incubate recovered animals at 15 °C, 20 °C, and 25 °C to test for temperature effects in silencing.

6. Dilute the worm solution to recover single animals, and analyze the phenotype of the worm by either microscopy or biochemical methods.

B. Soaking liposome-embedded dsRNA or inverted repeat DNA constructs

1. Wash L4-stage hermaphrodites in 0.2 M sucrose and 0.1x phosphate-buffered saline.

2. Transfer the worms into 10 μl of the same buffer in a siliconized tube.

3. In another siliconized tube, vigorously mix 4 μl of dsRNA or inverted repeat DNA plasmid (3.8 μg/μl) and 1 μl liposome (Lipofectamine 2000; Invitrogen).

4. Add 15 worms to the RNA-liposome mixture; this results in a total volume of 15 μl and a final RNA concentration of 1 mg/ml.

5. After 24 h, transfer the worms to an agar plate with *E. coli* OP50, and culture until mid-adulthood on NG plates.

2.5.6
RNAi Feeding Protocol

Beside the application of exogenous dsRNA by microinjection and soaking with plain dsRNA, RNAi can be induced by feeding bacteria to produce dsRNA. The bacterial vectors can be designed to either directly express dsRNA or long hairpin RNA (lhRNA) from inverted DNA repeats, which contain the fragments of the target cDNA in consecutive sense and antisense orientation. During transcription of the inverted repeat sequence an RNA molecule is formed that is supposed to fold back into a hairpin-like structure by intramolecular hybridization (Figure 2.12 B and C). The resulting RNA is then effectively double-stranded. It has been shown that those lhRNAs are finally processed by Dicer to siRNAs that can target endogenous mRNA for cleavage (Diallo et al. 2003). The method of expressing dsRNA from a DNA or plasmid construct can encompass two variants.

1. The DNA sequence of interest is cloned into a cloning site of a bacterial expression vector that is flanked by opposing bacteriophage T7 promoters (Timmons and Fire 1998). The recombinant plasmid is transformed into a bacterial strain that is expressing a T7 polymerase under the control of an inducible (lac) promoter (BL21(DE3) (Novagen) (Timmons and Fire 1998) or the RNase III-deficient strain HT115(DE3) (Fraser et al. 2000) (Figure 2.12 A).

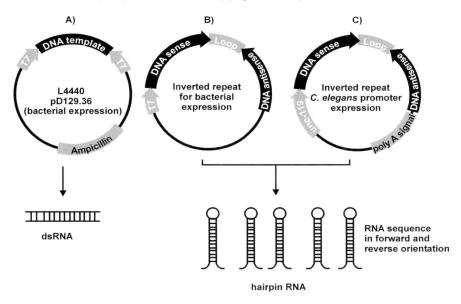

Fig. 2.12 Different methods to generate dsRNA-expressing DNA constructs for delivery to *C. elegans* by feeding. (A) Construct for bacterial T7-driven dsRNA expression in feeding bacteria. (B) Construct for bacterial T7-driven hairpin RNA expression in feeding bacteria. (C) Construct for hairpin RNA expression in *C. elegans* tissue using *C. elegans*-specific promoters.

2. A single T7 promoter can be used to express an inverted repeat of the DNA of interest. RNA transcribed from this sequence has the capacity to fold back into a hairpin double-stranded RNA (Figure 2.12 B).

3. Instead of T7-driven bacterial expression, one can fuse the inverted repeat or hairpin sequence to a *C. elegans*-specific promoter (Figure 2.12 C). This same inverted repeat sequence has been used in the feeding protocols, and is functional in eliciting RNAi (Timmons et al. 2001). Those inverted repeats are often separated by up to several hundred base pairs of nonrelated spacer sequence.

In order to enhance the cytosolic accumulation of the dsRNA, one can modify the inverted repeat by replacing the spacer sequence with intronic sequences. They are proposed to splice out when expressed in the worm, resulting in a perfect hairpin RNA with no spacer (Smith 2000). Likewise, one can introduce additional introns in the first repeat sequence.

One very important aspect for RNAi experiments in *C. elegans* is the developmental stage of the worm. The favorable stage will depend on the type of gene that must be silenced. In general, L1s are used for simplicity and to maximize exposure to dsRNA. Temperature can often have dramatic effects on phenotypes, so it is useful to test the exposure at different temperatures. In this respect, temperatures from 15 to 25 °C are recommended. RNAi produced by this protocol is inherited; for example, if L1s are treated until adulthood and removed to *E. coli* OP50, the progeny will inherit the phenotype.

2.5.7
DNA Templates for dsRNA Expression in Feeding *E. coli*

Expression of the dsRNA-producing DNA template as well as long inverted repeat DNAs is controlled by the very strong bacteriophage T7 promoter. The bacterial T7 RNA polymerase specifically recognizes this promoter. For expression of both kind of templates, it is necessary to deliver T7 RNA polymerase to the cells by either inducing expression of the polymerase or infecting the cells with a phage expressing the polymerase. The plasmid pPD129.36 (L4440) originally used and distributed by the Fire lab (ftp://www.ciwemb.edu/pub/FireLabInfo/FireLabVectors/) carries two opposing promoters for T7 RNA polymerase (Timmons and Fire 1998). It is designed for insertion of coding regions for genes to be targeted for interference.

In order to express dsRNA from the inserted DNA, the sequence is transferred to a bacterial strain expressing T7-RNA polymerase from an isopropyl-β-D-thio-galactoside (IPTG) inducible promoter.

Timmons and Fire used the *E. coli* strain BL21(DE3) in their initial development of bacteria-induced RNAi. However, the specific interference was limited, which was due to partially degraded dsRNA. To date, most *C. elegans* laboratories use *E. coli* HT115(DE3) (CGC strain #4597) as a suitable strain for the expression of T7-regulated inverted repeats. This strain carries the DE3 bacteriophage lambda lysogen. This lambda lysogen contains the *lacI* gene, the T7 RNA polymerase gene under control of the *lacUV5* promoter, and a small portion of the *lacZ* gene. This *lac* con-

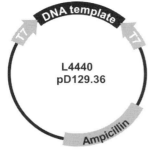

Multiple cloning site

```
                        ┌ Ncol
        ┌ BamHI         │        ┌ Nhel                    ┌ BamHI
        ↓               ↓        ↓                          ↓
5'.....GGATCCACCGGTTCCATGGCTAGCCACGTGACGCGTGGATCC.....3'
3'.....CCTAGGTGGCCAAGGTACCGATCGGTGCACTGCGCACCTAGG.....5'
```

Fig. 2.13 First vector described for feeding dsRNA to *C. elegans* from the Fire vector kit, 1999.

struct is inserted into the *int* gene, thus inactivating it. Disruption of the *int* gene prevents excision of the phage (i.e., lysis) in the absence of helper phage. The *lac* repressor represses expression of T7 RNA polymerase. Addition of the inducer IPTG allows expression of T7 RNA polymerase. Once sufficient T7 RNA polymerase is produced, it binds to the T7 promoter and transcribes the gene of interest. The disadvantage of this system is the leaky repression of the T7 polymerase. There is always some basal level expression of T7 RNA polymerase.

Beside the inducible T7 expression, HT115(DE3) has another key feature. It is RNase-III-deficient, which allows dsRNA expression without its degradation by the bacterial strain shown for other DE3 strains such as BL21(DE3). HT115(DE3) is able to produce high levels of specific dsRNA and trigger strong and gene-specific interference responses when fed to *C. elegans*. The *E. coli* HT115(DE3) strain grows on LB or 2XYT plates, and is tetracycline-resistant. This strain should be analyzed for expression by transforming in one of the plasmids from the Fire Vector Kit (1999) (Table 2.6) (e.g., pLT76) using standard CaCl$_2$. The strain has the following genotype: F-, mcrA, mcrB, IN(rrnD-rrnE)1, rnc14::Tn10(DE3 lysogen: lavUV5 promoter-T7 polymerase) (IPTG-inducible T7 polymerase) (RNase III minus). Prior to an actual feeding experiment, it can be grown in liquid in the presence of ampicillin alone, and then seeded onto NG medium plates containing 100 µg/ml ampicillin and 1 mM IPTG. This technique does not work well if the cells are old; therefore, the strain should be seeded onto IPTG-containing plates from a fresh overnight culture that was grown from a colony on an Amp/Tet plate. In order to optimize the production of dsRNA, the effects of varying the concentration of IPTG (0–1 mM), induction temperature (room temperature to 37 °C), induction time (2 h to overnight) and induction medium (LB versus Terrific) have been previously tested (Tenllado et al. 2003). Beside some minor variations, there was no significant difference in the production dsRNA when varying the parameters under test. For subsequent production of dsRNA, IPTG was added from 100 µM to 1 mM, and the culture (LB/TB medium) was incubated at 37 °C.

Tab. 2.6   Fire Lab Vector Kit 1999

Plasmid	Lig	Insert	Notes
pPD129.36	L4440	none/	Insert coding region of interest between T7 polymerase sites
pLT61.1	LT61	unc-22	Affected worms twitch strongly
pLT63.1	LT63	fem-1	Affected worms are female
pPD128	L4417	gfp	Interferes with gfp reporter constructs

2.5.8
DNA Templates for Hairpin RNA Expression

Although the cloning efficacy for long inverted repeats is very low, and the selection of cells is very time-consuming, this technique is especially valuable for long-term studies. These demand down-regulation rates of gene expression for longer time periods than are achievable by expression or exogenous application of dsRNA. For this purpose, the hairpin RNA is expressed under the control of *C. elegans*-specific promoters. Likewise, it can also be expressed under the control of the T7 promoter, as described for the Fire vector L4440.

2.5.8.1   *C. elegans* Promoters
The expression of hairpin or inverted repeat constructs relies on the correct choice of the promoter. Many promoters allow an ubiquitous expression of an inverted repeat such as the *let-858* promoter (Kelly et al. 1997) or the *ribosomal protein L28* promoter (Consortium 1998). In addition, there are many strong or weak tissue-specific promoters available to induce RNAi in a single cell type. The promoters presented here represent only a small selection of the tissue-specific promoters in *C. elegans*. Some promoters usually control expression of the *myo-3* (body wall muscle in late embryos and larvae) (Fire et al. 1998), *vit-2* (vitellogenin gene from adult hermaphrodite intestine) (MacMorris et al. 1994), and *unc-119* (neuron) genes (Maduro and Pilgrim 1995), which are considered strong transcriptional activators based on the fluorescence intensity of GFP that accumulates when under their regulation. The *myo-2* (pharyngeal muscle) and *snb-1* (nervous system) promoters are exceptionally strong. *snb-1* promoters have also been noted to elicit RNAi outside the nervous system.

Other promoters are specific for the developmental stage in the worm. There are promoters of somatic genes in larvae/adults such as the *unc-22* promoter, or maternal genes such as the *mex-3* and *mex-6* promoters. A variety of promoters have been identified in promoter trap screen, which are specific for the differentiation of other cell types during animal development (Hope 1991, 1994; Hope et al. 1996, 1998; Molin et al. 1999; Young and Hope 1993). There are many more tissue-specific promoters in *C. elegans* obtainable from the *C. elegans* webserver referred to above. The silencing of genes in defined tissues may depend on the choice of promoter used to drive the dsRNA expression.

Tab. 2.7 Selected *C. elegans* promoters.

Promoters	Tissue	Remarks
let-858	whole animal	– moderate to strong
myo-2	pharyngeal muscles	– exeptionally strong
		– constructs with this promoter can be toxic when present at high copy number
myo-3	body muscles	
	vulval ass. muscles	
	intestine ass. muscles	
rps-5, rpt-28	multiple tissues	– moderate to strong
hsp 16-41, hsp 16-2	heat/stress inducible	
snb-1	nervous system	– also outside the nervous system
		– exceptionally strong
unc-119	neurons	– strong
vit-2	intestine	– strong

2.5.8.2 **Inverted Repeat Constructs**

Different inverted repeat constructs are used for feeding protocols in *C. elegans*. Some contain only the plain exon-based gene fragment in a consecutive sense and antisense orientation, while others comprise a non-gene-related loop structure to separate the inverted repeat. Those loop sequences should not be homologous to any worm sequence, in order to prevent interference or an unspecific RNAi response. The most popular loop structures are intron-based, followed by antibiotic resistance genes such as the kanamycin resistance gene, or the bacterial tetracycline resistance gene (Peden 1983), and other non-related genes such as bacterial genes or GFP.

The use of antibiotic resistance genes is limited to plasmid vectors that do not contain the same resistance gene, since recombination events will prevent its amplifica-

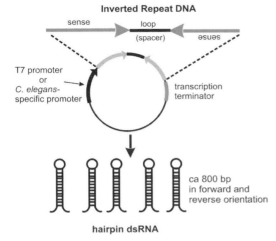

Fig. 2.14 Scheme of the hairpin RNA expression from an inverted repeat DNA construct.

tion. Transferring the inverted repeat insert into pBluescriptII KS+ will often help to enhance transformation and amplification efficiency in the feeding *E. coli* strain HT115(DE3) (Kamath et al. 2003). The following protocol is partially modified from the protocol published by Craig Hunter's group (http://mcb.harvard.edu/hunter/). The cloning of the inverted repeat is based on the PCR amplification of the respective cDNA sequence and ligation of the sense-loop-antisense sequence prior to its insertion into the expression vector. Since the protocol comprises several agarose gel purification steps, one should be sure to start with a reasonable amount of PCR product. Regular PCRs yield 1–4 μg of product, but this can be exceeded by using some improved Taq-polymerases (Long Expand Taq-polymerase, Roche; Takara), which additionally attach a single-stranded A overhang to the end of PCR product. This allows an insertion into T/A cloning vectors such as the TOPO™ cloning system from Invitrogen. This cloning system can definitely be recommended, but it is expensive. There is an alternative means of cloning the inverted repeat DNA, which is independent of the TOPO™ system. Instead of using a regular Taq polymerase, one can take a proofreading Taq polymerase such as Pfu-Taq (NEB) or Pwo-Taq (Roche) to generate blunt ends. The PCR product from this reaction can be processed as described in the following method. Subsequent phosphorylation of the final inverted repeat insert is necessary before ligation into a dephosphorylated blunt end restriction site of pBluescriptII KS+ (Stratagene) (see protocol B). Another method is to use restriction sites at both ends of the inverted repeat sequence to perform a directed cloning of the inverted repeat in the vector of choice (see Chapter 4, Protocol 34).

PROTOCOL 4

1. Generate a DNA fragment for the inverted repeat by PCR. The fragment size should be between 800 and 1500 bp. Add an *AvrII* site to the reverse primer.

   ▶ **Note**: To discriminate between the ligation products, one should choose an appropriate length of the DNA fragment. The DNA fragment should not be close to twice the length of the loop.

2. Amplify a 500-bp fragment for the loop by PCR with primers both containing *NheI* sites.

   ▶ **Note**: Make sure that the loop sequence is not homologous to something in the worm that will interfere with your experiments, because the loop sequence can actually produce an RNAi response.

3. Purify the PCR fragments on an 1–1.5% agarose gel to remove enzymes and residual primers. Extract the PCR product (for example using the QIAquick reagent from Qiagen).

4. Digest the DNA fragment with *AvrII* (1 u/μg DNA) and loop fragment with *NheI* (1 u/μg DNA) overnight.

5. Purify the PCR fragments on a 1–1.5% agarose gel to remove the restriction enzymes. Extract the PCR product.

6. Ligation: The ligation protocol depends on the supplier of the T4-ligase. The following protocol is based on the highly concentrated T4-ligase of NEB (A) or on the Rapid Ligation Kit protocol (Roche) (B). Both methods allow ligation at room temperature.

7. Pipette the following components in a microcentrifuge tube as depicted in Table 2.8. The sense-loop-antisense DNA ratio is 1:1:1 molecular weight ratio.

Tab. 2.8

(A) Method A (modified from Hunter lab)		Method B (Rapid Ligation Kit™, Roche)	
Gene fragment *Avril* (0.625 µg)	x µl	Gene fragment *Avril* (0.625 µg)	x µl
Loop fragment *Nhel* (0.13 µg)	x µl	Loop fragment *Nhel* (0.13 µg)	x µl
NEB 2 buffer (10x)	5 µl	Dilution buffer (5x) or NEB 2 buffer (10x)	3 µl
BSA (10 mg/ml)	5 µl	*AvrII* (10 u/µl)	1.5 µl
AvrII (10 u/µl)	1.5 µl	*NheI* (10 u/µl)	1.5 µl
NheI (10 u/µl)	1.5 µl	Ligation buffer (2x)	15 µl
ATP (10 mM)	5 µl	T4 ligase (8 u/µl)	1 µl
T4 ligase (8.8–20 u/µl) (NEB)	2 µl	Water	x µl
Water	x µl		
Total	50 µl	Total	20 µl

8. Incubate the ligation reactions, as shown in Table 2.9. Here, a PCR thermocycler can be used.

Tab. 2.9

Incubation cycles		
Step 1	20 °C	20 min
Step 2	37 °C	10 min
Step 3	repeat 1 and 2 for 10 times	

9. Separate the ligation products on a 1–1.5% agarose gel and extract the sense-loop-antisense fragment using the QIAquick gel extraction procedure (Qiagen).

▶ **Note**: If the product band is too weak repeat the reaction, collect several gel slices, and extract at once.

10. Elute in 20 µl 10 mM Tris-HCl, pH 8.5.

▶ **Note:** Trying to PCR amplify a gel-purified ligation product does usually not work.

11. Ligate the fragment into a vector, which can be used for T/A cloning like the pCR3.1-XL TOPO cloning vector (TOPO™ cloning system, Invitrogen) according to the TOPO™ system instructions (see Table 2.10).

Tab. 2.10

Ligation product	2.5 μl
pCR3.1-XL TOPO™	
or other TOPO™ cloning vectors	1 μl
Salt solution from the TOPO™ Kit	1 μl
Water	1.5 μl
Total volume	6 μl

12. Incubate for 30 min at room temperature.

13. Transform the 2 μl of ligated vector into *E. coli* TOP10 distributed with the TOPO™ cloning system or recA- cells such as SURE2 (Stratagene), as described in the TOPO™ protocol or in Sambrook and Russell (2001).

▶ **Note**: Using recombinase-deficient (recA-) *E. coli* strains such as SURE2 (Stratagene) reduces the recombination events, and therefore the disruption of the recombinant clones. It significantly reduces the amount of colonies to screen.

▶ **Note**: Some TOPO cloning vectors contain a kanamycin resistance gene cassette. They cannot be transformed into SURE2 cells, as the strain is kanamycin-resistant.

14. Plate the bacteria on selection LB/agar plates supplemented with the appropriate antibiotic.

▶ **Note**: Due to the high rate of recombination events, the number of growing colonies is very low. Besides, the percentage of positive clones that are bearing the inverted repeat is often below 1%. To increase the amount of positive recombinant colonies it is recommended to set up several ligation and transformation reactions for the same cloning procedure simultaneously. Likewise, the amount of DNA for the ligation must be increased for the protocol to work well (>500 ng for 750 bp gene fragment, and >250 ng for the 784 bp loop).

15. Pick the colonies and grow the bacteria in 1–3 ml LB medium + antibiotics.

▶ **Note**: For high-throughput DNA isolation, use either a pipetting robot such as the BioRobot system (Qiagen) or the 96-well turbo DNA isolation method described in Chapter 4.

16. Isolate the DNA (Qiagen Spinprep Kit, or see Chapter 4).

17. Digest with *EcoRI* to verify insert.

18. Separate the digestion products on a 1–1.5% TAE agarose gel, and extract the insert from the gel using a QiaQuick gel extraction Kit (Qiagen).

19. Sub-clone into the *EcoRI* site of pBluescript II KS+.

20. Transform into HT115 (DE3) (available from CGC) and select with 50 µg/ml ampicillin or kanamycin and 12.5 µg/ml tetracycline.

21. From glycerol stock or small culture of your plasmid in HT115(DE3) or BL21(DE3), inoculate 1 l culture of LB (Luria Bertani) media or TB (Terrific Broth) with 50 µg/ml ampicillin (or the more stable alternative, carbenicillin) and 12.5 µg/ml tetracycline. Incubate for 24 h at 37 °C, shaking until very dense.

▶ **Note**: Since the HT115(DE3) is in some way leaky for T7 polymerase expression, expression can be performed either in presence of 10 µM to 1 mM IPTG, or in absence of IPTG.

▶ **Note**: TB cultures can be grown much more densely than LB cultures.

22. Pellet culture at 4000 r.p.m. for 10 min.

23. Resuspend in 4 ml/g M9 medium + 15% sterile glycerol.

Tab. 2.11

M9 medium (10x)	
KH_2PO_4	30 g
K_2HPO_4	60 g
NaCl	50 g
Add Water	to 990 ml
$MgSO_4$*	1 ml

* Add sterile $MgSO_4$ after autoclaving the phosphate solution or filter sterilize the entire solution

24. Aliquot and freeze at –80 °C for future use.

▶ **Note**: Bacteria can be stored at 4 °C for at least 3 months.

The bacteria are now ready to be used for feeding the worms. Follow the protocol below and treat the animals. Worms fed *E. coli* engineered to express dsRNA from a worm gene can exhibit specific interference with the activity of the targeted gene. Both soaking and feeding appear to work with similar efficiency, but in all cases the effects are less potent than those obtained by direct microinjection. RNAi of neuronal genes is unlikely to be better than injection, which is usually poor. Only moderate RNAi has been observed of unc-119::GFP.

PROTOCOL 5

1. Prepare four wells of a 12-well plate containing NGM agar + 1 mM IPTG.

Tab. 2-12

NGM agar (cuture medium)			
Agar	17 g	After autoclaving add a sterile solution of:	
Peptone	2.5 g	$CaCl_2$ (1 M)	1 ml
NaCl	3 g	$MgSO_4$ (1 M)	1 ml
Cholesterol (5 mg/ml)	1 ml	KH_2PO_4/K_2HPO_4 (pH 6.0) (1M)	25 ml
Add Water	to 960–980 ml		

Optionally add IPTG to a final concentration of 100 μM to 1 mM

▶ **Note**: The plates can be prepared with or without IPTG (see protocol above). Do not use antibiotics for the plates.

2. Spot 100 μl bacteria onto 60 mm ∅ NG plates.

3. To spread the bacteria evenly, add a few glass beads (3–3.5 mm ∅) (Roth, Germany Cat# A557.1) to the plate and shake the plate in circles (it should dry very quickly). This will produce a thick lawn of bacteria.

4. Remove the glass beads by turning the plate upside down.

5. Incubate the plate for 24 h at room temperature to grow the bacteria and to induce dsRNA production.

6. Plates can be stored at 4 °C for some weeks.

7. Place 10–15 worms at stage L3–L4 in the first of the four wells for each gene, and incubate the worms for 72 h at 15 °C.

8. Remove three worms (these are now young adults) and place them individually on the three remaining wells for each gene.

9. Allow to lay embryos for 24 h at room temperature. The three worms are then removed ($t = 0$).

2.6
Mounting Animals for Microscopy

The protocol for mounting *C. elegans* onto slides for microscopy is a modified version of that published by Monica Driscoll (http://mbclserver.rutgers.edu/driscoll/). The protocol is suitable for upright microscopy. It cannot be used for inverse microscopy unless the coverslip is fixed to the slide with nail polish. Then, the animals cannot be recovered.

PROTOCOL 6

1. Before preparing the slides, make the following 5% agar solutions in water.

Tab. 2.13

A) For application of worms with a stick		B) For application of worms by pipetting	
Agar	5%	Agar	5%
Water		Sodium azide (NaN_3)	10–25 mM

▶ **Note:** Keep the solution on a heating block or in a waterbath at 65 °C.

2. Prepare two glass slides with labeling tape (for example, Fisher #11-880-5-D) taped over both ends to serve as spacers (0.4 mm).

3. Clean glass slides with water and ethanol.

▶ **Note:** Avoid lint and dirt on the slides.

4. (Method A) Prepare the agar pads on the slides as depicted in Figure 2.14.

5. Cover two slides with tape to generate a spacer.

6. Place a third clean slide between, and place a drop of warm agar onto the clean slide.

7. Cover the agar with another clean slide placed in a cross shape on top of the three slides. Press gently so the agar drop is flattened to a circle about until it reaches the thickness of the tape spacers (0.4 mm).

▶ **Note:** Avoid getting bubbles in the agar as the worms will be trapped in them.

8. When the agar is solid, gently pull out the taped slides.

9. Gently separate the remaining two slides so that the agar pad adheres to one of them.

▶ **Note:** Do not prepare the agar pad too soon in advance as it will dry out. Keeping the two clean slides in saran wrap without the spacer slides will prevent the drying for at least some hours.

10. Before mounting the live worms, place 1–2 µl of M9 containing 10–25 mM sodium azide (NaN_3) onto the center of the agar pad. NaN_3 will anesthetize the worms to prevent too much movement on the pad.

11. Transfer animals into the drop using a worm pick or an eyebrow hair fastened to a toothpick with wood glue or clear nail polish.

▶ **Note:** Getting animals stick to the eyebrow is a skill acquired with practice.

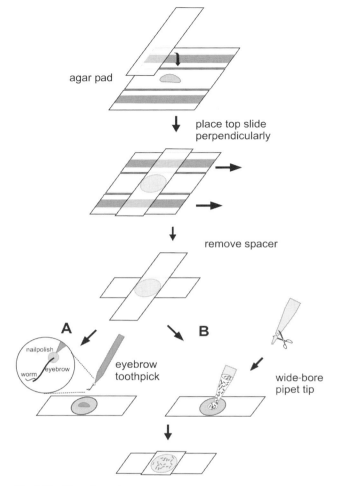

agar pad

place top slide
perpendicularly

remove spacer

A

nailpolish
worm eyebrow

eyebrow
toothpick

B

wide-bore
pipet tip

Fig. 2.15 Mounting scheme.

12. Move the hair or pick into the NaN_3 drop to let the animals float off.

13. (Method B) Prepare the agar pad with anesthetic. Include 10 mM NaN_3 in the agar instead of having it in the solution.

14. Spin down the worms in M9 medium in an Eppendorf tube at 1200–1500 r.p.m. at room temperature.

▶ **Note**: *C. elegans* is a very robust organism. It has been shown to survive even ultracentrifugation at 100 000 g, but rapid accelerations or decelerations must be avoided.

15. Discard the supernatant and resuspend the worms in 10 µl M9/10 mM NaN_3.

16. Transfer 2–3 µl of the solution to the agar pad with an Eppendorf pipet.

▶ **Note**: For easy transfer, use a cut pipet tip with a larger bore.

17. Gently cover the agar pads with coverslips. Most animals will lie on their sides.

▶ **Note**: Do not touch the coverslips as this will crush the worms.

18. Optional for inverse microscopy: Fix the coverslip to the slide by using nail polish around the rim.

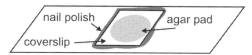

Fig. 2.16   Coverslip fixing for inverse microscopy.

▶ **Note**: The best nail polish is *Wet N' Wild*, as it will dry quickly and is less viscous than many others.

2.7
Genome Wide Screens

With the help of RNAi, major efforts have been undertaken to assign biological function to the known genes in high-throughput screens (Maeda et al. 2001). Genome-wide RNAi screens have recently become feasible with the generation of a library of bacterial strains that each produce dsRNA for an individual nematode gene. The current library contains 16 757 bacterial strains that will target ~86% of the 19 427 currently predicted genes of the *C. elegans* genome, and the loss-of-function phenotype when performing systemic RNAi on a genome-wide scale is estimated to be ~65% (Fraser et al. 2000; Kamath and Ahringer 2003). By this means, new insights may be obtained on genes involved in mitochondrial functions, lipid metabolism, and the structure of the genome itself. RNAi-based knock-out libraries have been established to simplify these studies even further. For reviews, see Castillo-Davis and Hartl (2003), Ashrafi et al. (2003), Gonczy et al. (2000), and Lee et al. (2003). Animals fed these bacteria induce RNAi in almost all tissues of the nematode (Kamath and Ahringer 2003; Kamath et al. 2003; Timmons and Fire 1998).

2.7.1
C. elegans RNAi Library

This *Caenorhabditis elegans* RNAi feeding library provided by MRC geneservice was constructed by Julie Ahringer's group at The Wellcome CRC Institute, University of Cambridge, Cambridge, England.

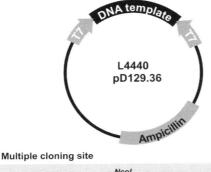

Multiple cloning site

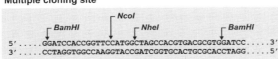

Fig. 2.17 First vector described for feeding dsRNA to *C. elegans* from the Fire vector kit 1999. This was used to generate the genome-wide RNAi feeding library for *C. elegans*.

C. elegans genomic fragments were PCR-amplified using Research Genetics Gene-Pairs, now distributed by Invitrogen (http://mp.invitrogen.com/) cloned into the *EcoRV* site of vector L4440 from Timmons and Fire (Timmons and Fire 1998) (Figure 2.17) and transformed into bacterial strain HT115(DE3) (Timmons et al. 2001) as described elsewhere (Fraser et al. 2000). The whole genome library consists of 16 757 bacterial strains, which cover 87% of *C. elegans* genes. The libraries are available by individual chromosome sets (I, II, III, IV,V and X), glycerol stocks of bacterial strains arrayed in 384-well plates, or as individual bacterial strains (clones). To order a single strain or whole libraries, go to the following website: http://www.hgmp.mrc.ac.uk/geneservice/reagents/products/rnai/index.shtm and fill out the respective website questionnaire (Figure 2.18).

Bacterial strains carry the GenePairs name, which usually correspond to a predicted *C. elegans* gene name. A current mapping of GenePair to gene can be found in Worm-Base (http://www.wormbase.org). GenePairs primer sequences are available at http://cmgm.stanford.edu/~kimlab/primers.12–22-99.html. There is also a "*C. elegans* Finder" tool available (http://www.hgmp.mrc.ac.uk/geneservice/reagents/tools/Celegans_Finder.shtml), which allows you to find the primer sequences for any given GenePair name in addition to the location of that bacterial strain in the HGMP 384-well plates. The complete *C. elegans* RNAi database can be downloaded. In this way, RNAi experiments can be carried out without the hassle of DNA cloning.

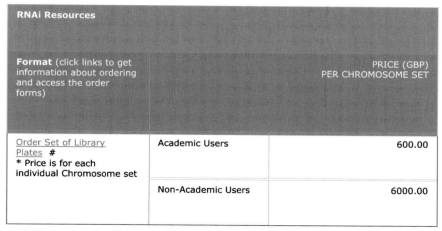

RNAi Resources		
Format (click links to get information about ordering and access the order forms)		PRICE (GBP) PER CHROMOSOME SET
Order Set of Library Plates **#** * Price is for each individual Chromosome set	Academic Users	600.00
	Non-Academic Users	6000.00

Fig. 2.18 Strain order sheet on the MRC geneservice website.

2.8
Selected Literature on *C. elegans* Research

BRENNER S **(1974)** The genetics of *Caenorhabditis elegans*. *Genetics* 77, 71–94.

MELLO CC, FIRE A **(1975)** DNA transformation. in *Methods in Cell Biology*. Caenorhabditis elegans: *Modern Biological Analysis of an Organism* (eds. Epstein HF, Shakes DC) pp. 451–482. Academic Press, San Diego.

HOPE IA **(1999)** *C. elegans*: A practical approach, Oxford University Press.

WOOD WB **(1988)** *The Nematode* Caenorhabditis elegans. Cold Spring Harbor Laboratory Press.

RIDDLE D et al. **(2003)** *C. elegans* II. Cold Spring Harbor Laboratory Press.

HOPE IA et al. **(1996)** The *C. elegans* expression pattern database: a beginning. *Trends Genet* 00, XX–XX.

2.9
Useful *C. elegans* Webpages

Nematode Net (official webpage of the nematode sequencing project)
ttp://www.nematode.net/index.php

Wormbase (very useful, contains many useful informations)
http://www.wormbase.org/

Other databases working with wormbase: RNAi Database
http://nematoda.bio.nyu.edu/

C. elegans Genetic Center (provides strains and mutants)
(http://biosci.umn.edu/CGC/CGChomepage.htm.)

C. elegans server (most comprehensive site with C. elegans links)
http://elegans.swmed.edu/

Worm Atlas (Morphology of C. elegans)
http://www.wormatlas.org/index.htm

Max Planck Institute Dresden (for genome wide screens on cell division)
http://worm-srv1.mpi-cbg.de/dbScreen/index.html

MRC Gene Service: Bacterial RNAi hairpin producing strains (genome wide libraries)
http://www.hgmp.mrc.ac.uk/geneservice/reagents/products/rnai/index.shtml

C. elegans protocols
http://cobweb.dartmouth.edu/~ambros/worms/index.html

2.10
References

ASHRAFI K, CHANG FY, WATTS JL, FRASER AG, KAMATH RS, AHRINGER J, RUVKUN G (**2003**) Genome-wide RNAi analysis of *Caenorhabditis elegans* fat regulatory genes. *Nature* 421: 268–272.

BETZ N (**2003**) RNAi in *Drosophila* S2 cells: effect of dsRNA size, concentration and exposure time. ENotes at http://www.promega.com/enotes/applications/ap0050_tabs.htm

BRENNER S (**1974**) The genetics of *Caenorhabditis elegans*. *Genetics* 77: 71–94.

BROVERMAN S, MACMORRIS M, BLUMENTHAL T (**1993**) Alteration of *Caenorhabditis elegans* gene expression by targeted transformation. *Proc Natl Acad Sci USA* 90: 4359–4363.

CASTILLO-DAVIS CI, HARTL DL (**2003**) Conservation, relocation and duplication in genome evolution. *Trends Genet* 19: 593–597.

CONSORTIUM CE (**1998**) Genome sequence of the nematode *C. elegans*: a platform for investigating biology. The *C. elegans* Sequencing Consortium. *Science* 282: 2012–2018.

DERNBURG AF, ZALEVSKY J, COLAIACOVO MP, VILLENEUVE AM (**2000**) Transgene-mediated cosuppression in the *C. elegans* germ line. *Genes Dev* 14: 1578–1583.

DIALLO M, ARENZ C, SCHMITZ K, SANDHOFF K, SCHEPERS U (**2003**) RNA interference: analyzing the function of glycoproteins and glycosylating proteins in mammalian cells. *Methods Enzymol* 363: 173–190.

FAGARD M, VAUCHERET H (**2000**) Systemic silencing signal(s). *Plant Mol Biol* 43: 285–293.

FEINBERG EH, HUNTER CP (**2003**) Transport of dsRNA into cells by the transmembrane protein SID-1. *Science* 301: 1545–1547.

FIRE A, XU SQ, MONTGOMERY MK, KOSTAS SA, DRIVER SE, MELLO CC (**1998**) Potent and specific genetic interference by double-stranded RNA in *Caenorhabditis elegans*. *Nature* 391: 806–811.

FRASER AG, KAMATH RS, ZIPPERLEN P, MARTINEZ-CAMPOS M, SOHRMANN M, AHRINGER J (**2000**) Functional genomic analysis of *C. elegans* chromosome I by systematic RNA interference. *Nature* 408: 325–330.

GONCZY P, ECHEVERRI C, OEGEMA K, COULSON A, JONES SJ, COPLEY RR, DUPERON J, OEGEMA J, BREHM M, CASSIN E, HANNAK E, KIRKHAM M, PICHLER S, FLOHRS K, GOESSEN A, LEIDEL S, ALLEAUME AM, MARTIN C, OZLU N, BORK P, HYMAN AA (**2000**) Functional genomic analysis of cell division in *C. elegans* using RNAi of genes on chromosome III. *Nature* 408: 331–336.

GRISHOK A, TABARA H, MELLO CC (**2000**) Genetic requirements for inheritance of RNAi in *C. elegans*. *Science* 287: 2494–2497.

Guo S, Kemphues KJ (**1995**) Par-1, a gene required for establishing polarity in *C. elegans* embryos, encodes a putative Ser/Thr kinase that is asymmetrically distributed. *Cell* 81: 611–620.

Hammond SM, Bernstein E, Beach D, Hannon GJ (**2000**) An RNA-directed nuclease mediates post-transcriptional gene silencing in *Drosophila* cells. *Nature* 404: 293–296.

Hope IA (**1991**) 'Promoter trapping' in *Caenorhabditis elegans*. *Development* 113: 399–408.

Hope IA (**1994**) PES-1 is expressed during early embryogenesis in *Caenorhabditis elegans* and has homology to the fork head family of transcription factors. *Development* 120: 505–514.

Hope IA (**1999**) C. elegans: *A practical approach*. Oxford University Press, Oxford.

Hope IA, Albertson DG, Martinelli SD, Lynch AS, Sonnhammer E, Durbin R (**1996**) The *C. elegans* expression pattern database: a beginning. *Trends Genet* 12: 370–371.

Hope IA, Arnold JM, McCarroll D, Jun G, Krupa AP, Herbert R (**1998**) Promoter trapping identifies real genes in *C. elegans*. *Mol Gen Genet* 260: 300–308.

Kamath RS, Ahringer J (**2003**) Genome-wide RNAi screening in *Caenorhabditis elegans*. *Methods* 30: 313–321.

Kamath RS, Fraser AG, Dong Y, Poulin G, Durbin R, Gotta M, Kanapin A, Le Bot N, Moreno S, Sohrmann M, Welchman DP, Zipperlen P, Ahringer J (**2003**) Systematic functional analysis of the *Caenorhabditis elegans* genome using RNAi. *Nature* 421: 231–237.

Kelly WG, Xu SQ, Montgomery MK, Fire A (**1997**) Distinct requirements for somatic and germline expression of a genera expressed *Caenorhabditis elegans* gene. *Genetics* 146: 227–238.

Lee SS, Lee RY, Fraser AG, Kamath RS, Ahringer J, Ruvkun G (**2003**) A systematic RNAi screen identifies a critical role for mitochondria in *C. elegans* longevity. *Nature Genet* 33: 40–48.

MacMorris M, Spieth J, Madej C, Lea K, Blumenthal T (**1994**) Analysis of the VPE sequences in the *Caenorhabditis elegans* vit-2 promoter with extrachromosomal tandem array-containing transgenic strains. *Mol Cell Biol* 14: 484–491.

Maduro M, Pilgrim D (**1995**) Identification and cloning of unc-119, a gene expressed in the *Caenorhabditis elegans* nervous system. *Genetics* 141: 977–988.

Maeda I, Kohara Y, Yamamoto M, Sugimoto A (**2001**) Large-scale analysis of gene function in *Caenorhabditis elegans* by high-throughput RNAi. *Curr Biol* 11: 171–176.

Mello C, Fire A (**1995**) DNA transformation. *Methods Cell Biol* 48: 451–482.

Mello CC, Draper BW, Krause M, Weintraub H, Priess JR (**1992**) The pie-1 and mex-1 genes and maternal control of blastomere identity in early *C. elegans* embryos. *Cell* 70: 163–176.

Milligan JF, Groebe D, Witherell GW, Uhlenbeck OC (**1987**) Oligoribonucleotide synthesis using T7 RNA polymerase and synthetic DNA templates. *Nucleic Acids Res* 15: 8783–8798.

Molin L, Schnabel H, Kaletta T, Feichtinger R, Hope IA, Schnabel R (**1999**) Complexity of developmental control: analysis of embryonic cell lineage specification in *Caenorhabditis elegans* using pes-1 as an early marker. *Genetics* 151: 131–141.

Palauqui JC, Elmayan T, Pollien JM, Vaucheret H (**1997**) Systemic acquired silencing: transgene-specific post-transcriptional silencing is transmitted by grafting from silenced stocks to non-silenced scions. *EMBO J* 16: 4738–4745.

Peden KW (**1983**) Revised sequence of the tetracycline-resistance gene of pBR322. *Gene* 22: 277–280.

Plasterk RHA, Ketting RF (**2000**) The silence of the genes. *Curr Opin Genet Dev* 10: 562–567.

Qiao L, Lissemore JL, Shu P, Smardon A, Gelber MB, Maine EM (**1995**) Enhancers of glp-1, a gene required for cell-signaling in *Caenorhabditis elegans*, define a set of genes required for germline development. *Genetics* 141: 551–569.

Riddle D (**2003**) *C. elegans* II. Cold Spring Harbor Press.

Roignant JY, Carre C, Mugat B, Szymczak D, Lepesant JA, Antoniewiski C (**2003**) Absence of transitive and systemic pathways allows cell-specific and isoform-specific RNAi in *Drosophila*. *Rn.a* 9: 299–308.

Sambrook J, Russell DW (**2001**) *Molecular Cloning: A Laboratory Manual*, 3rd edn. Cold Spring Harbor Press, New York.

SCHENBORN ET, MIERENDORF RC, JR. (1985) A novel transcription property of SP6 and T7 RNA polymerases: dependence on template structure. *Nucleic Acids Res* 13: 6223–6236.

SCHWARZ DS, HUTVAGNER G, DU T, XU Z, ARONIN N, ZAMORE PD (2003) Asymmetry in the assembly of the RNAi enzyme complex. *Cell* 115: 199–208.

SEYDOUX G, MELLO CC, PETTITT J, WOOD WB, PRIESS JR, FIRE A (1996) Repression of gene expression in the embryonic germ lineage of *C. elegans*. *Nature* 382: 713–716.

SIJEN T, FLEENOR J, SIMMER F, THIJSSEN KL, PARRISH S, TIMMONS L, PLASTERK RH, FIRE A (2001a) On the role of RNA amplification in dsRNA-triggered gene silencing. *Cell* 107: 465–476.

SIJEN T, FLEENOR J, SIMMER F, THIJSSEN KL, PARRISH S, TIMMONS L, PLASTERK RHA, FIRE A (2001b) On the role of RNA amplification in dsRNA-triggered gene silencing. *Cell* 107: 465–476.

SIMMER F, TIJSTERMAN M, PARRISH S, KOUSHIKA SP, NONET ML, FIRE A, AHRINGER J, PLASTERK RH (2002) Loss of the putative RNA-directed RNA polymerase RRF-3 makes *C. elegans* hypersensitive to RNAi. *Curr Biol* 12: 1317–1319.

SMARDON A, SPOERKE JM, STACEY SC, KLEIN ME, MACKIN N, MAINE EM (2000) EGO-1 is related to RNA-directed RNA polymerase and functions in germ-line development and RNA interference in *C. elegans*. *Curr Biol* 10: 169–178.

SMITH NA, SINGH SP, WANG MB, STOUTJEDIJK PA, GREEN AG, WATERHOUSE PM (2000) Total silencing by intron spliced hairpin RNAs. *Nature* 407: 319–320.

STEIN LD, BAO Z, BLASIAR D, BLUMENTHAL T, BRENT MR, CHEN N, CHINWALLA A, CLARKE L, CLEE C, COGHLAN A, COULSON A, D'EUSTACHIO P, FITCH DH, FULTON LA, FULTON RE, GRIFFITHS-JONES S, HARRIS TW, HILLIER LW, KAMATH R, KUWABARA PE, MARDIS ER, MARRA MA, MINER TL, MINX P, MULLIKIN JC, PLUMB RW, ROGERS J, SCHEIN JE, SOHRMANN M, SPIETH J, STAJICH JE, WEI C, WILLEY D, WILSON RK, DURBIN R, WATERSTON RH (2003) The Genome Sequence of

Caenorhabditis briggsae: A Platform for Comparative Genomics. *PLoS Biol* 1: E45.

STINCHCOMB DT, SHAW JE, CARR SH, HIRSH D (1985) Extrachromosomal DNA transformation of *Caenorhabditis elegans*. *Mol Cell Biol* 5: 3484–3496.

TABARA H, GRISHOK A, MELLO CC (1998) RNAi in *C. elegans*: soaking in the genome sequence. *Science* 282: 430–431.

TABARA H, SARKISSIAN M, KELLY WG, FLEENOR J, GRISHOK A, TIMMONS L, FIRE A, MELLO CC (1999) The rde-1 gene, RNA interference, and transposon silencing in *C. elegans*. *Cell* 99: 123–132.

TAVERNARAKIS N, WANG SL, DOROVKOV M, RYAZANOV A, DRISCOLL M (2000) Heritable and inducible genetic interference by double-stranded RNA encoded by transgenes. *Nature Genet* 24: 180–183.

TENLLADO F, MARTINEZ-GARCIA B, VARGAS M, DIAZ-RUIZ JR (2003) Crude extracts of bacterially expressed dsRNA can be used to protect plants against virus infections. *BMC Biotechnol* 3: 3.

TIMMONS L, COURT DL, FIRE A (2001) Ingestion of bacterially expressed dsRNAs can produce specific and potent genetic interference in *Caenorhabditis elegans*. *Gene* 263: 103–112.

TIMMONS L, FIRE A (1998) Specific interference by ingested dsRNA. *Nature* 395: 854.

TIMMONS L, TABARA H, MELLO CC, FIRE AZ (2003) Inducible systemic RNA silencing in *Caenorhabditis elegans*. *Mol Biol Cell* 14: 2972–2983.

WINSTON WM, MOLODOWITCH C, HUNTER CP (2002) Systemic RNAi in *C. elegans* requires the putative transmembrane protein SID-1. *Science* 295: 2456–2459.

WOOD WB (1988) *The Nematode* Caenorhabditis elegans. Cold Spring Harbor Press, Cold Spring Harbor.

YOUNG JM, HOPE IA (1993) Molecular markers of differentiation in *Caenorhabditis elegans* obtained by promoter trapping. *Dev Dyn* 196: 124–132.

ZAMORE PD, TUSCHL T, SHARP PA, BARTEL DP (2000) RNAi: Double-stranded RNA directs the ATP-dependent cleavage of mRNA at 21 to 23 nucleotide intervals. *Cell* 101: 25–33.

3
RNAi in *Drosophila*

3.1
Introduction

For almost a century, *Drosophila melanogaster* has been the most favored animal model to study developmental and cell biological processes, as it was predicted that some of the processes are conserved within a variety of organisms, including humans. Now that the sequencing of its genome has been completed, it has been shown that that more than 60 % of the genes identified in human diseases have counterparts in *Drosophila* (Rubin et al. 2000 b). The *Drosophila* genome is about 180 Mb in size, and so far has been shown to comprise 13 600 genes (published by Celera and the Berkeley *Drosophila* Genome Project (BDGP)), of which only approximately 20 % have been referred to in the literature, and only half of these have been characterized genetically (Adams et al. 2000; Celniker and Rubin 2003; Celniker et al. 2002; Hoskins et al. 2002). Recent studies from Renato Paro's group and others even show that the genome annotation is far from being complete. The validation of developmental profiling data by RT-PCR and in-situ hybridization substantially raised the number of genes that make-up the fly to approximately 17 000 (Hild et al. 2003). The genome sequence of *Drosophila melanogaster* (Adams et al. 2000; Myers et al. 2000) is available and annotated in the flybase (http://flybase.net/annot/). Thus, it is clear that a wealth of information remains to be mined from this model organism, not only on the function of those genes but also in the correlation to human gene function.

Shortly after its discovery in *C. elegans*, RNA interference (RNAi) was reported to function in *Drosophila*, when the injection of embryos with dsRNA successfully phenocopied characterized loss-of-function embryonic mutations (Kennerdell and Carthew 1998; Misquitta and Paterson 1999). The RNAi effect persists throughout development, and can be observed in the adult, although transmission of the RNAi effect to progeny has not been observed (Misquitta and Paterson 1999). To date, RNAi is most commonly applied by microinjection into embryos, larval instars, or pupae, depending on the stage at which gene function must be analyzed. Due to the lack of inheritable transmission, microinjected dsRNA only reveals transient interference with gene expression. Thus, in order to obtain heritable effects, transgenes expressing dsRNA must be introduced. Those transgenes usually comprise similar inverted repeat sequences such as those used in *C. elegans* which, when endogen-

RNA Interference in Practice: Principles, Basics, and Methods for Gene Silencing in C. elegans, Drosophila, and Mammals. Ute Schepers
Copyright © 2005 WILEY-VCH Verlag GmbH & Co. KGaA, Weinheim
ISBN: 3-527-31020-7

ously expressed, produce hairpin-loop RNA (Kennerdell and Carthew 2000). The expression of these heritable RNAi transgenes can be made conditional using binary expression systems (Figure 3.1; Wimmer 2003).

In contrast to *C. elegans*, there are a number of established *Drosophila* cell lines which facilitate the application of RNAi and offer many experimental advantages for biochemical studies. One of these is the popular Schneider S2 cell line (Schneider 1972), which was shown to be suitable for RNAi (Clemens et al. 2000). It is derived from a primary culture of late stage (20–24 h old) *Drosophila* embryos. Many features of the S2 cell line suggest that it is derived from a macrophage-like lineage. The use

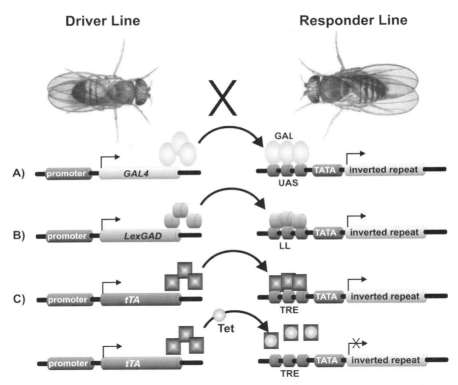

Fig. 3.1 Binary expression systems (modified from Wimmer 2003). **(A)** The most popular binary expression system is based on the yeast transactivator GAL4 and the corresponding upstream activation sequence (UAS) (Brand and Perrimon 1993). **(B)** In a second modified system, the activation domain of GAL4 has been fused to the DNA-binding domain of the bacterial LexA protein, resulting in the expression of the transactivator LexGAD. The corresponding activation sequence (LL) is suitable for the expression of distinct responders (Szuts and Bienz 2000). **(C)** The tetracycline-controlled transactivator tTA mediates gene expression by binding to the tTA-response element (TRE). Gene expression can be further controlled by the presence or absence of tetracycline in the fly food as a supplement (Bello et al. 1998). Tetracycline and tTA form a complex that prevents tTA from binding to its Tre responder, making it inactive to start transgene expression (Tet-off). Feeding tetracycline enables the fly to switch off the expression and allows a additional control beyond that by a tissue-specific promoter.

of dsRNA in S2 cells is technically simple, requiring only the production of dsRNA from a PCR product that has T7 RNA polymerase binding sites at each end (see Chapter 2). The uptake of dsRNA does not require any transfection, and is similar to the soaking method in *C. elegans*. The reason why those cells can take up dsRNA is unclear so far. The method is also quick, in that results of the protein "knock-out" experiment can be obtained within 2–3 days (Clemens et al. 2000). This contrasts sharply with the time necessary to produce selective gene "knock-outs" in mammalian cells (see Chapter 4). Likewise, the method appears to be highly reproducible.

One of the main benefits of this method is that it can be performed even in laboratories that are not specialized in *Drosophila* research. It can be used to perform a pre-analysis of a mammalian homologue before embarking into more complicated RNAi in mammalian cells.

3.2
Application of RNAi in *Drosophila*

The first RNAi experiments in *Drosophila* were carried out in S2 cells (Caplen et al. 2000; Clemens et al. 2000), but to date an increasing number of laboratories are using the whole organism to study protein function. In-vivo applications encompass injection or treatment with the Helios gene gun (Biorad) (Carthew 2003). The methods comprise either the exogenous application of linear dsRNA or the endogenous expression of dsRNA from inverted repeat DNA constructs, as already described for *C. elegans* (see Chapter 2). Since *Drosophila* does not exhibit a systemic RNA silencing mechanism as described for *C. elegans* (Timmons et al. 2003; Winston et al. 2002) and plants (Fagard and Vaucheret 2000; Palauqui et al. 1997), the induction of an RNAi phenotype in the whole fly is more difficult, and can be circumvented by using fly embryos in the early stage of their development. In contrast to other animals, the *Drosophila* embryo goes through a series of mitotic divisions without cleavage of the cytoplasm; this results in a syncytium in which many nuclei are present (Ashburner 1989) (Figure 3.2). This means that the embryo remains as a single cell during the first nine mitotic cycles of embryogenesis.

Hence, the syncytial blastoderm is an excellent target for the application of RNAi, as injected dsRNA can diffuse throughout the whole cytoplasm of later developing cells. Beside maternal genes, which can be silenced within a short period after injection, genes that are expressed during development of the fly can also be affected as the dsRNA remains in the cytoplasm during cell division for many hours. However, as shown for other organisms and cultured cells, the RNAi phenotype diminishes with time – apparently due to the dilution and degradation of the dsRNA. While the RNAi activity is more or less constant during embryogenesis, it decreases to only 20–10% in the adult fly (Carthew 2003).

As described in Chapter 2 for *C. elegans*, most of the RNAi experiments in *Drosophila* or S2 cell culture are carried out using larger dsRNA expressed from T7 promoter flanked linear DNA or derived from inverted repeat DNA constructs as long hairpin RNAs (lhRNAs).

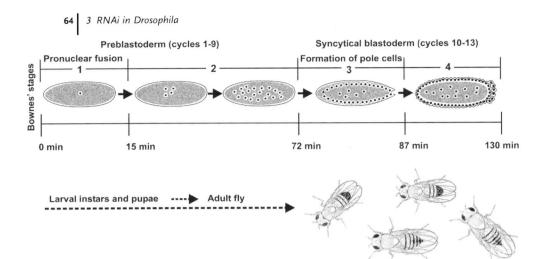

Fig. 3.2 Timeline of the first stages of *Drosophila* development after egg deposition (modified from Foe et al. 1993). Indicated are the Bownes' stages and the mitotic cycles of the embryonic cell.

3.2.1
dsRNA from Linear DNA Templates

For rapid and transient RNAi experiments the exogenous application of dsRNA is preferred, as it can be delivered directly into the cytosol of the cells or the embryo, where it immediately induces RNAi.

dsRNA synthesis can be performed by two methods. One method comprises the cloning of the chosen sequence into the multiple cloning site of a commercially available plasmid (pBluescript, Stratagene) that will be flanked with either T7, Sp6, or T3 RNA polymerase recognition sites (promoters) (Figure 3.3 A and B).

The second method, which makes the cloning step obsolete, is based on the amplification of the sequence using oligonucleotide primers with an additional T7, T3, or Sp6 recognition sequence at the 5′-end of the primer (Figure 3.3 C).

This allows the generation of several different dsRNAs in a very short time. DNA templates can be amplified from total mRNA using RT-PCR. However, the RT-PCR products are ready for direct use in dsRNA production, but cannot be reused for further reactions.

3.2.2
dsRNA from Inverted Repeat DNA

Instead of T7-driven transcription of linear DNA, one can generate dsRNA by transcription of an inverted repeat DNA. This method allows the endogenous expression of dsRNA and the production of stable RNAi phenotypes throughout the developmental stages of the fly and in cell culture. Such inverted repeats contain the desired

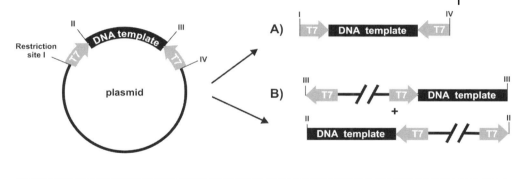

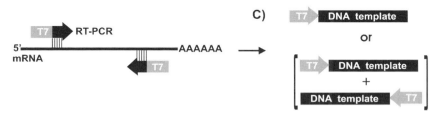

Fig. 3.3 Generation of the DNA template for dsRNA production.
(A) Restriction digest of a plasmid at the restriction sites I and IV
reveals a template flanked by two T7 promoters. (B) Restriction digest
at either site III or II reveals a template with only one T7 promoter
for the separate production of the sense and antisense strand.
(C) Template production by PCR.

sequence in a consecutive sense and antisense orientation. During transcription of
the inverted repeat sequence, an RNA molecule is formed that is supposed to fold
back into a hairpin-like structure by intramolecular hybridization (Figure 3.4).

The resulting RNA is then effectively double-stranded. It has been shown that
those long hairpin RNAs (lhRNAs) are finally processed by Dicer to siRNAs that can
target endogenous mRNA for cleavage (Zamore et al. 2000). To facilitate their transcription, those inverted repeats are often separated by up to several hundred base
pairs of nonrelated and non-palindromic spacer sequence.

Inverted repeats can be modified by replacement of the spacer sequence with intronic sequences to enhance the cytosolic accumulation of the dsRNA. They are proposed to splice out, when expressed in the fly resulting in a perfect hairpin RNA
with no spacer (Kalidas and Smith 2002; Smith 2000). Likewise, one can introduce
additional introns in the first repeat sequence.

Although the cloning efficacy for long inverted repeats is very low, and selection of
the cells is very time-consuming, this technique is especially valuable for long-term
studies in cells, which demand down-regulation rates of gene expression to longer
time periods than achievable by application of exogenous dsRNA, and further in
adult flies, in which RNAi activity from exogenous dsRNA is already diminished.

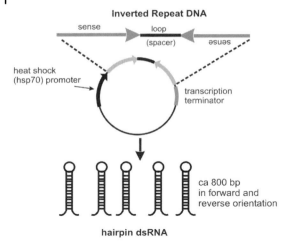

Fig. 3.4 Scheme of the hairpin RNA expression from an inverted repeat DNA construct.

For this purpose, the hairpin RNA is expressed as a transgene (Fortier and Belote 2000; Kennerdell and Carthew 2000; Lam and Thummel 2000; Lee and Carthew 2003; Martinek and Young 2000). To date, many methods have been employed to drive the expression of genetic elements in *Drosophila*, one of which is the ubiquitous and inducible expression driven by a heat shock promoter (hsp70 TATA). The gene can then be turned on at a specific point in development by heat shocking the transgenic animal (Blochlinger et al. 1991; Gonzalez-Reyes and Morata 1990; Ish-Horowicz and Pinchin 1987; Ish-Horowicz et al. 1989; Schneuwly et al. 1987; Stein-grimsson et al. 1991; Struhl 1985). Limitations of this system are basal levels of expression and induction of phenocopies (Petersen 1990; Petersen and Mitchell 1987). Beside this technique, the transcription can be driven by a tissue-specific promoter that allows the restricted expression in a defined subset of cells (Parkhurst et al. 1990; Zuker et al. 1988), but is limited to fully characterized promoters.

Nowadays, most *Drosophila* laboratories use an elaborate system for inducible and tissue-specific transgene expression in *Drosophila* that is based on the GAL4/UAS binary expression system (Brand and Perrimon 1993).

GAL4 is a transcription factor from yeast that can induce the transcription of a transgene in *Drosophila* without affecting any endogenous gene expression. To generate transgenic lines expressing GAL4 in numerous cell- and tissue-specific patterns, the GAL4 gene is usually inserted into the genome by P-element transposition (Brand et al. 1994; Brand and Perrimon 1993; Phelps and Brand 1998), thereby driving GAL4 expression from numerous different genomic enhancers. Since the transposition of such mobile elements happens more or less randomly, there are many insertions among the transformed animals showing an inducible and tissue-specific expression that is controlled by a distal promoter or enhancer. An additional mobilization of the transposable element can highly enhance the inducibility and the tissue specificity of the transformants. The inverted repeat will then be cloned into a second transformation vector behind several copies of a yeast GAL4 binding element

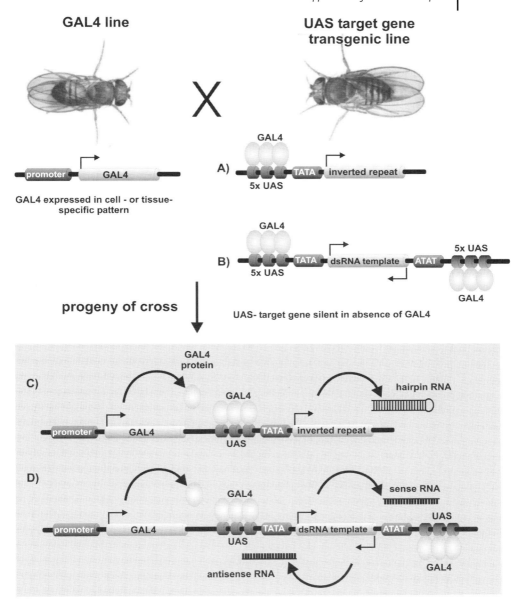

GAL4 line

UAS target gene transgenic line

X

A) GAL4 | 5x UAS | TATA | inverted repeat

GAL4 expressed in cell - or tissue-specific pattern

B) GAL4 | 5x UAS | TATA | dsRNA template | ATAT | 5x UAS | GAL4

UAS- target gene silent in absence of GAL4

progeny of cross

C) GAL4 protein | promoter | GAL4 | UAS | TATA | inverted repeat | hairpin RNA

D) promoter | GAL4 | UAS | TATA | dsRNA template | ATAT | UAS | GAL4 | sense RNA | antisense RNA

GAL4 drives expression of UAS target gene in cell- or tissue specific pattern

Fig. 3.5 Schematic view of the ectopic gene expression system using yeast GAL4/UAS recognition. *Drosophila* lines expressing the GAL4 transcription factor in specific tissues are either crossed with lines carrying an inverted repeat (A, C) (Lee and Carthew 2003) downstream of an UAS/GAL4 recognition site, or with lines that encode a dsRNA template DNA flanked by opposing UAS sites (B, D) (Giordano et al. 2002).

UAS (UAS = upstream activating enhancer sites) and a promoter (for example heat shock promoter). This transgene is also introduced into the *Drosophila* germ line by P-element-mediated transformation (Spradling and Rubin 1982) (Figure 3.5)

The target gene is silent in the absence of GAL4. To activate the target gene in a cell- or tissue-specific pattern, flies carrying the target (UAS-inverted repeat) are crossed to flies expressing GAL4. In the progeny of this cross, it is possible to activate UAS-inverted repeat in cells where GAL4 is expressed and to observe the effect of the directed expression of the inverted repeat (Brand and Perrimon 1993).

Taking advantage of the large number of existing GAL4 driver lines (Yang et al. 1995), this technique allows the tissue-specific reduction of gene function. A large variety of GAL4 driver lines are available in many established *Drosophila* laboratories, or at the Bloomington *Drosophila* Stock Center (http://flystocks.bio.indiana.edu/). The transformation vector pUAST (Fig 3.6C) is available at almost every *Drosophila* laboratory, especially from the Perrimon lab at Harvard (http://genetics.med.harvard.edu/~perrimon/), and the Brand lab (http://www.welc.cam.ac.uk/~brandlab/). More information about the cloning, application, and modification of the GAL4/UAS system can be obtained from the *Drosophila* literature and resources (Brand and Perrimon 1993; Kennerdell and Carthew 2000; Lee and Carthew 2003; Phelps and Brand 1998; Sulivan et al. 2000).

Currently, many similar systems have been developed for the inducible and tissue-specific endogenous expression of double-stranded RNA (dsRNA) molecules (Enerly et al. 2003; Giordano et al. 2002; Kalidas and Smith 2002; Nagel et al. 2002; Piccin et al. 2001; Reichhart et al. 2002; Schmid et al. 2002; van Roessel and Brand 2002; Van Roessel et al. 2002). As shown in Fig 3.6E, some have included a splicing competent intron in the UAS responder plasmid to facilitate the formation of the hairpin RNA (Kalidas and Smith 2002; Lee and Carthew 2003; Reichhart et al. 2002). Others are coupling the hairpin RNA expression with GFP expression that provides a convenient way to identify area of interference simultaneously visualized by green fluorescence (Nagel et al. 2002). In pUdsGFP, a second, UAS-GFP responder was inserted downstream of the position for the UAS-splice-activated hairpin (Duffy 2002; Nagel et al. 2002) (Figure 3.7).

3.2.3
Inducible Expression in *Drosophila* Cell Lines

Although the UAS/GAL4 system is widely used in whole flies, it is not well established in *Drosophila* cell lines. The great demand for *Drosophila* cell lines which stably express an RNAi phenotype forced the development of constitutive GAL4 drivers for *Drosophila* cell culture. The co-transfection of UAS responders with constitutive drivers, such as Act5C-GAL4 or arm-GAL4, now allows the expression of hairpin RNA in a variety of cell lines (Johnson et al. 2000; Klueg et al. 2002). Additionally, GAL4 expression controlled by a metallothionein promoter, can be induced by copper (Johnson et al. 2000). Unlike in whole flies, the RNAi phenotype can be analyzed within days after transfection.

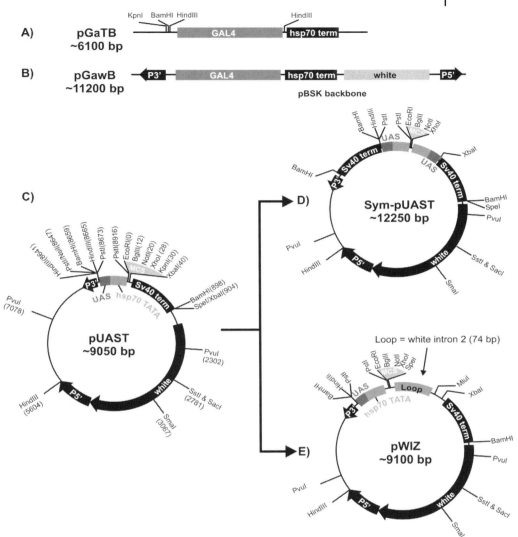

Fig. 3.6 Vectors for directed endogenous RNAi in *Drosophila*. The vectors pGaTB (A), pGawB (Brand and Perrimon 1993) (B), pUAST (C) (all three; see Brand and Perrimon 1993), and the pUAST derivatives Sym-pUAST (Giordano et al. 2002) and pWIZ (Lee and Carthew 2003) are illustrated. To target GAL4 expression to specific cells, tissue-specific promoters can be subcloned upstream of GAL4 into the unique *BamHI* site of pGaTB (A). pGawB is an enhancer detection vector that directs expression of GAL4 in a genomic integration site-dependent fashion (B). pUAST is designed to direct GAL4-dependent transcription of a gene of choice. The sequence is subcloned into a polylinker downstream of five tandemly arrayed, optimized GAL4 binding sites (C). In Sym-pUAST (D) opposing UAS sites are driving bidirectional expression toward the dsRNA template. The pWIZ vector (E) developed by the Carthew lab contains a 74-bp second intron of the white gene serving as a loop sequence to produce inverted repeat DNA constructs. The sense strand is cloned upstream and the antisense strand downstream of the intron. The intron bears splice sites at their ends that allow splicing in heterologous tissues.

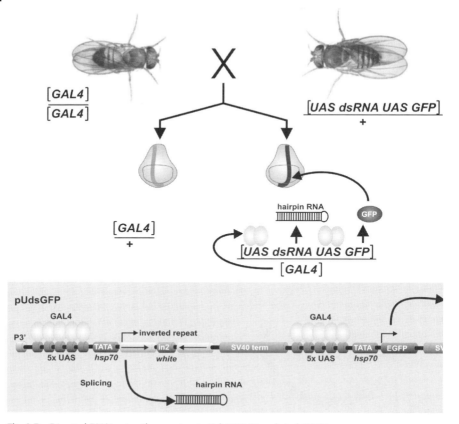

Fig. 3.7 Directed RNAi using the construct pUdsGFP (Nagel et al. 2002) modified from (Duffy 2002). The inverted repeat sequence is separated by an intron that will be spliced to form a snapback hairpin RNA. Such a snapback hairpin is sufficient to induce RNAi in a specific tissue. A second UAS site is cloned upstream of the open reading frame of an enhanced green fluorescent protein (EGFP). When crossed to a GAL4 driver, both the hairpin transcript for the inverted repeat sequence and the enhanced GFP marker are expressed in the corresponding GAL4 pattern.

3.2.4
Limitations

As in *C. elegans*, most of the methods exhibiting the same advantages and disadvantages when applied to *Drosophila*. Nevertheless, there are some slight differences in the procedures. Depending on the aim of the study, one must choose between the well-established techniques. In order to gain the best results, the following limitations of each procedure must be regarded.

A) Exogenous dsRNA

Advantages:
- Very rapid
- Application by simply soaking the dsRNA (only in S2 culture)
- Simultaneous application of more than one dsRNA (only in S2 culture)
- Allows high-throughput phenotype screening in S2 cells
- Transient injection in embryos allows a rapid phenotyping in the embryo
- Allows reduced phenotypes to be made in adult flies
- Can be easily performed by novices in the *Drosophila* field

Limitations:
- No knock-out phenotypes in adult flies
- No long-lasting RNAi in cells and flies
- Sometimes different phenotypes after injection due to partitioning of the dsRNA at one side of the embryo
- Does not allow tissue-specific RNAi

B) Endogenously expressed hairpin RNAs

Advantages:
- Knock-out phenotypes in adult flies are possible
- Allows long-lasting RNAi in cells and flies
- The use of the GAL4/UAS system with inverted repeats allows tissue-specific and inducible RNAi
- Cell lines and *Drosophila* lines can be stored for later experiments

Limitations:
- Time consuming and very laborious
- Scientists need quite extensive training to perform successful injections of embryos
- A fully equipped fly facility is needed

3.3
dsRNA Synthesis

As described in Chapter 2 for *C. elegans*, most of the RNAi experiments in *Drosophila* or S2 cell culture are carried out using larger dsRNA expressed from T7 promoter flanked linear DNA or derived from inverted repeat DNA constructs as long hairpin RNAs (lhRNAs). Some reports have suggested a length of 700–800 bp for optimal RNAi activity, while others have set a range between 200 and 2000 bp. The dsRNA can be made from cDNA or genomic DNA templates, as long as most of the dsRNA corresponds to exon regions. To avoid an unspecific silencing of other genes or off-gene targeting by the sense strand it is beneficial to use dsRNAs corresponding to untranslated regions (UTRs) such as the 3′-UTR, which encodes unique sequences. The following protocol describes the basic procedure, according to the protocols in

Chapter 2. For more information about the evaluation of the target sequence and the construction of the linear DNA templates for the generation of dsRNA, follow the comments in Chapter 2, Sections 2.2 and 2.4.2.

3.3.1
In-vitro dsRNA Transcription

To date, many companies offer RNA production kits (Ambion, Roche, Stratagene, Promega, etc.), which reveal RNA or dsRNA with similar qualities. Here, we describe the procedure using the T7 RiboMAX_ System or the new variant T7 RiboMAX™ Express RNAi System. Both systems are widely used, and have been shown capable of synthesizing milligram amounts of RNA that can be converted to dsRNA by simple annealing, nuclease digestion, and precipitation steps.

The T7 RiboMAX™ Express RNAi System is used to generate dsRNAs in the size range of 180 to 1000 bp. The yield of dsRNA is dependent on the first six bases or the GC content of the DNA-template, but can reach 1–2 mg dsRNA per ml (see Section 2.4.2 and Table 2.1). Meanwhile, it is very important for a successful reaction to use highly purified DNA template, which can either be obtained by phenol:chloroform extraction (Sambrook and Russell 2001) or agarose gel extraction (QIAquick, Qiagen, Germany). After transcription, the resulting RNA strands are annealed to form dsRNA or siRNA, and the remaining single-stranded RNAs and the DNA template are removed by nuclease digestion. The dsRNA or siRNA is then purified by isopropanol precipitation, and can be introduced into the organism of choice for RNAi applications (for a more detailed description, see the Promega manual for the T7 Ribomax™).

PROTOCOL 7

Vector Template Method

1. Prepare 1–8 μl of a DNA solution with a final DNA amount of 1 μg for a dual-opposed promoter PCR product or either 1 μg per separate reaction or 1 μg each in a combined reaction (2 μg total) for separate single-promoter templates.

2. Add the following reaction component into a DNase- and especially RNase-free microcentrifuge tube, and fill up to a final volume of 20 μl (for T7 Ribo-MAX™ Express RNAi System) or 100 μl (for T7 RiboMAX™ System). Mix by gently flicking the tube (see Table 3.1).

▶ **Note**: It is important that no RNase is present in the DNA. If contamination of RNase is suspected, treat the DNA with proteinase K (100 μg/ml) and SDS (0.5%) in 50 mM Tris-HCl (pH 7.5) and 5 mM $CaCl_2$ for 30 min at 37 °C.

Tab. 3.1

RiboMAX™Express T7		*RiboMAX™ T7/Sp6/T3*	
T7 Reaction components	Reaction	T7 Reaction components	Reaction
RiboMAX™Express T7 Buffer (2x)	10 µl	RiboMAX™T7 Transcription Buffer (5x)	20 µl
Linear DNA template (1–2 µg)	1–8 µl	rNTPs (25 mM)	30 µl
Enzyme Mix T7 Express	2 µl	Linear DNA template (1–10 µg)	1–40 µl
Nuclease-free water	0–7 µl	Enzyme Mix T7 Express	10 µl
		Nuclease-free water	0–39 µl
Total volume	20 µl	Total volume	100 µl

RiboMAX™T7 can be replaced by RiboMAX™Sp6 and RiboMAX™T3

3. Incubate at 37 °C for 30 min (for T7 RiboMAX™ Express RNAi System) or overnight (for T7 RiboMAX™ System).

▶ **Note:** In contrast to Promega Notes on T7 RiboMAX™, a dramatic increase in yield for almost all templates can be observed when incubating for longer than 6 h.

▶ **Note:** Incubation at 42 °C may improve the yield of dsRNA for transcripts containing secondary structure, which is often due to the GC-rich templates. The use of separate single-promoter templates in separate transcription reactions has also been observed to increase yield of targets.

4. The reaction can be monitored measuring the viscosity. As the yield increases, the reaction mixture turns into a gelatinous translucent pellet.

▶ **Note:** If the sample is too viscous for the annealing step, add a few µl of water or annealing buffer (Table 3.2).

Tab. 3.2

Annealing buffer I (1x)		*Annealing buffer II (10x)*	
Potassium acetate	100 mM	Tris-HCl, pH 8.0	100 mM
HEPES-KOH, pH 7.4	30 mM	EDTA, pH 8.0	10 mM
Magnesium acetate	2 mM	NaCl	1 M

5. After RNA synthesis, anneal both RNA strands, mix equal volumes of complementary RNA reactions, and incubate at 70 °C for 10 min.

6. Spin the samples very briefly! Then cool to room temperature very slowly.

▶ **Note:** It is essential to cool the reaction mixture very slowly. This can be done by heating the samples in a thermoblock (waterbath), switching off the thermoblock (waterbath heater), and waiting until the temperature of the metal block (water) reaches ambient temperature.

7. Dilute the supplied RNase Solution 1:200 by adding 1 µl RNase Solution to 199 µl nuclease-free water. Add 1 µl freshly diluted RNase Solution and 1 µl RQ1 RNase-Free DNase (for T7 Ribomax™ add 1 µl RQ1 for 1 µg DNA) per 20 µl (100 µl) reaction volume, and incubate for 30 min at 37 °C. This will remove any remaining single-stranded RNA and the template DNA, leaving double-stranded RNA.

8. Add 0.1 volume of 3 M sodium acetate (pH 5.2) and 1 volume of isopropanol or 2.5 volumes of 100% ethanol. Mix and place on ice for 1 h. The reaction will appear cloudy at this stage. Spin at top speed in a microcentrifuge for 10 min at 4 °C.

9. Carefully aspirate the supernatant, and wash the pellet with 0.5 ml of ice-cold 70% ethanol, removing all ethanol following the wash. Air-dry the pellet for 15 min at room temperature, and resuspend the RNA sample in 100 µl injection buffer (Table 3.3) with a final concentration of at least 1–2 mg/ml.

Tab. 3.3   For more information on injection buffers (see Spradling 1986).

Injection buffer (Spradling 1986)	
KCl	5 mM
NaH$_2$PO$_4$ (pH 7.8)	10 mM

Adjust a 100 mM stock of NaH$_2$PO$_4$ to pH 7.8 using NaOH

▶ **Note**: The sodium phosphate content can be reduced to as little as 100 µM. A low salt concentration will give better results in injections, but will reduce the stability of the dsRNA.

10. Aliquot the dsRNA and store at –20 °C or –70 °C.

11. Alternatively, further purify dsRNA following precipitation using a G25 micro spin column following the manufacturer's instructions (Amersham Biosciences, Cat.# 27–5325-01). This will remove any remaining rNTPs and allow accurate quantitation by absorbance at 260 nm.

▶ **Note**: Do not process more than an initial 40 µl reaction volume per spin column. A loss of yield can be expected following G25 purification (approximately 66% recovery).

▶ **Note**: Do not use those columns with water. The purification will also result in a desalting of the solution and water will decrease the dsRNA annealing efficiency.

12. Prepare a 1:100 to 1:300 dilution of the dsRNA and measure the concentration at 260 nm (OD (260) = 1 is equivalent to 40 µg RNA/ml).

13. Dilute 1 µl of dsRNA in 50–100 µl of 1x TAE buffer or nuclease-free water (DEPC water) and use 50–500 ng per lane and the respective amount of 10x DNA agarose loading dye (Table 3.4).

Tab. 3.4

Gel loading buffer (10x)	
Ficoll	25%
EDTA, pH 8.0	1 mM
Bromophenol blue	0.25%

14. Analyze the quality of the dsRNA on a 1–2% TAE-agarose gel using a DNA size marker (for example 1KB plus, Invitrogen).

▶ **Note**: Double-stranded RNA usually migrates more slowly than double-stranded DNA. Use 1–5 µl of diluted dsRNA per lane (dilute at least 1:50 with Nuclease-Free Water) or use 50–500 ng per lane.

15. Staining of the gel can be performed using 0.5 mg/ml ethidium bromide in 1x TAE buffer (Sambrook and Russell 2001).

▶ **Note**: Ethidium bromide (0.1 µg/ml) can be added to the gel, but in some cases it can interfere with the migration of the dsRNA and the resolution of the bands.

PCR Template Method

1. Choose primer sequences that will amplify the region you want to act as the dsRNA template.

2. The 5′-end of each primer must correspond to a minimal (A) or elongated (B) T7 promoter sequence (as depicted in Figure 3.8), plus the at least 21 nt of the target sequence Thus, each primer will be approximately 39+ nt long.

A) 5'-**TAATACGACTCACTATAG**-3' **Minimal T7 sequence**

B) 5´-GGATCC**TAATACGACTCACTATAG**-3' **Elongated T7 sequence**

Fig. 3.8 Extra bases added to the minimal T7 promoter (A) often increase the yield by allowing more efficient T7 polymerase binding and initiation (B)

▶ **Note**: For reactions containing two primers that both have T7 promoter sequences, a primer concentration of 100–500 nM is recommended for the PCR amplification. Higher concentrations may result in significant primer–dimer formation.

3. Perform a 50 µl PCR reaction with T7-linked primers and a suitable template.

> ▶ **Note**: Amplification strategies using primers containing T7 promoter se-
> quences may include an initial 5–10 cycles at an annealing temperature ap-
> proximately 5 °C above the melting temperature of the gene-specific se-
> quences, followed by 20–35 cycles of annealing approximately 5 °C above
> the melting temperature of the entire primer, including the T7 promoter.

4. Purify the PCR product either by agarose gel electrophoresis or by phenol:
   chloroform extraction and ethanol precipitation (Sambrook and Russell
   2001).

5. Dissolve in TE buffer and measure concentration spectrophotometrically.

6. Perform an RNA synthesis reaction as described in the previous protocol.

7. Finally, dsRNA should be analyzed on an 1.5% agarose gel.

3.3.2
Inverted Repeat DNA

Different inverted repeat constructs are used for RNAi in *Drosophila* embryos. Some
contain just the plain exon based gene fragment in a consecutive sense and antisense orientation, while others comprise a splice-activated intron loop structure to
separate the inverted repeat. Those inverted repeat constructs are eventually expressing an RNA precursor that snaps back to form a hairpin RNA. The intron loop will
be spliced and the hairpin sequence will only contain exonic regions and is released
into the cytosol. As described in Chapter 2 for *C. elegans*, the cloning of such inverted
repeat constructs is very laborious and time-consuming due to the recombination
events that take place during replication of the constructs. In vivo, large DNA palindromes are intrinsically unstable sequences (Leach 1994). Inverted repeats may initiate genetic rearrangements by formation of hairpin secondary structures that block
DNA polymerases or are processed by structure-specific endonucleases. The inverted
repeat base-pairing results in cruciform structures, which have proved difficult to detect in bacteria, suggesting that they are destroyed. Besides, shorter fragments of the
inverted repeats and the plasmid vector can often be isolated undermining this hypothesis. It has been shown that sbcCD, an exonuclease of *E. coli*, is responsible for
the processing and cleavage of large palindromic DNA sequences in *E. coli* (Davison
and Leach 1994; Leach 1994), thus preventing the replication of long palindromes.
SbcCD is cleaving cruciforms in duplex DNA followed by RecA-independent single-
strand annealing at the flanking direct repeats, generating a deletion.

There are two possibilities of preventing the degradation. One is the insertion of a
spacer between the inverted repeat, since it has been shown that inverted repeats
with an interruption of the pairing at the center are less likely to form cruciform
structures than perfect pairing inverted repeats (Bzymek and Lovett 2001; Sinden
et al. 1991; Zheng et al. 1991). The other possibility is the use of nuclease (sbcCD) or
recombinase (recA,B,J) -deficient *E. coli* strains such as SURE II (Stratagene),
JM105, JM103, or CE200 (ATCC, at http://www.atcc.org) (Table 3.5).

Tab. 3.5 A list of *E. coli* strains that facilitate the inverted repeat cloning (Hanahan 1983; Wyman et al. 1985; Yanisch-Perron et al. 1985).

Strains	Genotype	References	Source
SURE II	e14– (McrA–) ? (mcrCB-hsdSMR-mrr)171 endA1 Sup E44 thi-1 gyrA96 relA1 lac recB recJ sbcC umuC::Tn5 (Kanʳ) uvrC [F′ proAB lacIqZ? M15 Tn10 (Tetʳ) Amy Camʳ	http://www.stratagene.com/manuals/200238.pdf	Stratagene
JM103	F′ traD36 proA+ proB+ lacIq delta(lacZ)M15 delta(pro-lac) supE hsdR endA1 sbcB15 sbcC thi-1 rpsL lambda-	Hanahan (1983)	ATCC
JM105	F′ traD36 proA+ proB+ lacIq delta(lacZ)M15 delta(pro-lac) hsdR4 sbcB15 sbC? rpsL thi endA1 lambda-	Yanisch-Perron et al. (1985)	ATCC
CES200	F- delta(gpt-proA)62 thr-34::Tn10 lacY1 ara-14 galK2 xyl-5 mtl-1 leuB6 hisG4 argE3 hsdR mcrB rac- sbcB15 recB21 recC22 rpsL31 rfbD1 kdgK51 thi-1 tsx-33 lambda-	Wyman et al. (1985)	ATCC

The SURE II competent cells were designed to facilitate cloning of inverted repeat DNA by removing genes involved in the rearrangement and deletion of these DNAs such as the UV repair system (*uvrC*) and the SOS repair pathway (*umuC*) genes. This results in a 10- to 20-fold increase in the stability of DNA containing long inverted repeats. Furthermore, mutations in the sbcC and RecA,B,J genes involved in recombination events greatly increase stability of inverted repeats. The combination of *recB* and *recJ* mutations confers a recombination-deficient phenotype to the SURE cells that greatly reduces homologous recombination, similar to a mutation in the *recA* gene.

Even though those strains increase the potential to recover a clone, the recombination and cruciform formation is not completely abolished. For more information, see Bzymek and Lovett (2001).

The cloning of the inverted repeat is based on the PCR amplification of the respective cDNA sequence and ligation of the sense-intronic loop-antisense sequence prior to its insertion into the expression vector. Since the protocol comprises several agarose gel purification steps, one should make sure of starting with a reasonable amount of PCR product. Regular PCRs yield 1–4 µg of PCR product, but this can be exceeded by the use of a few improved Taq-polymerases (Long Expand Taq-polymerase, Roche; Takara). The PCR comprises the use of restriction sites at both ends of the inverted repeat sequence to perform a directed cloning of the inverted repeat in the pUAST vector (Figure 3.9). The PCR product from this reaction can be processed as described in the following protocol.

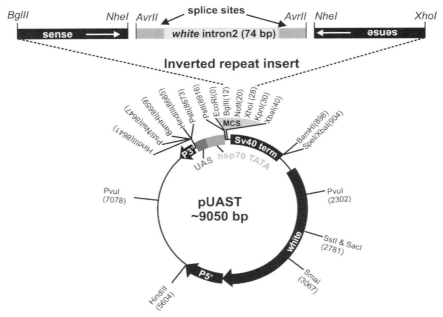

Fig. 3.9 Cloning strategy of the inverted repeat DNA encoding the splice activated snapback hairpin RNA into the transformation vector pUAST (Brand and Perrimon 1993). In this example the inverted repeat is inserted into the *BglII* and *XhoI* site of the vector. However, every other restriction site of the multiple cloning site (MCS) can be used. Using only one restriction site will decrease the number of positive clones because of the cruciform DNA formation (Bzymek and Lovett 2001).

PROTOCOL 8

1. Generate a DNA fragment for the inverted repeat by PCR. The fragment size should be between 800 and 1500 bp. Add the following restriction sites to the 5′-end of the gene specific primer (Table 3.6).

Tab. 3.6 Primer sequences for the PCR amplification of inverted repeat components.

	AvrII
white **forward**	5′-GAA**CCTAGG**TGAGTTTCTATTCGCAGTCGG-3′
	AvrII
white **reverse**	5′-GCA**CCTAGG**CTGAGTTTCAAATTGGTAATT-3′
	BglII
sense forward	5′-GTAA**AGATCT**▬▬(N18–21)▬▬
	XhoI
antisense forward	5′-GTAA**CTCGAG**▬▬(N18–21)▬▬
	NheI
sense/antisense reverse	5′-GTAA**GCTAGC**▬▬(N18–21)▬▬

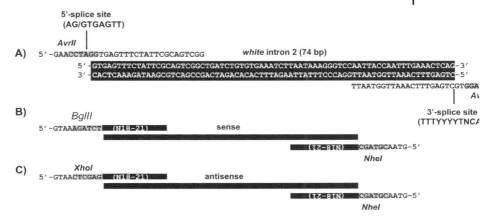

Fig. 3.10 Schematic view of the primers used to generate the white intron 2 loop with the splice sites (A), the sense strand (B), and the antisense strand of the inverted repeat (C). The primers contain the respective restriction sites at their 5′-end that allow the directed cloning.

▶ **Note**: Any other restriction site from the MCS will be suitable (see respective sequence).

▶ **Note**: Do not use proofreading Taq polymerase that generates mainly blunt-end PCR products. This will allow self-ligation of the fragments during PCR cloning.

2. Amplify the white intron 2 for the loop by PCR with primers both containing *AvrII* sites.

3. Purify the PCR fragments on a 1–1.5% agarose gel to remove enzymes and residual primers. Extract the PCR product (for example using the QIAquick reagent from Qiagen).

4. Digest the DNA fragment with *NheI* (1 U/μg DNA) and the loop fragment with *AvrII* (1 u/μg DNA) overnight at 37 °C.

▶ **Note**: Using a Taq polymerase generating 3′-T overhangs will prevent ligation of the other ends of the sense and antisense fragments.

5. Purify the PCR fragments on a 1–1.5% agarose gel to remove the restriction enzymes. Extract the PCR products.

6. Ligation: The ligation protocol depends on the supplier of the T4-ligase. The following protocol is based on the highly concentrated T4-ligase of NEB (A) or on the Rapid Ligation Kit protocol (Roche) (B). Both methods allow RT-ligation.

7. Pipette the following components in a microcentrifuge tube as depicted in Table 3.7.

Tab. 3.7

A) Method A		B) Method B (Rapid Ligation Kit™, Roche)	
Gene fragment *NheI* (0.625 µg)	x µl	Gene fragment *NheI* (0.625 µg)	x µl
Loop fragment *AvrII* (0.13 µg)	x µl	Loop fragment *AvrII* (0.13 µg)	x µl
NEB 2 buffer (10x)	5 µl	Dilution buffer (5x) or NEB 2 buffer (10x)	3 µl
BSA (10 mg/ml)	5 µl	*AvrII* (10 u/µl)	1.5 µl
AvrII (10 u/µl)	1.5 µl	*NheI* (10 u/µl)	1.5 µl
NheI (10 u/µl)	1.5 µl	Ligation buffer (2x)	15 µl
ATP (10 mM)	5 µl	T4 ligase (8 u/µl)	1 µl
T4 ligase (8.8–20 u/µl) (NEB)	2 µl	Water	x µl
Water	x µl		
Total	50 µl	Total	20 µl

8. Incubate the ligation reactions as shown in Table 3.8. A PCR thermocycler can be used for this.

Tab. 3.8

Incubation cycles		
Step 1	20 °C	20 min
Step 2	37 °C	10 min
Step 3	repeat 1 and 2 for 10 times	

9. Separate the ligation products on a 1–1.5% agarose gel and extract the sense-loop-antisense fragment using the QIAquick gel extraction procedure (Qiagen).

▶ **Note**: If the product band is too weak repeat the reaction, collect several gel slices, and extract at once.

10. Elute in 20 µl 10 mM Tris-HCl pH 8.5.

▶ **Note**: Trying to PCR-amplify a gel-purified ligation product does usually not work.

11. Digest both, the final sense-white-antisense fragment and the vector pUAST, with *BglII* and *XhoI* (1 U/µg DNA) for 2 h at 37 °C to allow directional cloning.

▶ **Note:** If it is necessary to use only one restriction site to insert the inverted repeat into the vector, dephosphorylate the pUAST vector using calf intestinal or alkaline phosphatase (1 µl/µg) DNA in the digestion buffer for 15–60 min at 37 °C to prevent re-ligation of the vector.

12. Purify the PCR and vector fragments on a 1–1.5% agarose gel to remove the enzymes. Extract the PCR products.

13. Ligate the fragment into the *BglII* site pUAST-vector (Table 3.9).

Tab. 3.9

Rapid Ligation Kit™, Roche

Insert ligation product	x µl
pUAST (*Bglll* and *Xhol* digested)	x µl
Dilution buffer (5x) or NEB 2 buffer (10x)	2 µl
Ligation buffer (2x)	10 µl
T4 ligase (8 u/µl)	1 µl
Fill up with water	

Total	20 µl

Insert / vector molar ratio should be 8 : 1

14. Incubate for 30 min at room temperature.

15. Transform the 2–8 µl of ligated vector into *E. coli* SURE II cells (Stratagene) as described by Sambrook and Russell (2001).

16. Plate the bacteria on selection LB/agar plates supplemented with the appropriate antibiotic.

▶ **Note**: Due to the high rate of recombination events, the number of growing colonies is very low. Besides, the percentage of positive clones that are bearing the inverted repeat is often below 1%. To increase the amount of positive recombinant colonies it is recommended to set up several ligation and transformation reactions for the same cloning procedure simultaneously.

Likewise, the amount of DNA for the ligation must be increased for the protocol to work well (>500 ng for the 750-bp gene fragment and >250 ng for the 784-bp loop).

17. Pick the colonies and grow the bacteria in 1–3 ml LB media + antibiotics.

▶ **Note**: For high-throughput DNA isolation use either a pipetting robot such as the BioRobot system (Qiagen), or the 96-well turbo DNA isolation method described in Chapter 4.

18. Isolate the DNA (Qiagen Spinprep Kit, or see Chapter 4).

19. Digest with the appropriate restriction enzymes to verify the direction of the intron.

20. Separate the digestion products on a 1–1.5% TAE agarose gel extract the insert from the gel using QiaQuick gel extraction Kit (Qiagen).

21. Precipitate the DNA with 0.1 vol of 3 M sodium acetate and 2.5 vol of 100% ethanol.

22. Freeze at –80 °C for future use, or dilute in injection buffer (Table 3.10).

Tab. 3.10 For more information on injection buffers (see Spradling 1986).

Injection buffer (Spradling 1986)	
KCl	5 mM
NaH$_2$PO$_4$ (pH 7.8)	10 mM

Adjust a 100 mM stock of NaH$_2$PO$_4$ to pH 7.8 using NaOH

3.4
Injections

3.4.1
Injection Services

There are two ways to apply RNAi to fly embryos. The first requires the equipment and a special training to inject the embryo. Since injection procedures are established in the majority of the *Drosophila* laboratories, this should be a minor problem. However, before embarking on the technique, novices in *Drosophila* research should bear in mind that establishing the injection procedure is very time-consuming and requires extensive training. Another way to obtain transgenic flies is to use commercial injection services, as are offered from the EMBL, Heidelberg, Germany (http://www.embl-heidelberg.de/~voie/). At the EMBL, they inject the desired inverted repeat DNA construct into "w1118" flies using "delta 2–3" as helper DNA. Based on EMBL facility statistics, a standard injection of a 12-kb construct yields 40–60 surviving larvae, with a transformation efficiency of 10–20%. The inverted repeat DNA must be purified and particle-free. Each injection cycle per construct requires a lyophilized or dried mixture of 10 µg inverted repeat DNA and 3 µg of delta 2–3 helper DNA. The price for a routine injection is 300 Euro.

Other transgenic services exist in the USA. The Duke University Non-Mammalian Model Systems Flyshop (http://www.biology.duke.edu/model-system/services.htm) provides P-element transformation, dsRNA injection, embryo sorting, and inverse PCR services. Further Genetic Services Inc. (http://www.geneticservices.com) even provides cloning of the transgenes, injection, genetic screening, and stock creation.

3.4.2
Injection Method

There are several well-established protocols for the injection of *Drosophila* embryos, which are currently in use. The most detailed protocols are those described by Richard Carthew and colleagues (http://www.biochem.northwestern.edu/carthew/manual/RNAi_Protocol.html) and Bruce Patterson and colleagues [Misquitta, 2000 #4206]. A prerequisite for the procedure is access to a working fly facility. Novices in *Drosophila* research should study the basic protocols for injection that can be found in common *Drosophila* handbooks, as listed at the end of this chapter. The procedure is relatively straightforward, but requires a significant amount of training. This especially includes the preparation and handling of the injection needles and the embryos. There are two ways to inject the embryos. One is to use dechorionated embryos, while the other is based on the injection of the dsRNA or inverted repeats through the chorion to minimize problems with embryo desiccation. In any case, embryos must be kept wet throughout the whole injection procedure.

3.4.3
DsRNA or Inverted Repeat DNA Preparation

PROTOCOL 9

1. Dilute dsRNA or inverted repeat DNA in a suitable amount of injection buffer (Table 3.9) to receive a final concentration of 1–2 mg/ml.

▶ **Note**: Use at least 100 µl to prevent precipitation of the dsRNA or the DNA.

2. Before starting the injection, all floating particles must be removed from the dsRNA or inverted repeat plasmid solution by centrifugation.

3. For injection, mix the dsRNA or inverted repeat DNA with a dye.

▶ **Note**: For this purpose, one can use either tetramethylrhodamine dextran or a regular food dye. Both are non-toxic to the embryos.

4. Prepare a solution of tetramethylrhodamine dextran (20 mg/ml of 70 000 MW) and filter-sterilize. Dilute the stock to a final concentration of 1 µg/µl with sterile injection buffer.

5. Mix 4.5 µl of dsRNA or inverted repeat DNA solution with either 0.5 µl of food dye or tetramethylrhodamine dextran.

3.4.4
Embryo Collection

The injection procedure requires several days of preparation. The collection of flies (usually 2–4 days old ry506, w1118, or yw67 c) and synchronization of the embryos takes some time. The flies are collected either in collection cages or small plastic beakers with air holes at room temperature. The eggs are collected either on agar-fruit juice (Table 3.11A) or agar molasses plates (Table 3.11B) placed underneath the plastic beaker.

Tab. 3.11

A) *Agar-fruit juice plates*		B) *Agar-molasses plates*	
Bacto-agar	3 g	Bacto-agar	4.4 g
Water	50 ml	Molasses	18 ml
Fruit juice (grapefruit)	50 ml	Water	111 ml
Methyl-*p*-hydroxy benzoate	1 ml	Methyl-*p*-hydroxy benzoate	1 ml
(10% solution in ethanol)		(10% solution in ethanol)	
Glac. acetic acid	1 ml	Glac. acetic acid	1 ml
		Propionic acid	0.63 ml

Heat agar in 50 ml in a microwave oven. Add prewarmed fruit juice and the other components and pore the plates.

Autoclave agar and molasses dissolved in the water (including a stir bar). After cooling down to 50 °C add the other components and pore the plates.

PROTOCOL 10

1. Dab the agar fruit juice plate with yeast paste (a viscous mixture of dry yeast with water) to facilitate egg-laying.

2. Place agar fruit juice plate underneath the plastic beaker filled with 2- to 4-day-old flies, and collect the eggs (Figure 3.11).

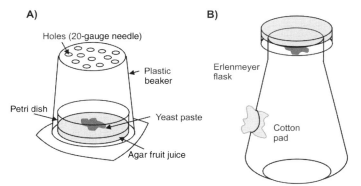

Fig. 3.11 Egg collection cage. (A) A very inexpensive method to collect the eggs is the use of a disposable plastic beaker containing holes punched into the top with a 20-gauge needle. (B) Another method is to use an Erlenmeyer flask with a cotton-filled aeration hole at the side and the egg collecting plate on the top.

▶ **Note**: Replace the plate every 30–60 min to synchronize the eggs.

▶ **Note**: It is important to inject the eggs within the same hour of collection before cellularization occurs.

3. Transfer the eggs into the wash basket containing a nylon mesh (Figure 3.12), using a wet brush and wash with tap water (at room temperature).

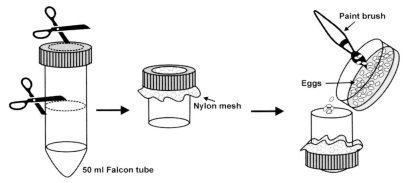

Fig. 3.12   Preparation of the wash basket or egg strainer from a 50-ml Falcon tube. The lid can be either cut with a hot scalpel or punctured with a cork-borer. The lid should be screwed back to hold a 120-μm Nitex Nylon mesh in place. The eggs are collected with a small wet paint brush and transferred to the strainer.

A) Dechorionation Method

- Remove the eggshell (chorion) by placing the wash basket into a small Petri dish containing 50% bleach (final concentration: 2.5% hypochlorite) for 1–3 min, gently swirling the strainer.

- Rinse the eggs with tap water until they have reached neutral pH.

- ▶ **Note**: Another more gentle method to dechorionate is to transfer the washed eggs onto a small strip of a double-sided tape. The embryos can be rolled on the tape gently from one side to the other, using forceps. The embryo can be move to an agar strip by simply touching it with the forceps. It will stick by itself.

- Remove the embryos with an artist's paint brush to a thin agar fruit juice strip (~5 × 10 × 2 mm cut from an agar plate) and line up the embryos, with their anterior end facing out from the agar.

- ▶ **Note**: Surface tension from the moisture of the agar-fruit juice plates holds the embryos in place, so that they can be oriented and lined up easily.

- ▶ **Note**: A very detailed protocol how to line up the embryos can be downloaded from the Carthew lab web page (http://www.biochem.northwestern.edu/carthew/manual/RNAi_Protocol.html)

- Carefully pick up the line of embryos with a rectangular coverslip that is covered with double-stick tape (a nontoxic variant is 3M, type 415) such that the posterior end is not sticking to the tape (Figure 3.13A)

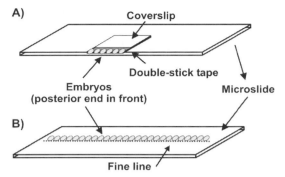

Fig. 3.13 Line-up of the embryos a described in the dechorionation method (A) or for injection through the chorion (B).

- Attach the coverslip to a microslide.

- Desiccate the embryos in a desiccator for 3–15 min.

▶ **Note**: Desiccation will relieve the turgor pressure that is sometimes explosively released upon micropipette penetration through the vitelline membrane.

- Cover the embryos with carbon halocarbon oil.

B) Chorion Injection Method

- After washing the eggs (step 4), remove the eggs from the strainer with a fine brush to the injection slide.

- Line up the 50–60 embryos on the slide, with their posterior ends perpendicular to the length of the slide, as depicted in Fig 3.13B.

▶ **Note**: Draw a fine line on the microscope slide, using a diamond pencil.

- Keep the embryos wet, by brushing them with water during lining.

▶ **Note**: If the embryos start to dry on the outside, the chorion will harden and will break the needles.

- Keep the slides in a moist chamber and inject within an hour.

- Do not inject the embryo directly into the posterior end of the chorion, as it is too hard and the needle will break. Use a position slightly off-center.

4. Prepare micropipettes from borosilicate glass capillaries with a glass filament (World Precision Instruments TWF100–4) using a needle puller and bevel or break a tip to a diameter of 0.5 to 2.5 μm.

▶ **Note**: For a detailed protocol to bevel the tips of the pipette, see (http://www.biochem.northwestern.edu/carthew/manual/RNAi_Protocol.html)

▶ **Note**: The loaded tip of the needle can be broken by gently pushing it to the edge of the mounted coverslip or slide. The process has to be followed up with the microscope.

5. Place the freshly pulled needles horizontally imbedded in a small amount of clay fixed into a Petri dish.

6. Back-fill the pipette by pipetting a drop of dsRNA or inverted repeat plasmid in injection solution onto a piece of Parafilm and placing the back end of the pipette into the solution. The solution will be introduced by capillary forces.

▶ **Note**: One can also use Eppendorf microloader tips.

▶ **Note**: It is very important to remove all insoluble particles from the injection solution. Any particles will clog the pipette.

7. Align the micropipette perpendicular to the posterior end and inject the embryo (according to the instructions of the injector for example: the Transjector from Eppendorf). The volume of injection varies between 50 and 100 pl.

8. The posterior end is preferred for injection, as this is where the germ cells are eventually located.

9. After injection, place the slides into a moist chamber at 18–25 °C until the desired developmental stage.

10. Wash the embryos and fix them for antibody staining.

11. To rescue injected inverted repeat transgenic lines, apply a small amount of yeast paste to the slide and transfer the embryos to food vials.

3.5
Cell Lines

Nowadays, most investigators use *Drosophila* cell lines rather than whole embryos for a quick RNAi experiment. Many embryonic cell lines are available, and protocols even exist which allow the application of RNAi in primary cells. Cell lines and primary cells can be found by searching the bionet.*Drosophila* web page (http://www.bio.net:80hypermail/Dros/). Some often-used cell lines are listed in Table 3.12.

Beside the available cell types, cell lines can be differentiated from embryonic cells. They can comprise nerve, muscle, fat-body, chitin-secreting, and macrophage-like cells (possibly hemocytes) that appear in the first 24 h and mature over the next week. Tracheal, imaginal disc, a second stage of the macrophage-like, and a number of unidentified fibroblastic and epithelial cells appear in the second and third weeks, following a resumption of cell multiplication (Dubendorfer et al. 1975; Shields et al. 1975).

Tab. 3.12 Selected *Drosophila* cell lines, origins, and morphology (Echalier and Ohanessian 1970; Haars et al. 1980; Peel et al. 1990; Ui et al. 1994; Yanagawa et al. 1998).

Line	Origin	Reference	Characteristics
Schneider's Line S2 – (S2)	Dissociated embryos, near hatching (Oregon R)	Schneider (1972)	hemocyte-like gene expression, phagocytic, semi-adherent in colonies, round, granular cytoplasm
Schneider's S2 – (S2*)	Dissociated embryos, near hatching (Oregon R)		hemocyte-like gene expression, phagocytic, semi-adherent in colonies, round, granular cytoplasm
Schneider's S2 – (S2C)	Dissociated embryos, near hatching (Oregon R)		hemocyte-like gene expression, semi-adherent in colonies, round, granular cytoplasm
Schneider's S2 – (S2-R+)	Dissociated embryos, near hatching (Oregon R)	Yanagawa et al. (1998)	hemocyte-like gene expression, phagocytic, adherent, flat cells; Fz+ and Wg-responsive
Schneider's S2 – (DL2)	Dissociated embryos, near hatching (Oregon R)	peterc@ento.csiro.au	hemocyte-like gene expression, phagocytic, adherent monolayer of uniformly round, smooth cells
Schneider's Line S3	Dissociated embryos, near hatching (Oregon R)	Schneider (1972)	adherent, spindle-shaped cells, ecdysone responsive, grow in clumps
Kc (Kc 167)	Dissociated embryos, 8–12 h (F2 ebony x sepia)	Echalier and Ohanessian (1969)	hemocyte-like gene expression, phagocytic, uniformly round, clump in sheets, ecdysone responsive into adherent, bipolar spindle-shape cells
l(2)mbn	3rd instar larvae tumorous hemocytes, (l(2)malignant blood neoplasm)	Gateff et al. (1980)	larger cells, larger granular, complex cytoplasm, phagocytic, aneuploid, heterogenous size and shape
ML-DmBG2	Dissociated 3rd instar larvae brain and ventral ganglia (y v f mal)	Ui et al. (1994)	acetylcholine, HRP expression, neuronal-like processes
ML-DmBG6	Dissociated 3rd instar larvae brain and ventral ganglia (y v f mal)	Ui et al. (1994)	acetylcholine, HRP expression, neuronal-like processes
Clone 8	3rd instar larvae wing imaginal discs	Peel et al. (1990)	columnar epithelial, adherent, will form multiple layers, conserved signaling pathways

Most cell lines can be grown in M13 or Schneider's medium supplemented with 10% fetal calf serum. These may be commercially available (Shields and Sang M3 insect medium (Sigma; Shields and Sang 1970), DES medium or *Schneider's Drosophila* medium (Invitrogen), Schneider's medium (Sigma), or for serum-free culture Hyclone CCM-3), or can be prepared following the recipe of Schneider and Blumenthal (1978). For a more detailed discussion of cell lines and their cultivation, see Cherbas and Cherbas (1998).

3.6
Protocols

S2 cells grow at room temperature without CO_2 as a loose, semi-adherent monolayer in tissue culture flasks and in suspension in spinners and shaker flasks (Figure 3.14). They can be easily dislodged from the flask by blowing medium over the surface using a pipette. As the cells double every 24 h, they must be split 10-fold every three to four days in order to remain healthy and to ensure normal cell growth. Insect cells are density-sensitive; they die if too dense or too dilute. The following protocol is designed to initiate a cell culture from a frozen stock, and is modified from the Invitrogen S2 cell line protocol.

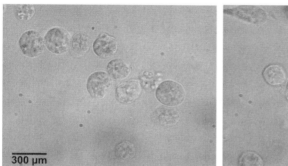
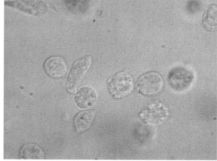

300 µm

Fig. 3.14 Images of S2 cell in culture. The morphology of the cells varies widely, as shown in the left image.

3.6.1
Thawing and Maintenance of S2 Cells

PROTOCOL 11

1. Remove the vial of cells from liquid nitrogen and thaw quickly at room temperature.

▶ **Note**: The vial of S2 cells should contain ~1×10^7 cells. Upon thawing, cells should have a viability of 60–70%.

2. Just before the cells are completely thawed, transfer the cells to a 25-cm² flask containing 5 ml of S2 insect cell medium (DES complete; *Schneider's Drosophila* (Invitrogen), supplemented with 10% heat-inactivated fetal calf serum).

▶ **Note**: Once the culture is established, cell viability should be >95%.

3. Incubate at 22–24 °C for 5 to 16 h (overnight).

4. Replace the medium to remove the DMSO left over from the freezing medium.

5. Incubate at 22–24 °C until cells reach a density of 6 to 20×10^6 cells/ml. This may take 3–4 days.

▶ **Note**: Cells will start to clump at a density of $\sim 5 \times 10^6$ cells/ml in serum-containing medium. This does not seem to affect growth. Clumps can be broken up during passage.

6. S2 cells should be subcultured to a final density of 2 to 4×10^6 cells/ml. Do not split cells below a density of 0.5×10^6/ml.

▶ **Note**: Do not passage the cells when they are not dense enough. This will stress the cells, and they will not leave the bottom of the flask with simple tapping and dislodging.

7. Tap the flask several times to dislodge the cells, and use a pipet to remove the remaining adherent cells from the bottom.

8. Briefly pipette the solution up and down to break up clumps of cells.

9. Split cells at a 1 : 5 dilution into new culture vessels.

▶ **Note**: The cells are dividing very rapidly. Split them regularly. If they are growing too densely they will start to differentiate into elongated cells.

10. Expand the cells from an early passage and freeze them.

▶ **Note**: Differentiation of the cells occurs when the culture ages. Do not keep cells in culture for longer than 2 months.

11. Perform RNAi experiments on S2 cells of a low passage number.

3.6.2
Freezing Protocol

PROTOCOL 12

1. When the cell density is between 1.0–2.0×10^7/ml in a 75-cm^2 flask, dislodge the cells from the flask.

▶ **Note**: There should be 12 ml of cell suspension.

2. Count a sample of cells in a hemocytometer to determine actual cells per ml, and the viability (95–99 %).

3. Pellet the cells by centrifuging at 1000 g for 2–3 min at +4 °C.

▶ **Note**: Be sure to reserve the medium after centrifuging cells; it will serve as the conditioned medium for freezing. Optimal recovery of S2 cells requires growth factors in the medium. Be sure to use conditioned medium in the freezing medium. In addition, fetal calf serum that has not been heat-inactivated will inhibit the growth of S2 cells.

4. Resuspend the cells in 10 ml phosphate-buffered saline, and pellet at 1000 g for 2–3 min.

▶ **Note**: Meanwhile prepare the freezing medium (Table 3.13).

Tab. 3.13 Freezing medium.

Freezing medium	
Schneider's *Drosophila* medium (conditioned) + 10% FCS	45%
DMSO	10%
FCS	10%

Conditioned = medium from a growing culture

5. Resuspend the cells at a density of 1.1×10^7/ml in freezing medium.

6. Aliquot 1 ml of the cell suspension per vial.

▶ **Note**: Repeat resuspending the cells to avoid a cell density gradient.

7. Freeze cells in a controlled-rate freezer (Nalgene cryo-bottle) to –80 °C, or place the vials into aluminum block cooled to 4 °C. Place the aluminum block in a styrofoam box and transfer the containers to –80 °C and hold for 24 h to allow the cells to freeze very slowly.

8. Transfer vials to liquid nitrogen for long-term storage.

3.7
RNAi in S2 Cells

Three methods have been reported for RNAi in S2 cells, including transfection with calcium-phosphate (Hammond et al. 2000), lipofection (Lipofectamine, Invitrogen or Superfect, Qiagen) (Caplen et al. 2000), and simple soaking of the naked dsRNA (Clemens et al. 2000). So far, for transient assays soaking has been shown to be the simplest and most efficient method to apply RNAi in S2 cells (Figure 3.15). Other cells, such as primary insect cells, rather require other methods (Biyasheva et al. 2001). It has been shown that primary cells can be transfected with hypertonic medium (Okada and Rechsteiner 1982), allowing pinocytosis of the dsRNA. Changing from a hypertonic to a hypotonic medium releases the dsRNA into the cytosol (Okada, 1982, # 4172; Carthew, 2003, #4173).

3.7.1
dsRNA Transfection Using the Calcium Phosphate Method

This calcium phosphate method is not the most favored, and for novices it may cause some difficulties. The method requires careful preparation of the transfection reagent, and the pH of the solution is crucial for successful transfection. For novices,

Control cells

dsRNA-treated cells

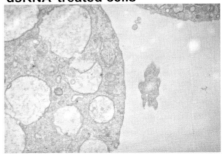

Fig. 3.15 Electron microscopy image of a dsRNA soaking experiment. Cell were incubated with 15 µg per 6-well dsRNA encoding clathrin heavy chain mRNA. Whilst the control cells show clathrin-coated vesicles and clathrin-coated invaginations (left image, see also magnifi-cation of the membrane), dsRNA-treated cells completely lack those structures (right image). If there is a need to pre-screen the phenotype of a mammalian homologue in *Drosophila*, it must be remembered that S2 cells show a different organelle structuring than mammalian cells.

it may be better to use commercially available transfection reagents such as the Phar-mingen transfection buffer set A+B (cat. # 554806) to ensure a good transfection effi-ciency. Transfection Buffer A is Grace's insect medium containing 10% fetal calf serum. Transfection Buffer B contains 25 mM HEPES, pH 7.1, 125 mM $CaCl_2$, and 140 mM NaCl. For the regular calcium phosphate method, refer to the established cell culture methods (Hammond et al. 2000)

3.7.2
dsRNA Soaking of S2 Cells

There are two ways of applying dsRNAs to S2 cell cultures to induce RNAi when using the soaking method. One can either plate the cells first, before adding the dsRNA, or one can coat the plate with dsRNA before plating the cells. Either of the methods works well.

PROTOCOL 13

Method 1:

1. Count the S2 cells and centrifuge them for 5 min at 1200 r.p.m. at room temperature.

2. Resuspend cells at $1–5 \times 10^6$ cells/ml in the appropriate media.

▶ **Note:** This step can be performed in either serum-containing or serum-free medium. However, using serum-containing media accelerates the attach-ment of the cells to the bottom of the dish.

3. Plate 1×10^6 cells per well of a 6-well tissue culture plate and let the cells adhere for 20 min.

4. Meanwhile, resuspend the dsRNA in DEPC water.

▶ **Note**: The dsRNA dilution should contain a concentration of at least 1 µg/µl.

5. Replace the medium with 1 ml serum-free S2 medium.

6. Dilute the dsRNA in annealing buffer (Table 3.14) to a final concentration of 1–2 mg/ml.

Tab. 3.14 Annealing buffers.

Annealing buffer I (1x)		Annealing buffer II (10x)	
Potassium acetate	100 mM	Tris-HCl, pH 8.0	100 mM
HEPES-KOH, pH 7.4	30 mM	EDTA, pH 8.0	10 mM
Magnesium acetate	2 mM	NaCl	1 M

7. Add 10–70 µg dsRNA per well to the cells in a dropwise fashion while rotating the plate.

8. Incubate for 30 min at room temperature.

▶ **Note**: Slow rotation of the 6-well plate on a tangential shaker during the incubation will keep the cell layer from drying, and will evenly distribute the dsRNA.

9. Add 2 ml serum-containing medium (10% fetal calf serum).

10. Incubate at 24 °C for 3–4 days.

▶ **Note**: The incubation time will depend on the turnover of the protein to be analyzed.

Method 2:

1. Dilute dsRNA in water or annealing buffer (see Table 3.14).

2. Add 10–70 µg dsRNA per well of a 6-well tissue culture plate and cover the entire bottom of the dish.

3. Count the S2 cells and centrifuge them for 5 min at 1200 r.p.m. at room temperature.

4. Resuspend cells at $1–5 \times 10^6$ cells/ml in serum-free medium.

5. Plate 1×10^6 cells per well of a 6-well tissue culture plate.

6. Incubate dsRNA with cells at room temperature for 30 min.

7. Add 2 ml serum-containing medium (10% fetal calf serum) per well.

8. Incubate for 3–4 days and analyze.

For the application of RNAi in other sizes of tissue culture plates, the number of cells, volumes and dsRNA concentrations must be adjusted.

3.8
High-Throughput Screens

The availability of sequence data from the complete genome and the easy application of RNAi in S2 cell cultures allows the design of a functional genomic approach to many cell biological processes.

This goal was already the focal point of researchers at Harvard University (*Drosophila* RNAi Screening Center, http://134.174.160.115/RNAi_index.html), who have generated a set of 21 000 dsRNAs covering every annotated gene in the *Drosophila* genome (http://flyrnai.org and http://134.174.160.115/RNAi_index.html). They have developed high-throughput RNAi screens using *Drosophila S2* cells cultured in 384-well plates covered with an array of gene-specific dsRNAs. After incubation, the phenotype of the cells is analyzed mainly by fully automated microscopic imaging (Figure 3.16).

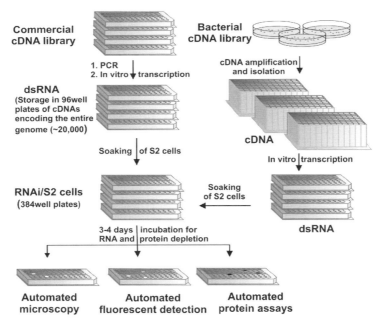

Fig. 3.16 Schematic view of a set-up for high-throughput RNAi screens in *Drosophila* cell lines modified from the DRSC (http://134.174.160.115/ RNAi_ index.html). The screening can be started with prepared *Drosophila* DNA libraries, or with bacterial libraries. The phenotyping depends on the respective phenotyping assay used.

In order to generate those genome-wide screens, dsRNAs are transcribed from a *Drosophila* cDNA library, and are eventually aliquoted into unique wells (~10–30 µl per well) of fifty-six, 384-well assay plates. Only minimal amounts of dsRNA are necessary (25–75 nM, or ~0.2 µg of 500-nt length) to completely degrade the endogenous mRNA.

Finally, the cells are plated directly into the dsRNA-containing assay plate wells in serum-free medium, but this is later changed to full medium. The resulting phenotypes are analyzed in a fully automated fashion depending on the respective cell-based assay. Data acquisition from the 384-well plates is achieved using a very sophisticated automated microscopy system, the so-called "autoscope". Recently, several reports have been made of genome-wide RNAi in *Drosophila*, using the methods described above (Boutros et al. 2004; Kiger et al. 2003; Lum et al. 2003). Beside the participating groups at Harvard, other groups can ask to screen the library by submitting a research proposal. To initiate such a screening, check the application webpage of the *Drosophila* RNAi Screening Center at Harvard Medical School (http://134.174.160.115/RNAi_index.html).

3.8.1
Drosophila RNAi Library

In order to facilitate RNAi experiments, or to set up your own RNAi screen, genomic or cDNA *Drosophila* libraries can be purchased from several sources. This can save time, especially for the cloning of respective templates. A *Drosophila* RNAi library provided by MRC geneservice was constructed together with Cyclacel Limited (http://www.cyclacel.com) at The Wellcome CRC Institute, University of Cambridge, Cambridge, UK. The same MRC geneservice (http://www.hgmp.mrc.ac.uk/geneservice/reagents/products/) as described in Chapter 2 for *C. elegans* offers a collection of 13 600 RNAi constructs for *Drosophila* species covering ~90% of its genome (Figure 3.17) The whole genome/mRNA pool is arrayed in 142 96-well microtiter plates.

Drosophila genomic/cDNA fragments were PCR-amplified using gene-specific primers with dual T7 promoter sequences ready for RNA synthesis using an in-vitro transcription reaction. The fragment sizes ranged between 300 and 1000 bp. A database of genes and primer sequences on the MRC website allows the search for genes by name or plate ids. The PCR products (~10 µl of each) are provided as individual 96-well plates or as a complete set of 142 plates. A Finder Tool (http://www.hgmp.mrc.ac.uk/geneservice/reagents/tools/DrosRNAi_Finder.shtml) enables the user to translate from the MRC Plate ID to the database.

For other resources of *Drosophila* genomic DNA, ESTs, or even cDNAs, refer to the BDGP web site (http://www.fruitfly.org). This is an abundant resource of almost all genomic clones, cDNAs and ESTs, that is constantly updated (Rubin et al. 2000a; Stapleton et al. 2002). The cDNA listed are available, and can be purchased. Another source for *Drosophila* clones is Open Biosystems (http://www.openbiosystems.com/Drosophila_rnai_collection.php?), who provide the *Drosophila* RNAi collection version 1.0, a collection of dsDNA constructs for generation of dsRNA.

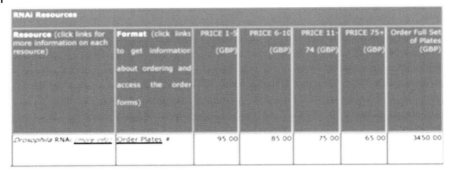

RNAi Resources						
Resource (click links for more information on each resource)	**Format** (click links to get information about ordering and access the order forms)	PRICE 1-5 (GBP)	PRICE 6-10 (GBP)	PRICE 11-74 (GBP)	PRICE 75+ (GBP)	Order Full Set of Plates (GBP)
Drosophila RNAi (more info)	Order Plates #	95.00	85.00	75.00	65.00	3450.00

Fig. 3.17 Strain order sheet on the MRC geneservice website.

3.9
Useful Webpages for *Drosophila* Research

- The most useful web page on *Drosophila* research: http://flybase.bio.indiana.edu/

- Online atlas and database of the *Drosophila* nervous system: http://www.flybrain.org/

- The Berkeley *Drosophila* Genome Project: DNAs, cDNA, ESTs, and annotations: http://www.fruitfly.org

- A basic introduction to *Drosophila*: http://www.ceolas.org/fly/intro.html

- A cyberspace guide to *Drosophila* genes and their roles in development: http://sdb.bio.purdue.edu/fly/aimain/1aahome.htm

- A daily updated list of interesting RNAi publications: http://www.orbigen.com/RNAi_Orbigen.html

- Very detailed protocols on injection of embryos: http://www.biochem.northwestern.edu/carthew/

Other web pages that are interesting are mentioned in the respective chapters in the text.

3.10
Books and Literature on *Drosophila*

Many biology and developmental biology textbooks contain chapters on *Drosophila*. These include:

- *The Making of a Fly* (Peter Lawrence; Blackwell Scientific, 1992). An easy and simple introduction to *Drosophila* development.

- *Drosophila* (Brian Shorrocks; Ginn & Co, London, 1972). General biology of the fly, with chapters on laboratory and field ecology.

- The Development of *Drosophila melanogaster* (ed. Bate & Martinez-Arias; Cold Spring Harbor Press, 1993). A major summary of current knowledge on most developmental aspects of *Drosophila* (not suitable for novices).

- Biology of *Drosophila* (ed. M. Demerec; Cold Spring Harbor Press, 1994). A descriptive biology of the fly.

- The Embryonic Development of *Drosophila melanogaster* (Campos-Ortega & Hartenstein; Springer-Verlag, 1985). A detailed book on embryonic development.

- *Drosophila*, a Practical Approach (ed. DB Roberts; IRL Press, 1986). An edited book covering laboratory techniques and molecular biology.

- *Drosophila*, a Laboratory Manual (Michael Ashburner; Cold Spring Harbor Press, 1989). A "recipe book" for many techniques.

- *Drosophila* Protocols (ed. W Sullivan, M Ashburner, RS Hawley, Cold Spring Harbor Press, 2000). An edited book covering laboratory techniques and molecular biology. This contains state-of-the-art protocols.

- *Drosophila* melanogaster: Practical Uses in Cell and Molecular Biology (ed. Goldstein & Fryberg; Academic Press 1994). Excellent coverage of cell biological techniques in *Drosophila* with both recipes and explanations. This is Volume 44 in the *Methods in Cell Biology* series.

3.11
References

ADAMS MD, CELNIKER SE, HOLT RA, EVANS CA, et al. (**2000**) The genome sequence of *Drosophila* melanogaster. *Science* 287: 2185–2195.

ASHBURNER M (**1989**) Drosophila, *a Laboratory Manual*. Cold Spring Harbor Press, Cold Spring Harbor, NY, USA.

BELLO B, RESENDEZ-PEREZ D, GEHRING WJ (**1998**) Spatial and temporal targeting of gene expression in *Drosophila* by means of a tetra-cycline-dependent transactivator system. *Development* 125: 2193–2202.

BIYASHEVA A, DO TV, LU Y, VASKOVA M, ANDRES AJ (**2001**) Glue secretion in the *Drosophila* salivary gland: A model for steroid-regulated exocytosis [Review]. *Dev Biol* 231: 234–251.

BLOCHLINGER K, JAN LY, JAN YN (**1991**) Transformation of sensory organ identity by ectopic expression of Cut in *Drosophila*. *Genes Dev* 5: 1124–1135.

BOUTROS M, KIGER AA, ARMKNECHT S, KERR K, HILD M, KOCH B, HAAS SA, CONSORTIUM HF, PARO R, PERRIMON N (**2004**) Genome-wide RNAi analysis of growth and viability in *Drosophila* cells. *Science* 303: 832–835.

BRAND AH, MANOUKIAN AS, PERRIMON N (**1994**) Ectopic expression in *Drosophila*. *Methods Cell Biol* 44: 635–654.

BRAND AH, PERRIMON N (**1993**) Targeted gene expression as a means of altering cell fates and generating dominant phenotypes. *Development* 118: 401–415.

BZYMEK M, LOVETT ST (**2001**) Evidence for two mechanisms of palindrome-stimulated deletion in *Escherichia coli*: single-strand annealing and replication slipped mispairing. *Genetics* 158: 527–540.

CAPLEN NJ, FLEENOR J, FIRE A, MORGAN RA (**2000**) dsRNA-mediated gene silencing in cultured *Drosophila* cells: a tissue culture model for the analysis of RNA interference. *Gene* 252: 95–105.

CARTHEW RW (**2003**) *RNAi applications in* Drosophila melanogaster, 11th edn. Cold Spring Harbor Press, Cold Spring Harbor, New York.

CELNIKER SE, RUBIN GM (**2003**) The *Drosophila melanogaster* genome. *Annu Rev Genomics Hum Genet* 4: 89–117.

CELNIKER SE, WHEELER DA, KRONMILLER B, CARLSON JW, et al. (**2002**) Finishing a whole-genome shotgun: release 3 of the *Drosophila melanogaster* euchromatic genome sequence. *Genome Biol* 3: RESEARCH0079.

CHERBAS L, CHERBAS P (**1998**) *Cell Culture*, 2nd edn. IRL Press at Oxford University Press, UK, Oxford.

CLEMENS JC, WORBY CA, SIMONSON-LEFF N, MUDA M, MAEHAMA T, HEMMINGS BA, DIXON JE (**2000**) Use of double-stranded RNA interference in *Drosophila* cell lines to dissect signal transduction pathways. *Proc Natl Acad Sci USA* 97: 6499–6503.

DAVISON A, LEACH DR (**1994**) The effects of nucleotide sequence changes on DNA secondary structure formation in *Escherichia coli* are consistent with cruciform extrusion in vivo. *Genetics* 137: 361–368.

DUBENDORFER A, SHIELDS G, SANG JH (**1975**) Development and differentiation in vitro of *Drosophila* imaginal disc cells from dissociated early embryos. *J Embryol Exp Morphol* 33: 487–498.

DUFFY JB (**2002**) GAL4 system in *Drosophila*: a fly geneticist's Swiss army knife. *Genesis* 34: 1–15.

ECHALIER G, OHANESSIAN A (**1970**) In vitro culture of *Drosophila melanogaster* embryonic cells. *In Vitro* 6: 162–172.

ENERLY E, LARSSON J, LAMBERTSSON A (**2003**) Silencing the *Drosophila* ribosomal protein L14 gene using targeted RNA interference causes distinct somatic anomalies. *Gene* 320: 41–48.

FAGARD M, VAUCHERET H (**2000**) Systemic silencing signal(s). *Plant Mol Biol* 43: 285–293.

FOE VE, ODELL GM, EDGAR BA (**1993**) *Mitosis and morphogenesis in the* Drosophila embryo. Cold Spring Harbor Press, Cold Spring Harbor, New York.

FORTIER E, BELOTE JM (**2000**) Temperature-dependent gene silencing by an expressed inverted repeat in *Drosophila*. *Genesis* 26: 240–244.

GIORDANO E, RENDINA R, PELUSO I, FURIA M (**2002**) RNAi triggered by symmetrically transcribed transgenes in *Drosophila melanogaster*. *Genetics* 160: 637–648.

GONZALEZ-REYES A, MORATA G (**1990**) The developmental effect of overexpressing a Ubx product in *Drosophila* embryos is dependent on its interactions with other homeotic products. *Cell* 61: 515–522.

HAARS R, ZENTGRAF H, GATEFF E, BAUTZ FA (**1980**) Evidence for endogenous reovirus-like particles in a tissue culture cell line from *Drosophila melanogaster*. *Virology* 101: 124–130.

HAMMOND SM, BERNSTEIN E, BEACH D, HANNON GJ (**2000**) An RNA-directed nuclease mediates post-transcriptional gene silencing in *Drosophila* cells. *Nature* 404: 293–296.

HANAHAN D (1983) Studies on transformation of *Escherichia coli* with plasmids. *J Mol Biol* 166: 557–580.

HILD M, BECKMANN B, HAAS SA, KOCH B, SOLOVYEV V, BUSOLD C, FELLENBERG K, BOUTROS M, VINGRON M, SAUER F, HOHEISEL JD, PARO R (**2003**) An integrated gene annotation and transcriptional profiling approach towards the full gene content of the *Drosophila* genome. *Genome Biol* 5: R3.

HOSKINS RA, SMITH CD, CARLSON JW, CARVALHO AB, HALPERN A, KAMINKER JS, KENNEDY C, MUNGALL CJ, SULLIVAN BA, SUTTON GG, YASUHARA JC, WAKIMOTO BT, MYERS EW, CELNIKER SE, RUBIN GM, KARPEN GH (**2002**) Heterochromatic sequences in a *Drosophila* whole-genome shotgun assembly. *Genome Biol* 3: RESEARCH0085.

ISH-HOROWICZ D, PINCHIN SM (**1987**) Pattern abnormalities induced by ectopic expression of the *Drosophila* gene hairy are associated with repression of ftz transcription. *Cell* 51: 405–415.

ISH-HOROWICZ D, PINCHIN SM, INGHAM PW, GYURKOVICS HG (**1989**) Autocatalytic ftz activation and metameric instability induced by ectopic ftz expression. *Cell* 57: 223–232.

JOHNSON RL, MILENKOVIC L, SCOTT MP (**2000**) In vivo functions of the patched protein: requirement of the C terminus for target gene inactivation but not Hedgehog sequestration. *Mol Cell* 6: 467–478.

KALIDAS S, SMITH DP (**2002**) Novel genomic cDNA hybrids produce effective RNA interference in adult *Drosophila*. *Neuron* 33: 177–184.

KENNERDELL JR, CARTHEW RW (**1998**) Use of dsRNA-mediated genetic interference to demonstrate that frizzled and frizzled 2 act in the wingless pathway. *Cell* 95: 1017–1026.

KENNERDELL JR, CARTHEW RW (**2000**) Heritable gene silencing in *Drosophila* using double-stranded RNA. *Nature Biotechnol* 18: 896–898.

KIGER A, BAUM B, JONES S, JONES M, COULSON A, ECHEVERRI C, PERRIMON N (**2003**) A functional genomic analysis of cell morphology using RNA interference. *J Biol* 2: 27.

KLUEG KM, ALVARADO D, MUSKAVITCH MA, DUFFY JB (**2002**) Creation of a GAL4/UAS-coupled inducible gene expression system for use in *Drosophila* cultured cell lines. *Genesis* 34: 119–122.

LAM G, THUMMEL CS (**2000**) Inducible expression of double-stranded RNA directs specific genetic interference in *Drosophila*. *Curr Biol* 10: 957–963.

LEACH DR (**1994**) Long DNA palindromes, cruciform structures, genetic instability and secondary structure repair. *BioEssays* 16: 893–900.

LEE YS, CARTHEW RW (**2003**) Making a better RNAi vector for *Drosophila*: use of intron spacers. *Methods* 30: 322–329.

LUM L, YAO S, MOZER B, ROVESCALLI A, VON KESSLER D, NIRENBERG M, BEACHY PA (**2003**) Identification of Hedgehog pathway components by RNAi in *Drosophila* cultured cells. *Science* 299: 2039–2045.

MARTINEK S, YOUNG MW (**2000**) Specific genetic interference with behavioral rhythms in *Drosophila* by expression of inverted repeats. *Genetics* 156: 1717–1725.

MISQUITTA L, PATERSON BM (**1999**) Targeted disruption of gene function in *Drosophila* by RNA interference (RNA-i): A role for nautilus in embryonic somatic muscle formation. *Proc Natl Acad Sci USA* 96: 1451–1456.

MYERS EW, SUTTON GG, DELCHER AL, DEW IM, et al. (**2000**) A whole-genome assembly of *Drosophila*. *Science* 287: 2196–2204.

NAGEL AC, MAIER D, PREISS A (**2002**) Green fluorescent protein as a convenient and versatile marker for studies on functional genomics in *Drosophila*. *Dev Genes Evolution* 212: 93–98.

OKADA CY, RECHSTEINER M (**1982**) Introduction of macromolecules into cultured mammalian cells by osmotic lysis of pinocytic vesicles. *Cell* 29: 33–41.

PALAUQUI JC, ELMAYAN T, POLLIEN JM, VAUCHERET H (**1997**) Systemic acquired silencing: transgene-specific post-transcriptional silencing is transmitted by grafting from silenced stocks to non-silenced scions. *EMBO J* 16: 4738–4745.

PARKHURST SM, BOPP D, ISH-HOROWICZ D (**1990**) X:A ratio, the primary sex-determining signal in *Drosophila*, is transduced by helix-loop-helix proteins. *Cell* 63: 1179–1191.

PEEL DJ, JOHNSON SA, MILNER MJ (**1990**) The ultrastructure of imaginal disc cells in primary cultures and during cell aggregation in continuous cell lines. *Tissue Cell* 22: 749–758.

PETERSEN NS (**1990**) Effects of heat and chemical stress on development. *Adv Genet* 28: 275–296.

PETERSEN NS, MITCHELL HK (**1987**) The induction of a multiple wing hair phenocopy by heat shock in mutant heterozygotes. *Dev Biol* 121: 335–341.

PHELPS CB, BRAND AH (**1998**) Ectopic gene expression in *Drosophila* using GAL4 system. *Methods* 14: 367–379.

PICCIN A, SALAMEH A, BENNA C, SANDRELLI F, MAZZOTTA G, ZORDAN M, ROSATO E, KYRIACOU CP, COSTA R (**2001**) Efficient and heritable functional knock-out of an adult phenotype in *Drosophila* using a GAL4-driven hairpin RNA incorporating a heterologous spacer. *Nucleic Acids Res* 29: E55.

REICHHART JM, LIGOXYGAKIS P, NAITZA S, WOERFEL G, IMLER JL, GUBB D (**2002**) Splice-activated UAS hairpin vector gives complete RNAi knockout of single or double target transcripts in *Drosophila melanogaster*. *Genesis* 34: 160–164.

RUBIN GM, HONG L, BROKSTEIN P, EVANS-HOLM M, FRISE E, STAPLETON M, HARVEY DA (**2000a**) A *Drosophila* complementary DNA resource. *Science* 287: 2222–2224.

RUBIN GM, YANDELL MD, WORTMAN JR, GABOR MIKLOS GL, et al. (**2000b**) Comparative genomics of the eukaryotes. *Science* 287: 2204–2215.

SAMBROOK J, RUSSELL DW (**2001**) *Molecular Cloning: A Laboratory Manual*, 3rd edn. Cold Spring Harbor Press, New York.

SCHMID A, SCHINDELHOLZ B, ZINN K (**2002**) Combinatorial RNAi: a method for evaluating the functions of gene families in *Drosophila*. *Trends Neurosci* 25: 71–74.

SCHNEIDER I (**1972**) Cell lines derived from late embryonic stages of *Drosophila melanogaster*. *J Embryol Exp Morphol* 27: 353–365.

SCHNEIDER I, BLUMENTHAL AB (**1978**) *Drosophila cell and tissue culture*. Academic Press-cher, London.

SCHNEUWLY S, KLEMENZ R, GEHRING WJ (**1987**) Redesigning the body plan of *Drosophila* by ectopic expression of the homoeotic gene Antennapedia. *Nature* 325: 816–818.

SHIELDS G, DUBENDORFER A, SANG JH (**1975**) Differentiation in vitro of larval cell types from early embryonic cells of *Drosophila melanogaster*. *J Embryol Exp Morphol* 33: 159–175.

SHIELDS G, SANG JH (**1970**) Characteristics of five cell types appearing during in vitro culture of embryonic material from *Drosophila melanogaster*. *J Embryol Exp Morphol* 23: 53–69.

SINDEN RR, ZHENG GX, BRANKAMP RG, ALLEN KN (**1991**) On the deletion of inverted repeated DNA in *Escherichia coli*: effects of length, thermal stability, and cruciform formation in vivo. *Genetics* 129: 991–1005.

SMITH NA, SINGH SP, WANG MB, STOUTJEDIJK PA, GREEN AG, WATERHOUSE PM (**2000**) Total silencing by intron spliced hairpin RNAs. *Nature* 407: 319–320.

SPRADLING AC (**1986**) *P-element-mediated transformation*. IRL Press, Oxford.

SPRADLING AC, RUBIN GM (**1982**) Transposition of cloned P elements into *Drosophila* germline chromosomes. *Science* 218: 341–347.

STAPLETON M, LIAO G, BROKSTEIN P, HONG L, CARNINCI P, SHIRAKI T, HAYASHIZAKI Y, CHAMPE M, PACLEB J, WAN K, YU C, CARLSON J, GEORGE R, CELNIKER S, RUBIN GM (**2002**) The *Drosophila* gene collection: identification of putative full-length cDNAs for 70 % of *D. melanogaster* genes. *Genome Res* 12: 1294–1300.

STEINGRIMSSON E, PIGNONI F, LIAW GJ, LENGYEL JA (**1991**) Dual role of the *Drosophila* pattern gene tailless in embryonic termini. *Science* 254: 418–421.

STRUHL G (**1985**) Near-reciprocal phenotypes caused by inactivation or indiscriminate expression of the *Drosophila* segmentation gene ftz. *Nature* 318: 677–680.

SULLIVAN W, ASHBURNER M, HAWLEY RS (**2000**) *Drosophila* Protocols. Cold Spring Harbor Press, NY, USA.

SZUTS D, BIENZ M (**2000**) LexA chimeras reveal the function of *Drosophila* Fos as a context-dependent transcriptional activator. *Proc Natl Acad Sci USA* 97: 5351–5356.

TIMMONS L, TABARA H, MELLO CC, FIRE AZ (**2003**) Inducible systemic RNA silencing in *Caenorhabditis elegans*. *Mol Biol Cell* 14: 2972–2983.

UI K, NISHIHARA S, SAKUMA M, TOGASHI S, UEDA R, MIYATA Y, MIYAKE T (**1994**) Newly established cell lines from *Drosophila* larval CNS express neural specific characteristics. *In Vitro Cell Dev Biol Anim* 30A: 209–216.

VAN ROESSEL P, BRAND AH (**2002**) Imaging into the future: visualizing gene expression and protein interactions with fluorescent proteins. *Nature Cell Biol* 4: E15–E20.

VAN ROESSEL P, HAYWARD NM, BARROS CS, BRAND AH (**2002**) Two-color GFP imaging demonstrates cell-autonomy of GAL4-driven RNA interference in *Drosophila*. *Genesis* 34: 170–173.

WIMMER EA (**2003**) Innovations: applications of insect transgenesis. *Nature Rev Genet* 4: 225–232.

WINSTON WM, MOLODOWITCH C, HUNTER CP (**2002**) Systemic RNAi in *C. elegans* requires the putative transmembrane protein SID-1. *Science* 295: 2456–2459.

WYMAN AR, WOLFE LB, BOTSTEIN D (**1985**) Propagation of some human DNA sequences in bacteriophage lambda vectors requires mutant *Escherichia coli* hosts. *Proc Natl Acad Sci USA* 82: 2880–2884.

YANAGAWA S, LEE JS, ISHIMOTO A (**1998**) Identification and characterization of a novel line of *Drosophila* Schneider S2 cells that respond to wingless signaling. *J Biol Chem* 273: 32353–32359.

YANG MY, ARMSTRONG JD, VILINSKY I, STRAUSFELD NJ, KAISER K (**1995**) Subdivision of the *Drosophila* mushroom bodies by enhancer-trap expression patterns. *Neuron* 15: 45–54.

YANISCH-PERRON C, VIEIRA J, MESSING J (**1985**) Improved M13 phage cloning vectors and host strains: nucleotide sequences of the M13mp18 and pUC19 vectors. *Gene* 33: 103–119.

ZAMORE PD, TUSCHL T, SHARP PA, BARTEL DP (**2000**) RNAi: Double-stranded RNA directs the ATP-dependent cleavage of mRNA at 21 to 23 nucleotide intervals. *Cell* 101: 25–33.

ZHENG GX, KOCHEL T, HOEPFNER RW, TIMMONS SE, SINDEN RR (**1991**) Torsionally tuned cruciform and Z-DNA probes for measuring unrestrained supercoiling at specific sites in DNA of living cells. *J Mol Biol* 221: 107–122.

ZUKER CS, MISMER D, HARDY R, RUBIN GM (**1988**) Ectopic expression of a minor *Drosophila* opsin in the major photoreceptor cell class: distinguishing the role of primary receptor and cellular context. *Cell* 53: 475–482.

4
RNAi in Mammals

4.1
Introduction

The interesting features of RNAi which were demonstrated in *C. elegans* and *Drosophila* led to many investigations being conducted that focused on the ultimate adaptation of this technique in mammalian and human cell lines. Initially, the application of RNAi in mammalian cells seemed impossible, mainly due to the fact that mammals evolved a different anti-dsRNA protection mechanism. The introduction of dsRNA into mammalian cell lines results in an interferon-like response, part of which involves the activation of dsRNA-responsive protein kinase R (PKR) – an enzyme which forms part of the mammalian defense machinery against viruses (Clemens 1997; Clemens and Elia 1997). PKR phosphorylates and inactivates the translation factor EIF2α, leading to the global suppression of protein biosynthesis and subsequently to programmed cell death (apoptosis). Furthermore, interferon induces and activates the 2'-5'-oligoadenylate synthase and RNaseL, which eventually triggers the non-specific degradation of mRNA (Figure 4.1).

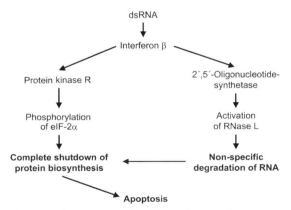

Fig. 4.1 Schematic view of the mammalian interferon response.

RNA Interference in Practice: Principles, Basics, and Methods for Gene Silencing in C. elegans, Drosophila, and Mammals. Ute Schepers
Copyright © 2005 WILEY-VCH Verlag GmbH & Co. KGaA, Weinheim
ISBN: 3-527-31020-7

Despite these arguments that RNAi would not function in mammalian cells, several independent groups proved the existence of mammalian RNAi pathways by the introduction of dsRNA or vectors producing dsRNA in cell lines which lacked the interferon machinery, such as mice oocytes or mice embryonic cancer cell lines (Billy et al. 2001; Wianny and Zernicka-Goetz 2000). These findings suggest either that embryonic cells have a more effective RNAi response, or that the interferon response to dsRNA is absent or not yet fully developed. However, in most somatic mammalian cells this approach provokes a strong cytotoxic response.

The decisive breakthrough in applying this new RNAi technique to the field of mammalian functional genomics was made in the studies by Tuschl and colleagues (Elbashir et al. 2001a). These authors found that transiently applied siRNAs of 21–23 nt are able to trigger the RNAi machinery in cultured mammalian cells, without initiating the programmed cell death response. The finding that transfection of synthetic 21 nt-long siRNA duplexes induces RNAi rather than the interferon response created a new hype in the field (Elbashir et al. 2001a; Harborth et al. 2001). However, recent experiments have shown the up-regulation of interferon-stimulated genes, such as the Jak-Stat pathway upon transfection with siRNAs (Sledz et al. 2003).

Further studies by Tuschl and coworkers showed that dsRNAs shorter than 21 bp and longer than 25 bp are inefficient in initiating RNAi (Elbashir et al. 2001c), as well as siRNAs with blunt ends. Only short dsRNAs with a 2-nt 3'-overhang, which resembles the naturally active products of Dicer, are efficient mediators of RNAi. With this technology even somatic primary neurons have been successfully treated to produce knock-down RNAi phenotypes (Krichevsky and Kosik 2002).

Currently, several methods are available for generating siRNAs for RNAi studies:

- chemical synthesis
- in-vitro transcription
- digestion of long dsRNA by recombinant Dicer or other RNase III family enzymes, ribozymes
- endogenous expression as short hairpin RNAs (shRNAs) or long hairpin RNAs (lhRNA) from expression plasmids or viral vectors
- endogenous expression from PCR-derived siRNA expression cassettes
- endogenous expression from allosterically regulated ribozymes
- commercially available siRNA/shRNA/lhRNA libraries

The first of these methods is mainly transient, and involves the in-vitro preparation of siRNAs. The uptake of these siRNAs to cultured cells can be mediated by lipid transfection reagents, electroporation, or other methods and generally result in a "knock-down" of the target gene expression by up to 90% (maximum). Other methods are based on the introduction of DNA-vectors which endogenously express siRNAs as short hairpin RNAs or dsRNA (shRNAs) as long hairpin RNAs (lhRNAs). Each of these methods has both advantages and disadvantages.

4.2
Transient RNAi in Cell Culture

4.2.1
Chemical Synthesis and Modifications of siRNAs

This aim of this chapter is to help those who wish to embark upon the synthesis of RNA to meet their needs in terms of modified siRNAs. Basically, the chapter is an overview of, or introduction to, modern RNA chemistry and state of the art technology, rather than serves as a detailed description of the different syntheses. In this respect it should guide novices in the field to the currently available literature, and help them through the first organizational steps. The synthesis of RNA requires a good knowledge of chemistry and the corresponding laboratory equipment, and if one or both of those prerequisites are absent, it might be best to purchase customized siRNAs from a commercial source, even though this is a very expensive process.

As the market for DNA oligonucleotides has expanded dramatically during the past decade, the solid-phase chemistry of DNA using phosphoramidites has become extensively optimized. However, the situation with RNA synthesis is much less efficient.

The ability to routinely synthesize RNA has become increasingly important as RNA has moved into the focus of research. The great hype derived from the discovery that small duplexes of synthetic RNA oligonucleotides can induce RNA interference in mammals has led to an improvement in their chemical synthesis. In contrast to DNA, RNA possesses an additional hydroxyl group at the 2'-position of the each ribose building block, which destabilizes RNA under the basic conditions (pH >12 at 25 °C) generally present in DNA synthesis. Hence, the most difficult step in RNA synthesis is the simultaneous protection of the 5'- and the 2'-hydroxyl groups during solid-phase chemistry. Consequently, an additional 2'-OH-protecting group requires the following features: Quantitative removal at the end of the synthesis, and stability under all reaction conditions. The 2'-OH-protecting groups can be divided into different subgroups, including acid- (Scaringe et al. 1998), photo- (Ohtsuka et al. 1974; Pitsch et al. 1999; Schwartz et al. 1995), and fluoride-labile groups (Beaucage and Caruthers 1996; Ogilvie et al. 1974). To date, most of the combinations used to protect the 5'-OH and 2'-OH functions are based on groups such as 5'-O-dimethoxytrityl (DMT) and the fluoride-labile 2'-O-tert-butyldimethylsilyl (TBDMS) protection, or a combination of 5'-O-DMT with 2'-O-[1-(2-fluorophenyl)-4-methoxypiperidin-4-yl] (FPMP), though this technology still faces many hurdles and as yet has not led to yields suitable for commercial success. Thus, the aim was to develop novel protecting groups at the 5'- and 2'- positions of RNA-ribonucleotides that would allow much higher yields of nucleoside coupling. During the past few years, two novel strategies have been reported that are equally well established in siRNA and RNA production. One of these strategies is based on the introduction of a 2'-O-[(triisopropylsilyl)oxy]methyl (TOM) group to protect the 2'-OH position in the phosphoramidites (Figure 4.2)

The 2'-O-TOM protection group was developed by Stefan Pitsch (Pitsch et al. 1999, 2001) and Xeragon, which was recently acquired by Qiagen (http://www.qiagen.

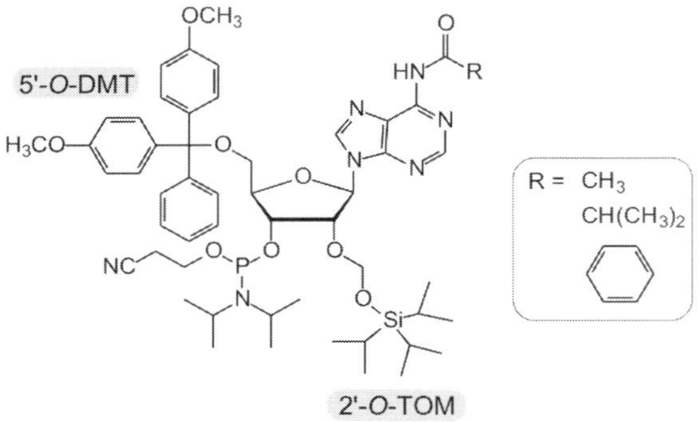

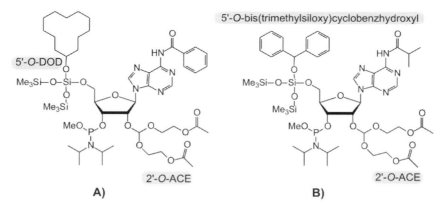

Fig. 4.2 Structure of a 5'-O-DMT- and 2'-TOM-protected adenosine phosphoramidite.

com). Qiagen is now one of the main providers of high-quality siRNA, and is using this technology for their RNA oligo production.

More recently, a novel combination of 5'-/2'-protection groups has been reported that is based on an acid-labile 2'-orthoester (bis(2-acetoxyethoxy)methyl (ACE) orthoester) and a 5'-silyl ether group (bis(trimethylsiloxy)cyclododecyloxysilyl (DOD) ether) (Scaringe 2001). The 2'-protection can be easily removed via mild acid-catalyzed hydrolysis, while the 5'-protection remains stable under these conditions. Eventually, the 5'-protection can be eliminated with fluoride ions under neutral pH. Both groups are stable during regular nucleoside phosphoramidite chemistry on solid support (Matteucci and Caruthers 1981; Matteucci and Caruthers 1992), and this can be accomplished on rebuild DNA synthesizers. High-quality siRNAs synthesized using this method are produced by Dharmacon (http://www.dharmacon.com). The structures of the protected and functionalized ribonucleoside phosphoramidites currently in use are illustrated in Figure 4.3.

Fig. 4.3 Structure of 5'-O-DOD- (A) and 5'-bis(trimethylsiloxy)-cyclobenzhydroxyl- (B) and 2'-O-ACE-protected adenosine phosphoramidites.

RNA oligonucleotides are synthesized in a stepwise fashion using the 3'- to 5'-nucleotide addition reaction cycle illustrated in Figure 4.4. Since 5'-DOD deprotection is performed with fluoride ions, which would destroy the control pore glass (CPG) usually used in DNA synthesis, the use of CPG as a solid support is prohibited. Instead, one can use aminomethyl polystyrene (Amersham, Dharmacon), which will then be loaded with the ribonucleoside succinate of base 1. For the first base, one can use a 5'-O-DMT-protection that facilitates the quantification of the polystyrene loading. Deprotection with 3% dichloroacetic acid or trichloroacetic acid in the presence of catalytic amounts of 4-dimethylaminopyridine in dichloromethane releases the DMT cation, which can be quantified by a simple and sensitive method for the quantitative determination of free amino groups. This method is based on a modification of a method originally described for free amino acids on solid supports (Ngo 1986).

The released DMT cation, which strongly absorbs at 498 nm ($\varepsilon(498) = 70\ 000$), is then determined spectrophotometrically (Gaur et al. 1989; Gaur 1989).

The synthesis is fully automated and comparable to that of DNA synthesis. The first nucleoside at the 3'-end of the chain is covalently attached to a solid support. After capping the 2'-OH position by esterification with acetic anhydride, the 5'-position is removed with trichloroacetic acid in dichloromethane to allow the coupling of the next nucleotide at the 5'-position and elongation of the chain in 3' → 5'-direction. In each cycle the chain is elongated by sequential addition of the respective 5'-O-DOD-2'-O-ACE-protected ribonucleoside phosphoramidite (Figure 4.4) to the solid support-bound oligonucleotide using S-ethyltetrazole as an activator. The support is washed, and any unreacted 5'-OH groups are capped with acetic anhydride to prevent the addition of other nucleoside phosphoramidites in the next cycles. The trivalent phosphorus atom (P(III)) at the 3'-position is then oxidized to the more stable pentavalent P(V). Finally, the 5'-DOD group is removed with triethylammonium fluoride ions (TEAHF) and the cycle is repeated to introduce the next nucleotide.

After addition of the last nucleotide, the methyl protecting groups on the phosphates are removed with 1 M disodium-2-carbamoyl-2-cyanoethylene-1,1-dithiolate trihydrate (S_2Na_2) in DMF (Dahl et al. 1990), and the support is washed with water.

Treatment with 40% methylamine (Reddy et al. 1994; Wincott et al. 1995) (10 min/55 °C) releases the RNA oligonucleotides from the support, while simultaneously deprotecting the exocyclic amines and removing the acetyl groups from the 2'-ACE protection. The resulting 2-ethyl-hydroxyl substituents are less electron-withdrawing than the acetylated full 2'-ACE, and this results in a more labile 2'-protection group to acid-catalyzed hydrolysis (http://www.dharmacon.com).

4.2.1.1 **Advantages**

The introduction of either of the novel protection groups, the 2'-TOM as well as the 5'-DOD (or Sil)-2'-ACE, has greatly improved RNA oligonucleotide chemistry, with coupling yields that are comparable to that for DNA. In particular, the 5'-DOD (or Sil)-2'-ACE chemistry allows a mild and fast acid deprotection. Since the 2'-ACE groups are very hydrophilic and readily solvated by water, a complete acid-

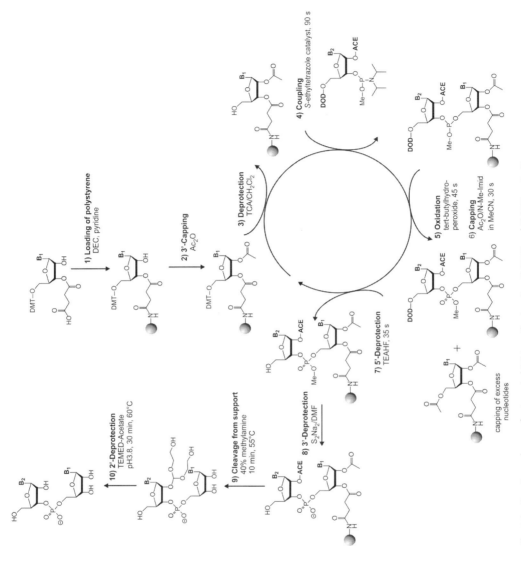

Fig. 4.4 Reaction scheme of automated RNA synthesis using 5'-DOD and 2'-ACE protection groups.

A)

B)

Fig. 4.5 Pictures of the commonly used DNA/RNA synthesizers:
(A) the Äkta Oligo Pilot from Amersham Biosciences; (B) the ABI 394
(or the updated version ABI 3400) from Applied Biosystems. To fit the
synthesis requirements, Dharmacon is offering a rebuilding service
of the ABI 394 (B) and the Äkta Oligo Pilot (A).

catalyzed hydrolysis can be achieved. The resulting 2′-ACE-protected oligonucleotides are water-soluble, which enables the routine handling of 2′-protected RNA in water.

4.2.1.2 Limitations
The 5′-DOD/2′-ACE technology requires a specialized DNA synthesizer, and several of these are available commercially. The choice may be between the Äkta Oligo Pilot from Amersham Biosciences (http://www.amershambiosciences.com) (Figure 4.5A) or the more commonly used Applied Biosystems DNA synthesizer ABI 394 (http://europe.appliedbiosystems.com). Both instruments have to be rebuilt to ensure that no glass compounds are present, and Dharmacon currently offers such a rebuilding service. This customization service includes remodeled valve blocks, new tubing, calibration, and final loading for direct use (http://www.dharmacon.com).

The respective 5′-O-DOD-2′-O-ACE phosphoramidites and all other chemicals used in the synthesis can be purchased directly from Dharmacon.

4.2.2
Custom Synthesis of siRNA Oligos

In the first successful attempt to trigger RNAi in mammalian cells, short double-stranded RNAs were used to mimic the siRNA intermediates of the RNAi pathway (Elbashir 2001a). However, in order to be recognized by RISC and thus efficiently mediate RNAi, the siRNA generated in vitro must meet a certain set of requirements.

Short dsRNAs of 21–23 bp have been shown to be the most efficient species. Short RNAs of less than 21 or more than 25 bp have been shown to be significantly less efficient, and species of more than 30 bp are large enough to trigger the interferon response. Exogenous siRNAs typically consist of a double-stranded region of 19 bp and 2 nt 3′-overhangs that have been shown to be crucial in the recognition of siRNAs by

RNaseIII enzymes. As described by Tuschl and coworkers, 3′-UU overhangs or their DNA counterparts, 3′-TT, both of which enhance the stability of siRNAs, are generally preferred over other sequences (Elbashir et al. 2001 a). Efficient gene silencing can be achieved through the use of either a duplex siRNA oligo that consists of a sense and antisense strand paired in a manner to have a 2-nt overhang or the single antisense siRNA 5′-phosphate strand.

Experiments with chemically modified siRNAs have shown that minor modifications of the sense strand, such as the addition of fluorescent probes, are generally tolerated, whereas small changes in the antisense strand lead to a dramatic loss of efficiency (Chiu and Rana, 2002). If the antisense strand of the siRNAs is tagged with a bulky fluorescent FITC group at the 3′-OH position, the siRNAs usually display very poor activity.

A free 3′-OH group on the antisense strand seems to be required by some organisms, but can be dispensed with by others (Chiu and Rana, 2002). It is assumed that the free 3′-hydroxyl group is necessary if the antisense strand serves as a primer for an RNA-dependent RNA polymerase (RdRp) (Lipardi, 2001; Sijen, 2001). It is also suggested that any impairment of incorporation of the antisense strand into the active RISC* is due to the presence of a bulky functional group. It has been further shown that the 5′-phosphorylated synthetic RNA is more active than its hydroxylated form.

4.2.3
siRNA Design Rules

4.2.3.1 siRNA Strand Bias and Off-Gene Targeting
The target recognition process is highly sequence specific, but not all positions of a siRNA contribute equally to target recognition. Mismatches in the center of the siRNA duplex prevent target RNA cleavage (Elbashir et al. 2001 c). Other studies have suggested that the introduction of point mutations had only a moderate effect on the silencing potential, indicating that the silencing machinery does not require perfect sequence identity (Boutla et al. 2001; Holen et al. 2002). Modified or mismatched ribonucleotides incorporated at internal positions in the 5' or 3' half of the siRNA duplex, as defined by the antisense strand, indicated that the integrity of the 5' and not the 3' half of the siRNA structure is important for RNAi, highlighting the asymmetric nature of siRNA recognition for initiation of unwinding (Chiu and Rana 2003). Detailed mutation scans suggest that mispairing is more crucial for the first 10 nucleotides from the 5′-end since the 10th nucleotide seems to be important for RISC-mediated cleavage of the target mRNA (Chiu and Rana 2003; Elbashir et al. 2001 c). The effects of a mismatch are less severe in the 3′-terminus of the sequence (Amarzguioui et al. 2003). The low base-pairing stability at the 5′-end is essential for the antisense strand, but not for the sense strand (Khvorova et al. 2003), resulting in a strand bias of siRNAs. Mismatches or internal thermodynamic stability are not only responsible for the overall efficacy but also for a major side effect. It has been recently reported in expression profiling studies that not only the antisense but also the sense strand of an siRNA can direct gene silencing of nontargeted genes contain-

ing as few as 11 contiguous nucleotides of identity to the siRNA (Jackson et al. 2003). Such a silencing effect is called an "off-target" effect. In order to carry out an efficient RNAi experiment, off-target effects must be considered and the siRNA must be designed accordingly. The off-target effects can be circumvented by either evaluating the antisense sequence in a Smith–Waterman or BLAST search for possible targets, or by the application of the new design rules proposed by Zamore and colleagues (Schwarz et al. 2003) and Khvorova and co-workers at Dharmacon (Khvorova et al. 2003; Reynolds et al. 2004). siRNAs can be designed to function asymmetrically so that only the anti-sense strand enters the RNAi pathway.

Both groups have shown that siRNA structure can profoundly influence the entry of the antisense siRNA strand into the RNAi pathway. Furthermore, many reports have suggested that the structure of the siRNA duplex, rather than that of the target site, explains ineffective siRNAs in mRNA silencing. According to Zamore and co-workers, most of the inactive siRNAs can be made active by simply changing the terminal sequence of the oligo. An siRNA strand, whose 5′-end is more weakly bound to the complementary strand, more readily incorporates into RISC, which favors the siRNA strand whose 5′-end has a greater propensity to fray.

Thus, a sense strand with a A-U 5′-end is preferentially incorporated into RISC than the antisense strand with a G-C 5′-end, while the antisense strand retains RNAi activity (Figure 4.6). Asymmetry of strand incorporation is dramatically increased when a G-U wobble is introduced at the 5′-end of the antisense strand of the siRNA, suggesting that an siRNA end that is more loosely attached is more easily accessible for an ATP-dependent RNA helicase that unwinds siRNA duplexes (Schwarz et al. 2003). The results of studies performed by Zamore and colleagues are summarized in Figure 4.6, and provide some ideas as how to improve the siRNA efficacy by simple modifications (Schwarz et al. 2003).

Further, single nucleotide mismatches at positions 2–4 of each strand indicated that mismatches, but not G-U wobbles, of an siRNA strand influence the incorporation into RISC. Together with a G-U wobble at position 1 of the antisense 5′-end strand, those mismatches not only lead to the preferred antisense strand usage but also increase the efficacy of the siRNA. These observations also suggest a need to revise the current design rules for the construction of siRNAs. Besides those studies, Khvorova and colleagues evaluated hundreds of siRNAs for one target. The outcome of the study was that the thermodynamic stability of the siRNA corresponds with its efficacy, and this factor has now been introduced into recently developed algorithms, as described in the following sections.

4.2.3.2 Improvements in siRNA Stability

Attempts to improve the long-term activity of the siRNAs by stabilizing the RNA strands against nucleases through introduction of modified nucleotides such as phosphorothioates or modifications of the 2′-OH position have led to controversial results. Remarkably, modifications at the 2′-OH position increase the persistence of RNAi, indicating that the 2′-OHs are not essential for either for recognition of siRNAs or for catalytic RNase activity of RISC (Chiu and Rana 2003). With exception of certain allyl-modifications, the activity of the siRNA is only marginally reduced

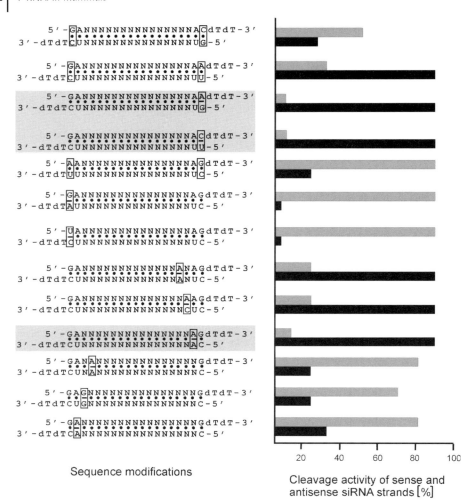

Sequence modifications

Cleavage activity of sense and antisense siRNA strands [%]

Fig. 4.6 Cleavage activity and off-gene target activity of siRNAs and asymmetric strand modifications.

(Amarzguioui et al. 2003). These findings are contradictory to the first reports from Tuschl and coworkers, who showed that the substitution of one or both siRNA strands by 2′-deoxy or 2′-O-methyl oligonucleotides abolished RNAi (Boutla et al. 2001; Elbashir et al. 2001c), even though siRNAs tolerate the replacement of the 3′-overhangs by 2′-deoxyoligonucleotides (3′-UU-3′-TT). The use of phosphorothioate is well tolerated, while causing only a marginal reduction; however, this chemical modification severely increases cytotoxicity.

Besides, Invitrogen now offers new Stealth™ siRNAs that are chemically modified siRNAs to overcome the limitations of traditional siRNAs such as the induction of the interferon response (Sledz et al. 2003). The chemical modifications concern the backbone of the ribonucleotide, and enhance its stability. More information on the chemical modifications of these Stealth™ siRNAs will be published shortly.

Even though many reports have been made showing that mRNA secondary structure has no significant impact on gene silencing, some experimental data suggest the opposite. According to its sequence, mRNA can adopt secondary structure that interferes with the binding of the RISC-bound template strand to the target mRNA, and this lowers the silencing efficiency of the chosen siRNA sequence. Regions of high secondary structure can be detected either experimentally or by using computer programs that consider base-pairing energies and spatial relationships of the individual nucleobases. A comparative study of optimized antisense oligonucleotides designed to work by an RNA interference mechanism to oligonucleotides revealed that the potency and maximal efficacy of optimized RNase H-dependent oligonucleotides and siRNAs are comparable. In an experimental approach, sequences of high binding affinity can be localized by hybridization of the target mRNA with DNA-oligomers corresponding to the antisense-strands of the siRNA sequences to be tested. The resulting DNA-RNA hybrids are eventually digested by RNase H that specifically degrades hybrid double-strands (Vickers et al. 2003). Degradation is only possible at the sites where the template DNA binds to the mRNA, thereby revealing suitable target sequences that are more easily accessible or free from protein-binding sites.

4.2.3.3 siRNA Design: Novel Modifications of the "Tuschl Rules"
There are many rules to design the most efficient siRNA for a certain target. However, these are still mainly based on experience from siRNA validation experiments rather than on useful mathematic algorithms. The newest generation of those algorithms, which are much more reliable than their ancestors, are now available on the web. They guarantee an siRNA silencing efficiency of at least 70–90%. Nevertheless, many bioinformatics groups are still developing new algorithms that might allow an exact prediction of the most efficient siRNA/mRNA target site in the near future.

Exogenous siRNAs typically consist of a double-stranded region of 19 bp and 2 nt 3'-overhangs. Ideally, the sequence of the target mRNA should comprise the 23-bp, AA(N19)TT with N is any nucleotide, as depicted in Figure 4.7A. The 3'-UU overhangs – or their DNA counterparts, 3'-TT – to stabilize the siRNAs against ribonucleases are generally preferred over other sequences (Elbashir et al. 2001a). Less favorable but still efficient are sequences such as NA(N19)TT (Figure 4.7B) or NA(N21) (Figure 4.7B), where the sequence of the sense siRNA corresponds to (N19)TT or N21 (position 3 to 23 of the 23-nt motif).

The last two nucleotides of the N21 sequence are converted to TT to facilitate synthesis. The antisense siRNA is synthesized as the complement to position 1 to 21 of the 23-nt motif. Since position 1 of the 23-nt motif is not necessary for the silencing effect, the 3'- nucleotide residue of the antisense siRNA can be changed to 3'-T. However, the +2 position of the 23-nt motif should always be complementary to the target sequence. Tuschl and coworkers found that functional siRNAs can be designed corresponding to the target motif NAR(N17)YNN, (R is purine (A, G) and Y is pyrimidine (C, U)). This rule also facilitates the design of small hairpin RNAs (shRNAs) corresponding to the siRNA. The respective 21-nt sense and antisense siRNAs therefore begin with a purine nucleotide, which is absolutely necessary for expression from pol III promoters, as described later in this chapter (Figure 4.7C). Recently, a new algorithm was published which suggests the motif (N4)A(N6)T(N2)H(N5)W(N2), (N is any nucleotide, H is not G, W is A but not G or C), and this is introduced into the following design protocol (Reynolds et al. 2004) (Figure 4.7D).

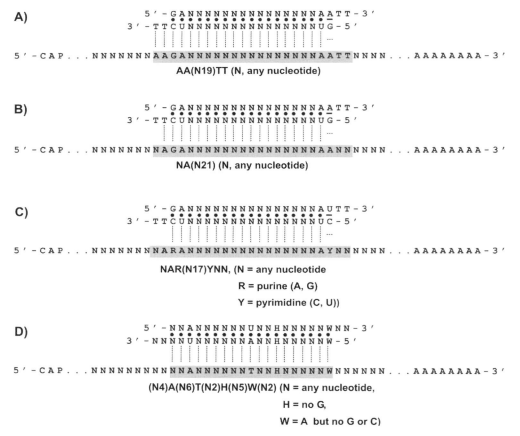

Fig. 4.7 Schematic view of the mRNA target site for efficient siRNAs. (A) The optimal sequence motif for siRNA design is perfectly matching AA(N19)TT (N, any nucleotide). If no suitable sequences are found, one can use either the NA(N21) motif (B), where the sequence of the sense siRNA corresponds to (N19)TT or N21 (position 3 to 23 of the 23-nt motif). The last two nucleotides of the N21 sequence are converted to TT to facilitate synthesis. (C) Functional siRNAs can be designed corresponding to the target motif NAR(N17)YNN, (R is purine (A, G) and Y is pyrimidine (C, U)). (D) Recently, a new algorithm was published that suggest the motif (N4)A(N6)T(N2)H(N5)W(N2), (N is any nucleotide, H is not G, W is A but not G or C). A little less efficient but also functional is a target sequence of N19 flanked by non-pairing 5'-AA and 3'-TT (not shown).

For the further design of the siRNAs one should follow the guidelines summarized below:

PROTOCOL 14

1. Choose a sequence in the coding region of the mRNA with a G/C content of 45–50% (Elbashir et al. 2001 a), or which fits the following criteria suggested by Khvorova and coworkers (Reynolds et al. 2004) (Table 4.1).

Tab. 4.1

Criterion I	G/C content = 30–52%
Criterion II	3 or more A/U bp at positions 15–19 (sense strand)
Criterion III	Hairpin prediction (no internal repeats or palindroms) T_m <20 °C
Criterion IV	A at position 19 (sense strand)
Criterion V	A at position 3 (sense strand)
Criterion VI	T at position 10 (sense strand)
Criterion VII	No G or C at position 19 (sense strand)
Criterion VIII	No G at position 13 (sense strand)

Adapted from Reynolds et al. (2004)

▶ **Note:** If the target region does not begin with AA, choose a 23-nt region of the coding region of the gene to determine the % GC.

▶ **Note:** An siRNA with 50% GC content will in most cases work as well, while higher ratios result in a decrease of silencing activity.

2. Avoid multiple base repeats such as more than three guanosines, cytosines, or more than four adenosines in a row.

▶ **Note:** Poly G and poly C sequences can hyperstack and therefore form agglomerates that potentially interfere in the siRNA silencing mechanism, while poly A can interrupt shRNA synthesis by premature transcription termination (see later in this chapter).

3. Avoid sequences from the 5′-UTR, as they can contain protein binding sites that block target recognition and start with the search at least 100 nt from the AUG start codon.

▶ **Note:** 3′-UTRs, however, are suitable target sequences and maybe very unique preventing unwanted knock-down of conserved genes.

4. Perform a comprehensive homology search (BLAST (nucleotide BLASTN) at http://www.ncbi.nlm.nih.gov/) or Smith–Waterman algorithm (http://www.paralign.org/)) with the sense sequence of the selected siRNA against EST libraries to prevent multiple gene targeting.

▶ **Note:** BLAST is easy to use for most scientists as it is easily accessible and guarantees 100% success (see Section 4.2.3.4; Homology search BLAST or Smith–Waterman algorithm).

5. To reduce the risk of unwanted off-gene targeting with the sense strand as a RNAi guiding strand, perform the same homology search with the antisense strand against EST libraries or sequences listed in the NCBI database.

6. Synthesize sense siRNA with a dTdT 3' end to generate equal 3'-ends of the sense and antisense strands.

▶ **Note:** The modification of the siRNA sense overhang to 3'-TT is not affecting silencing efficacy. Therefore, an AA(N19) or NA(N19) target sequence is sufficient for the design of a 21-mer siRNA.

7. Simultaneous knock-down of more than on gene is possible using the respective siRNAs in equal concentrations. A simultaneous knock-down of related genes can be achieved using siRNA encompassing conserved mRNA domains (see Table 4.2).

8. Design two additional siRNAs to control for specificity of the silencing effect of different siRNAs.

4.2.3.4 Homology Search by BLAST, FASTA, or Smith–Waterman Algorithm

As described above and in the previous chapters, off-target effects and non-specific gene silencing are derived from siRNAs that are not exclusively homologous to the mRNA of interest. In order to avoid those side effects it is recommended to perform extensive homology searches with all listed exon-derived sequences. There are some algorithms available that compare the sequence of interest. The BLAST program (Altschul et al. 1990, 1997) is probably the most widely used sequence matching method today due primarily to its availability on the on the National Center for Bioinformatics (NCBI) server (http://www.ncbi.nih.gov), and to its speed. A second search program, FASTA (Pearson and Lipman 1988), is also available on the NCBI server and is commonly used, although it is slower than BLAST because it is more sensitive. BlastN (N for nucleotide) requires at least seven contiguous matching bases to score as a homologous sequence. It is easy to use and will reveal the desired results in a short period of time. The Smith–Waterman (S-W) algorithm (Smith and Waterman 1981) is an exhaustive search, which is a pair-wise sequence alignment method. It is based on Bellman's dynamic programming algorithm to find the best local alignment, and is therefore the most sensitive and also the most historic of the three methods. It reports hits that can be missed by either BLAST or FASTA (Brenner et al. 1998; Hubbard et al. 1998; Janaki and Joshi 2003; Rognes and Seeberg 1998, 2000), and is more suitable for the design of siRNAs as it permits a defined search of local homologous regions smaller than seven bases within an siRNA. This can be especially important if the siRNA is designed with regard to the recently developed algorithms by Khvorova and Schwarz and colleagues. As described above,

Tab. 4.2 siRNA requirements for a variety of targets and different knock-down experiments

Gene specific knock-down

Target	Requirements
Unique genes	• siRNA sequence from ORF • BlastN with sense and antisense strand • Exclusion of homology (>12 nt) with other mRNAs
Mutated mRNAs (oncogenes)	• Introduction of mutation within the 5–10 nt position of the antisense strand • Validation of the siRNA for silencing of wild-type mRNA
One specific splice variant	• Longer transcripts: siRNA from the 5′ non-spliced region • Shorter transcripts: siRNA from 3′-UTR and co-transfection of the full ORF of the longer transcript
One gene family member	• siRNA from the 3′-UTR • Exclusion of homology (>12 nt)
One siRNA for different species (i. e. human, mouse, *Drosophila*)	• siRNA encodes highly homologous mRNA sequence mainly from 3′-end

Simultaneaous knock-down of many genes

2 different non-related genes	• siRNA sequence of different genes from ORFs • Transfection of equal concentrations
Genes from a gene family	• One siRNA encoding conserved mRNA sequence
All splice variants of an mRNA	• siRNA of the mRNA region encoded by all transcripts
Genes encoding a conserved domain	• One siRNA encoding conserved mRNA sequence
One siRNA for different species (i. e. human, mouse, *Drosophila*)	• siRNA encoding highly homologous mRNA sequence mainly from 3′-end

those design rules allow the introduction of single mismatches within the sequence, which would be not tolerated by the BLAST search.

Due to the demand for fast and sensitive searches, much effort has been made to produce fast implementations of the Smith–Waterman search (SSEARCH) (Janaki and Joshi 2003; Rognes and Seeberg 2000). To speed up database searching with the Smith–Waterman algorithm, several approaches – using heuristic searches of the S-W algorithm such as ParAlign (http://www.paralign.org/) (Rognes and Seeberg 2000) or the parallel search of Smith–Waterman algorithm (SSEARCH) and FASTA programs (Pearson 1991) (http://fasta.bioch.virginia.edu/fasta/home.html and http://www2.igh.cnrs.fr/bin/fasta-guess.cgi) – have been developed and can be performed on the respective servers that have a graphical web interface (Table 4.3).

4.2.3.5 **Troubleshooting**

If the knock-down effect is less efficient than predicted, it is recommended that another siRNA be used from a different region of the mRNA. A moderate efficacy of a

Tab. 4.3

Algorithm	Web site
BLAST	http://www.ncbi.nih.gov
FASTA	http://fasta.bioch.virginia.edu/fasta/home.html
Smith-Waterman algorithm	
• SSEARCH/FASTA	http://fasta.bioch.virginia.edu/fasta/home.html
	http://www2.igh.cnrs.fr/bin/fasta-guess.cgi
• Paralign	http://www.paralign.org/

certain siRNA can also be due to sequencing errors of the gene, database errors, or common polymorphisms. Therefore, it is important to check the sequence of the respective mRNA before embarking on the design of the siRNAs.

4.2.3.6 siRNA Design Programs and Algorithms

There are several siRNA design and search engines available on the Internet to facilitate the production of highly efficient siRNAs for silencing experiments. The most valuable target finder tool is now available for everybody in academic research, and is found at the Whitehead Institute for Biomedical Research (http://jura.wi.mit.edu/ pubint/http://iona.wi.mit.edu/siRNAext/) (Yuan and Lewitter 2003). This tool was developed by the local bioinformatics research group, and was one of the first to design siRNAs. After registration, which is no problem for academic users, siRNAs can be designed by a variety of options. One can decide whether to search the sequence as suggested by Tuschl and coworkers, such as AA(N19)TT or NA(N21) (Elbashir et al. 2001 c), or by the recently published pattern (N4)A(N6)T(N2)H(N5)W(N2) (N = any nucleotide, H = no G, W = A but no G or C), from Khvorova and coworkers (Reynolds et al. 2004). Currently, the program is updated with the newest data and algorithms published in the literature (Khvorova et al. 2003; Reynolds et al. 2004; Schwarz et al. 2003), and allows the exclusion of single polymorphic sites from the siRNA.

After the design, the appropriate sequences are searched in a database for homologous sequences using BLAST against the NCBI UniGene database. Finally, a built-in sort function permits the ranking of the predicted siRNAs either by specificity or by other criteria, such as thermodynamic values, BLAST results, %GC.

Nowadays, many companies offer similar or perhaps less advanced design tools as the Whitehead designer. One of these is Invitrogen design tool (http://rnaidesigner.- invitrogen.com/sirna/), which also regards thermodynamic properties and 5'- and 3'- end base composition. In addition, the Invitrogen design tool offers the design of so-called Stealth™ siRNAs that are chemically modified to overcome the limitations of traditional siRNAs such as the induction of the interferon response (Sledz et al. 2003). More information on the chemical modifications of those Stealth™ siRNAs will be published shortly. Another very helpful, but still crude, program is the siRNA Target Finder provided by Ambion (http://www.ambion.com/techlib/misc/siRNA_ finder.html). This allows a search of your entire sequence for appropriate siRNA sites based on the "Tuschl" rules. Additionally, together with Ambion, Cenix

BioScience (http://www.cenix-bioscience.com) has developed a novel algorithm that is based on a more stringent analysis of each siRNA sequence to maximize target specificity and that represents a major improvement over the "Tuschl" rules, as it can also determine siRNA with a high efficacy at low concentrations (<10 nM). As it has been shown that the specificity of the RNAi effect is increased with decreased amounts of siRNAs (Jackson et al. 2003; Semizarov et al. 2003), this feature is of common interest. This algorithm also considers enhanced flexibility of the 5′-end and the internal stability profile of the siRNAs as described above (Khvorova et al. 2003; Reynolds et al. 2004; Schwarz et al. 2003). As a special service Ambion provides the custom synthesizes of the siRNAs pre-designed using the algorithm developed by Cenix BioScience. Beside Ambion and Cenix BioScience, Qiagen is offering the same service on their siRNAs using their newly developed algorithm, which is similar to the Whitehead design tool with one exception. In order to avoid off-target effects, the sequence is searched for homology by the Smith–Waterman algorithm, which is more reliable than the BLAST search by the Whitehead design tool.

4.2.3.7 Preparation of siRNA Duplexes

siRNAs have emerged as powerful RNAi reagents, not only for functional genomics (Elbashir et al. 2001 a, b; Harborth et al. 2001; Lewis et al. 2002; Paddison and Hannon 2003), but also for the inhibition of viral propagation (Gitlin et al. 2002; Jacque et al. 2002; Jiang and Milner 2002; Randall et al. 2003). Although chemically synthesized siRNAs are used in most laboratories for RNAi experiments, they are still very expensive compared to their DNA counterparts. However, enzymatic synthesis (described later in this chapter) can still not compete with the excellent yield and purity of the chemical process, and usually does not allow modifications. As there is no 100% guarantee that the chosen siRNA will work with high efficiency, it is always recommended to validate more than one siRNA for one target mRNA. Beside Qiagen and Dharmacon there are several companies that provide services on custom synthesis of siRNAs as duplexes or single RNA strands. Some of the main providers are listed in the Table 4.4. The quality of the oligos is usually good, and does not really vary between the companies. However not all of the companies offer custom modification such as a fluorescent label or thio- or thioate-modifications.

Although siRNA duplexes are relatively stable compared to single-stranded RNA, storage should be at –20 °C, and repetitive freeze–thaw cycles should be avoided by

Tab. 4.4

Dharmacon	http://www.dharmacon.com
Qiagen	http://www.qiagen.com
Proligo	http://www.proligo.com
Ambion	http://www.ambion.com
MWG Biotech	http://www.mwgdna.com
Eurogentec	http://www.eurogentec.com
Invitrogen	http://www.invitrogen.com

storing as aliquots. Usually, siRNAs are deprotected and desalted, but they can also be obtained as purified duplexes after reversed-phase HPLC (RP-HPLC) or polyacrylamide gel electrophoresis (PAGE).

4.2.4
Enzymatic Synthesis of siRNAs

Beside the chemical synthesis, siRNAs can also be prepared by in-vitro transcription, which makes the RNAi experiments less expensive and allows a faster synthesis of the siRNAs. Some kits are available commercially and are easy to use for the enzymatic synthesis. One such kit is the T7 RiboMAX™ Express RNAi System (Promega); this is based on the T7 Ribomax™ system, as described in Chapter 2 for the synthesis of long dsRNA. It has been shown to be suitable for the synthesis of siRNAs. siRNAs were obtained by in-vitro transcription of short DNA-oligonucleotides encoding the appropriate 21-nt sequences (G(N17)CTT, N = any nucleotide) of the respective target genes containing a guanosine at the 5′-end and a 2-nt 3′-TT overhang. The DNA-templates also contain a T7 recognition sequence at the 5′-end of the siRNA according to the sense strand as a double-stranded recognition site for the T7 RNA polymerase. Since the enzyme usually requires GTP to efficiently initiate transcription, not any sequence can be chosen as an siRNA target. Yields equal to or greater than 500 μg/ml of siRNA are generally observed.

4.2.4.1 Designing DNA Oligonucleotides
In order to synthesize siRNAs in vitro using T7 RNA polymerase, the selected target mRNA sequence must be screened for the sequence 5′-G(N17)C-3′. For this, one can also use the Whitehead design tool (http://jura.wi.mit.edu/pubint/http://iona.wi.mit.edu/siRNAext/). It is recommended to select at least three different oligos, since the silencing efficiency can vary between different oligos. Enzymatic synthesis does not always allow the application of the current algorithms since T7-based synthesis requires a G in position 1 of both strands. Replacement by any other base reduces the synthesis yield (Milligan et al. 1987).

Optimized start sequences reveal the highest amount of dsRNA. The first six bases are crucial for yield and success of the transcription reaction. In addition to full-length RNAs, the reaction also yields large amounts of abortive initiation products. Variants in the +1 to +6 region of the promoter are transcribed with reduced efficiency. Transcription reaction conditions must be optimized in order to obtain milligram amounts (Table 4.5) (Milligan et al. 1987).

The optimal sequence for a transcription start is GGG followed by GG or GC. GGG should be followed by 17 to 22 gene-specific nucleotides. Avoiding A and T in

A)   5′-**TAATACGACTCACTATAG**-3′   Minimal T7 sequence

B)   5′-GGATCC**TAATACGACTCACTATAG**-3′   Elongated T7 sequence

Fig. 4.8 Extra bases added to the minimal T7 promoter (A) often increase the yield by allowing more efficient T7 polymerase binding and initiation (B).

Tab. 4.5 Yields of T7-based RNA transcription of oligonucleotides between 12 and 18 nt. The first six bases are crucial for the transcription of the full-length transcripts. Each reaction contained 50 nM template, and incubation was continued at 37 °C for 4 h (Milligan et al. 1987)

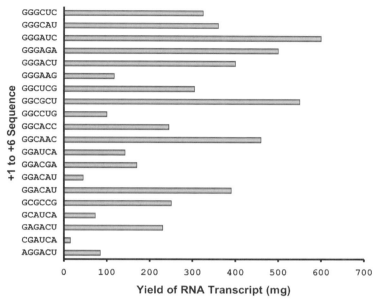

the first three base pairs would greatly enhance the yield, but decreases the silencing efficacy (Reynolds et al. 2004). The oligonucleotides consist of the target sequence plus a minimal or elongated T7 RNA polymerase promoter sequence (Figure 4.8A and B). Moreover, the elongated version of the T7 promoter may increase yield by allowing more efficient polymerase binding and initiation (Figure 4.8B), but is more expensive.

The target sequence can be either single-stranded or double-stranded, while the T7 promoter region is required to be double-stranded for correct binding of the polymerase (Figure 4.9).

It is recommended to order the sense and antisense template DNA oligonucleotides for each siRNA. The smallest scale synthesis (40 nmol or less/desalted) is sufficient for many transcriptions. The DNA oligonucleotides encoding either the sense or antisense strand (100 pmol each) are annealed with the T7 coding DNA oligonucleotide by boiling a 1:1 mixture and slowly cooling to room temperature, as depicted in Figure 4.10. The partially double-stranded DNA is transcribed according to the protocol described in the next section. Thereby, the yield of dsRNA is dependent on the first six bases or the GC content of the DNA-template but can reach 1–2 mg dsRNA/ml. Extending the incubation time during the initial transcription reaction from 30 min to 12 h and incubating at 42 °C instead of 37 °C, usually increases the yield. Following the reaction, the strands are treated with a DNAse/RNase mixture.

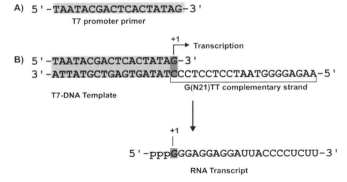

Fig. 4.9 (A) Sequence of the minimal T7 promoter, that must be introduced into the PCR primers. (B) Transcription of a T7 promoter flanked DNA sequence. The last base of the minimal T7 promoter sequence is also encoding the first RNA nucleotide of the transcript, which will contain a triple phosphate tail at the 5′-end.

Since it has been shown that the 5′-triple phosphate generated by the T7 polymerase is responsible for an elevated interferon response compared to synthetic siRNAs (Kim et al. 2004), it is recommended to treat the enzymatically produced siRNAs with phosphatase (CIP) to remove the 5′-phosphate. After transcription, the resulting RNA strands are annealed to form the siRNA, and the remaining single-stranded RNAs and the DNA template are removed by nuclease digestion. The siRNA is then purified by isopropanol precipitation, after which it may be introduced into the organism of choice for RNAi applications (for a more detailed description, see the Promega manual for T7 RiboMax™).

4.2.4.2 In-vitro dsRNA Transcription
At present, many companies offer RNA production kits (Ambion, Roche, Stratagene, Promega, etc.) that produce RNA or dsRNA in similar qualities. In this chapter, only a selection of methods will be described; all of these work well, and can even be applied without the use of special RNAi reagents. The methods that will be described in detail are based either on the T7 RiboMAX™ System and the new variant T7 RiboMAX™ Express RNAi System, or on the *Silencer*™ siRNA Construction Kit provided by Ambion. Both systems are widely used and have been shown to synthesize milligram amounts of siRNA.

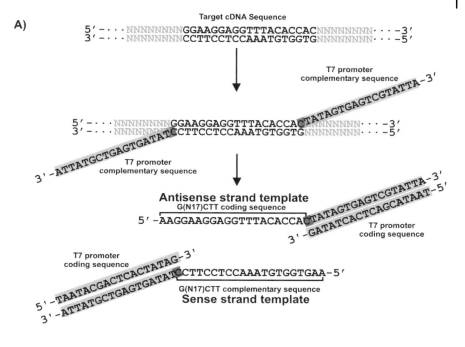

A)

B) T7 promoter coding strand

5'-**TAATACGACTCACTATAG**-3'

Antisense strand template

5'-AAGTGGTGTAAACCTCCTTC**CTATAGTGAGTCGTATTA**-3'

Sense strand template

5'-AAGGAAGGAGGTTTACACCA**CTATAGTGAGTCGTATTA**-3'

Fig. 4.10 Scheme of the DNA-template design for the T7-based production of siRNAs. (A) After choosing the siRNA region on the target mRNA, connect the T7 promoter complementary sequence to the 3'-end of the desired sequence. Add two additional adenine nucleotides to the 5'-end. Annealing of the T7 promoter coding sequence will result in a double-stranded T7 promoter, which is necessary for transcription (see Figure 4.9). The minimal T7 promoter can be exchanged by the elongated T7 promoter, as depicted in Figure 4.8 to possibly increase transcription efficiency (B) DNA sequence to order in the correct 5'→3' direction.

PROTOCOL 15

1. Combine equal amounts (100 pmol each) of either the sense or antisense DNA oligonucleotides with the T7 coding DNA oligonucleotide (see Figure 4.10).

2. Boil the mixture for 10 min and let it slowly cool to room temperature.

▶ **Note:** It is essential to cool the mixture very slowly. This can be done by heating the samples in a thermoblock (waterbath), switching off the thermoblock (waterbath heater) and waiting until the temperature of the metal block (water) reaches ambient.

3. Prepare 1–8 µl of a DNA solution with a final DNA amount of 1 µg.

4. Add the following reaction components (see Table 4.6) into a DNAse- and especially RNase-free microcentrifuge tube and fill up to a final volume of 20 µl (for T7 RiboMAX™ Express RNAi System) or 100 µl (for T7 RiboMAX System). Mix by gently flicking the tube.

Tab. 4.6

RiboMAX™Express T7		RiboMAX™ T7	
T7 Reaction components	**Reaction**	**T7 Reaction components**	**Reaction**
RiboMAX™Express T7 Buffer (2x)	10 µl	RiboMAX™ T7 Transcription Buffer (5x)	20 µl
DNA Oligonucleotide (20–100 pmol)	1–8 µl	rNTPs (25 mM)	30 µl
Enzyme Mix T7 Express	2 µl	DNA Oligonucleotide (20–100 pmol)	1–40 µl
Nuclease-free water	0–7 µl	Enzyme Mix T7 Express	10 µl
		Nuclease-free water	0–39 µl
Total volume	20 µl	Total volume	100 µl

5. Incubate at 37 °C for 30 min for the T7 RiboMAX™ Express RNAi System, or overnight for the T7 RiboMAX™ System).

▶ **Note:** In contrast to Promega Notes on T7 RiboMAX, a dramatic increase in yield for almost all templates can be observed when incubating for longer than 6 h.

▶ **Note:** Incubation at 42 °C may improve the yield of dsRNA for transcripts containing secondary structure.

6. The reaction can be monitored measuring the viscosity. As the yield increases, the reaction mixture turns into a gelatinous, translucent pellet.

▶ **Note:** If the sample is too viscous for the annealing step, add a few µl of water or annealing buffer (Table 4.7)

Tab. 4.7

Annealing buffer 1 (1x)		Annealing buffer 2 (10x)	
Sodium acetate	100 mM	Tris-HCl, pH 8.0	100 mM
HEPES-KOH pH 7.4	30 mM	EDTA, pH 8.0	10 mM
Magnesium acetate	2 mM	NaCl	1 mM

7. After RNA synthesis, anneal both RNA strands, mix equal volumes of complementary RNA reactions, and incubate at 70 °C for 10 min.

8. Spin the samples very briefly! Then cool to room temperature very slowly.

▶ **Note:** It is essential to cool the mixture very slowly. This can be done by heating the samples in a thermoblock (waterbath), switching off the thermoblock (waterbath heater), and waiting until the temperature of the metal block (water) reaches ambient.

9. Dilute the supplied RNase Solution 1:200 by adding 1 µl RNase Solution to 199 µl nuclease-free water. Add 1 µl freshly diluted RNase Solution and 1 µl RQ1 RNase-Free DNase (for T7 Ribomax™ add 1 µl RQ1 for 1 µg DNA) per 20 µl (100 µl) reaction volume, and incubate for 30 min at 37 °C. This will remove any remaining single-stranded RNA and the template DNA, leaving double-stranded RNA.

10. Add 0.1 volume of 3 M sodium acetate (pH 5.2) and 1 volume of isopropanol or 2.5 volumes of 100% ethanol. Mix and place on ice for 1 h. The reaction will appear cloudy at this stage. Spin at top speed in a microcentrifuge for 10 min at 4 °C

11. Carefully aspirate the supernatant, and wash the pellet with 0.5 ml of ice-cold 70% ethanol, removing all ethanol following the wash. Air-dry the pellet for 15 min at room temperature, and resuspend the RNA sample in 100 µl DEPC water or annealing buffer (Table 4.7).

12. Aliquot the dsRNA and store at –20 °C or –70 °C.

13. Alternatively, further purify siRNA following precipitation using a G25 micro spin column following the manufacturer's instructions (Amersham Biosciences, Cat.# 27–5325-01). This will remove any remaining rNTPs and allow accurate quantitation by absorbance at 260 nm.

▶ **Note:** Do not process more than an initial 40 µl reaction volume per spin column. A loss of yield can be expected following G25 purification (approximately 66% recovery).

▶ **Note:** Do not use these columns with water. The purification will also result in a desalting of the solution, and water will decrease the siRNA annealing efficiency.

14. Prepare a 1 : 100 to 1 : 300 dilution of the siRNA and measure the concentration at 260 nm (OD (260) = 1 is equivalent to 40 µg RNA/ml).

15. Dilute 1 µl of siRNA in 50–100 µl of 1x TAE buffer or nuclease-free water (DEPC water) and use 50–500 ng per lane and the respective amount of 10x DNA agarose loading dye (Table 4.8).

Tab. 4.8

Gel loading buffer (10x)	
Ficoll	25 %
EDTA, pH 8.0	1 mM
Bromophenol blue	0.25 %

16. Analyze the quality of the siRNA on a 3% TAE-agarose gel using a DNA size marker, or on a polyacrylamide gel as described later in the chapter.

Another method is based on the use of double-stranded DNA oligonucleotides (*Silencer*™ siRNA Construction Kit provided by Ambion) (Figure 4.11).

The design of the DNA oligonucleotides is similar to those described above. The main advantage of this method over that described above is that it allows the production of siRNAs that do not start with a G or GGG. Since the GGG is necessary for highly efficient transcription, it is introduced into the template but is not part of the siRNA sequence. It will later be removed from the desired siRNA by RNase digestion. Further, the 29-nt oligonucleotides must be designed to include an 8-base sequence complementary to the 5′ end of the T7 promoter primer, as depicted in Figure 4.12.

Both template DNA oligonucleotides are annealed with the T7 promoter oligonucleotide provided by Ambion in separate reactions. The 3′-ends of the hybridized DNA oligonucleotides are extended by Klenow fragment of DNA polymerase to generate double-stranded siRNA transcription templates. The siRNA strands are then transcribed by adding the T7 polymerase and nucleotides, and the RNA strands are annealed to generate dsRNA. It must be noted that the nucleotide mixture contains a modified UTP to enhance the siRNA efficiency. The dsRNA consists of 5′-terminal single-stranded leader sequences, a 19-nt target-specific dsRNA, and 3′-terminal UUs. The template DNA is digested with DNase. Further, the single-stranded sequence that is attached to bind the T7 promoter oligonucleotide is removed by digestion with a single-strand-specific ribonuclease (RNase T1), which does not affect the UU single-stranded overhang. The DNA template is removed at the same time by a deoxyribonuclease. After polishing with RNase, the duplex resembles the structure of natural siRNAs containing a 3′-UU overhang. The exact protocol can be found on the Ambion's website, and is supplied with the kit.

The main advantage of the enzymatically synthesized siRNAs is the lower price and the amount of time to acquire a decent amount of substance. The most effective siRNAs can reduce target gene expression by more than 90%. As for the synthetic siRNA, the key feature for high efficiency is correct design of the sequence. It is

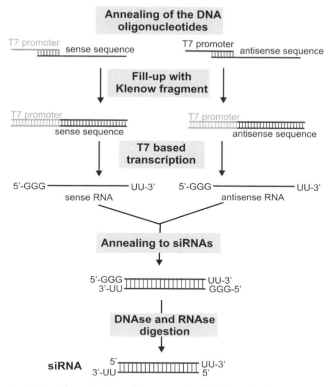

Fig. 4.11 Schematic view of the enzymatic siRNA production using the *Silencer*™ siRNA Construction Kit provided by Ambion.

recommended to use the Tuschl rules or the algorithm designed by Khvorova or Zamore and co-workers (Elbashir et al. 2001a; Khvorova et al. 2003; Reynolds et al. 2004; Schwarz et al. 2003). Nevertheless, one should test at least 3–5 siRNAs per gene and additional control siRNAs, which can encode housekeeping genes such GAPDH or β-actin.

The main disadvantage is that the enzymatically synthesized siRNAs can induce interferon response, as recently reported (Kim et al. 2004). Compared to their synthetic counterparts, the enzymatically synthesized siRNAs contain a triple phosphate residue at their 5'-end. Removal of the triple phosphate residue with calf intestinal phosphatase (CIP) or RNaseT1/CIP prevents induction of the interferon response. Based on these findings, an improved method for T7-based in-vitro transcription has been developed by Rossi and co-workers. Incorporation of 3'-AA allows for the dephosphorylation and removal of the pppGGG (optimal start sequence for T7 polymerase) by RNaseT1 and CIP (Kim et al. 2004).

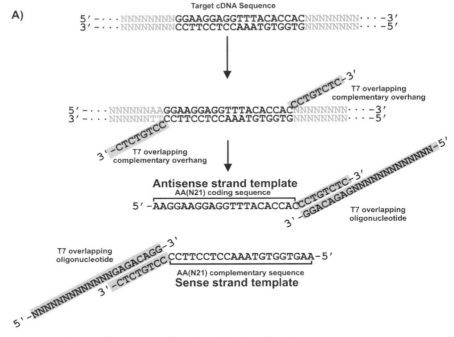

Fig. 4.12 Scheme of the DNA-template design for the T7-based production of siRNAs. (A) After choosing the siRNA region on the target DNA, connect the T7 promoter overlapping sequence 5′-CCTGTCTC-3′ to the 3′-end of the desired sequence. Add two additional adenine nucleotides to the 5′-end. Annealing of the T7 promoter oligonucleotide provided by Ambion and production of the double strand by a fill-in reaction with Klenow fragment allows the T7 transcription.
(B) DNA sequence to order in the correct 5′→3′ direction.

4.2.5
Processing of Long dsRNA

Another method for generating siRNAs is the enzymatic digestion of long dsRNA by ribonucleases III such as Dicer or RNaseIII from *E. coli* (Yang et al. 2002a). The digestion procedure can become a straightforward and inexpensive alternative, when recombinant protein expression is established in the laboratory. Both, the dsRNA and ribonuclease, can be obtained from recombinant precursor DNA. Long dsRNA fragments are generated by simple T7-based in-vitro transcription (as described in Chapter 2) using commercially available T7 RNA polymerase-based in-vitro transcription systems such as the T7 RiboMax™ System (Promega). So far, many groups have reported the recombinant expression of Dicer from mouse (Nicholson and Nicholson 2002), humans (re-hDicer) (Kawasaki et al. 2003; Myers et al. 2003; Provost et al. 2002; Zhang et al. 2002), and *Drosophila* (Hoffmann and Schepers, in preparation). Dependent on the species, Dicer is a polypeptide of about 215–250 kDa. Due

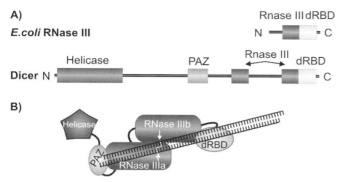

Fig. 4.13 Schematic view of the domain structure of Dicer and RNaseIII from *E. coli*. Dicer encompasses at least four domains: a helicase domain; a PAZ (Piwi-Argonoaute-Zwille) domain; two RNaseIII domains; and a dsRNA-binding domain (dRBD), whereas RNaseIII from *E. coli* consists only of one catalytic RNaseIII domain and a dsRNA binding domain. (B) Current proposed model of Dicer RNase III activity (W. Filipowicz, Keystone Symposium 2004).

to its high molecular weight, it must be expressed in eukaryotic cells such as insect cells or mammalian cells (Kawasaki et al. 2003; Myers et al. 2003; Provost et al. 2002). Since Dicer is the type of ribonuclease that is present in the RNAi pathway, the recombinant enzyme specifically generates the 21- to 23-nt siRNA products from dsRNA. Mg^{2+} is required for the digestion process, but ATP has been shown to be dispensable for human recombinant Dicer activity in vitro (Provost et al. 2002; Zhang et al. 2002).

In-vitro digestion with any form of recombinant Dicer leads to a pool of different siRNAs for one target RNA, which makes the selection of a target site and the design of siRNAs unnecessary. However, it is important to avoid homologous sequences within a target mRNA in a given gene family. Yields of more than 70% are obtained. Dicer-generated siRNAs (d-siRNAs) can be easily separated from the residual large dsRNA by a series of gel filtration chromatography or gel purification steps (Myers et al. 2003), making this method an alternative for the production of siRNAs.

4.2.5.1 Expression of Dicer in Hi5 Insect Cells

Expression of recombinant human or *Drosophila* Dicer can be performed in Hi5 insect cells using the Bac-to-Bac® baculovirus expression system (Invitrogen).

PROTOCOL 16

1. Amplify the open reading frame (ORF) of human (NCBI Acc. no. NM030621) or *Drosophila* Dicer 2 (NCBI Acc. no. AB073024) in an RT-PCR. Add the respective restriction sites to the forward and the reverse primer to allow the directed cloning of the PCR product (Figure 4.14).

Human Dicer Primers

Forward
5'-AGCTAG***CTCGAG***ATGGCAGGCCTGCAGCTCATGACCCCTGC-3'
      XhoI

Reverse
5'-CGAGCT***AAGCTT***TCAGCTATTGGGAACCTGAGGTTGATTAGC-3'
      HindIII

***Drosophila* Dicer 2 Primers**

Forward
5'-GTAGCT***ACTAGT***ATGGAAGATGTGGAAATCAAGCCT-3'
      SpeI

Reverse
5'-AGCTAGCTAGCT***GCGGCCGC***TTAGGCGTCGCATTTGCTTAGCTG-3'
      NotI

Fig. 4.14 RT-PCR primers for the amplification of human or *Drosophila* Dicer. Forward and reverse primers contain restriction sites for the directed ligation into the baculovirus shuttle vector (depicted in bold-italics).

▶ **Note:** Since the ORF is quite large (between 5168 and 5768 bp), the use of the "Titan One Tube" RT-PCR reaction system (Roche) is highly recommended, as it ensures high yield production even of longer products.

2. Digest the PCR product and the baculovirus shuttle vector pFastBac(HTc) (HT = His$_6$-Tag) (Invitrogen) with *SpeI* and *NotI* (*Drosophila* Dicer 2) or *SalI* and *HindIII* (Human Dicer) according to the enzyme supplier for 2 h at 37 °C.

▶ **Note:** The PCR fragment is quite large; therefore, cloning efficiency is often reduced. It helps to pre-clone the ORF into the pCR TOPO XL vector using the TOPO XL cloning system for the cloning of extra-large fragments (Invitrogen). The ORF is eventually excised by enzymatic digestion with the respective enzymes.

3. Ligate the ORF into
   • *Drosophila* Dicer 2: *SpeI* and *NotI* sites of pFastBac(HTc)
   • Human Dicer: *SalI* and *HindIII* sites of pFastBac(HTc)
   and amplify the DNA.

▶ **Note:** For human Dicer: The *XhoI* end of the fragment can be ligated into the *SalI* site of the vector.

4. To generate the recombinant baculovirus genome, carry out the transposition procedure described in the Bac-To-Bac® manual supplied with the Baculovirus expression system from Gibco-Life Technologies (now Invitrogen).

▶ **Note:** For novices in insect cell expression, this baculovirus expression system is so far the system of choice for the expression of recombinant Dicer because it is very easy to apply and allows the production of protein within 4–6 weeks after starting with the cloning of the shuttle vector.

5. Once the recombinant baculovirus genome (bacmid) is purified, it can be transfected into insect cells using simple calcium phosphate transfection, as described in many molecular biology textbooks.

▶ **Note:** To ensure highly efficient transfection, the use of a commercial transfection reagents is recommended, such as the Pharmingen Transfection Buffer A+B System (Pharmingen) (see the following modified protocol).

6. Plate 2×10^6 Sf9 insect cells into a well of a 6-well plate and allow the cells to attach (15 min at room temperature).

7. Make sure to use a positive and a negative control (e.g., bacmid DNA previously used by someone else, and only transfection reagent).

8. Remove the culture medium from the plate and add 333 µl of Pharmingen Transfection Buffer A.

▶ **Note:** Make sure that the plate is completely covered so that the cells do not dry out.

9. Mix 5 µg of the baculovirus genome with 330 µl of Transfection Buffer B and mix well by gently flicking the tube.

10. Incubate for 5 min at room temperature.

11. Add the transfection mixture B in a dropwise fashion to the cells, while rocking the plate to allow the even distribution of buffer B.

▶ **Note:** During this procedure a fine precipitate should form, making the solution slightly milky.

12. Incubate on a tangential shaker for at least 4 h at room temperature to keep the cells moist.

13. Remove the transfection reagent and replace with insect cell medium (Hink's TNM or Grace's medium supplemented with 10% fetal calf serum (FCS)).

▶ **Note:** The transfection can be monitored under the microscope. The cells should show crystals attached to the surface.

14. Incubate the cells for 3–5 days at 27 °C until they have lyzed.

15. Harvest the supernatant that contain the primary virus.

▶ **Note:** Store the virus at 4 °C and protected from light. Do not freeze the virus as this will result in a reduction of the titer.

16. The remaining cells on the plate may be lyzed in 0.2 ml of 1x Laemmli buffer and run on an SDS-PAGE (5 µl per lane) and eventually transferred onto a PVDF membrane to probe for expression of the recombinant protein.

17. If expression of the recombinant protein appears to be very poor or non-existent after the primary transfection, large-scale amplification of the virus may help (in most cases this is necessary).

18. To amplify on a large scale, inoculate 50 ml Sf9 cells (density 2×10^6 cells/ml in Hink's TNM medium) in a 250 ml Erlenmeyer flask with 50 µl of the virus (dilution 1:1000), and incubate the cells for 4–7 days in a tangential shaker (140 r.p.m.) at 27 °C. Loosen the cap during incubation.

19. Harvest 4 days later by transferring the culture to sterile centrifuge tubes and spinning off the cell debris at 2000 r.p.m. for 5–7 min. Transfer the supernatant into a new centrifuge tube and store at 4 °C, protected from light.

20. Infect Hi5 insect cells with the recombinant virus to produce the His_6-tagged Dicer protein according to common baculovirus protocols (Bac-To-Bac® System, Invitrogen).

▶ **Note:** Expression time depends on the quality and the titer of the virus. Determination of the titer is therefore highly recommended. The amount of protein should peak at 48 h after infection, before the cells start to be lyzed by the virus.

21. After 48 h, harvest the cells and lyze them by sonication on ice in a few ml of sonication buffer (Table 4.9).

▶ **Note:** The amount of the sonication buffer depends on the size of the respective centrifuge tubes being used.

Tab. 4.9 Buffers for purification of His_6-tagged Dicer under native conditions

Sonication buffer		Wash buffer		Elution buffer	
NaH_2PO_4 pH 8.0	50 mM	NaH_2PO_4 pH 8.0	50 mM	NaH_2PO_4 pH 8.0	50 mM
NaCl	300 mM	NaCl	300 mM	NaCl	300 mM
Imidazole	10 mM	Imidazole	20 mM	Imidazole	250 mM
Nonidet P-40	1%	Nonidet P-40	1%	Nonidet P-40	1%

22. Insoluble debris are removed by centrifugation at 10 000 g at 4 °C.

23. Incubate the supernatant with Ni-NTA (nitrilo triacetic acid) resin (Qiagen) on a rotator for 1 h at 4 °C

▶ **Note:** Use 1 ml of 50% Ni-NTA slurry per 4 ml of supernatant.

24. Wash the resin twice with 5 volumes of sonication buffer.

25. Wash the resin twice with 1 volume of wash buffer.

26. Elute the protein with 4×0.5 ml of elution buffer, and pool the active fractions.

27. Dialyze the protein against the digestion buffer (Table 4.10) containing 20% glycerol and 1% Nonidet-P40.

28. Determine the protein concentration by the method of Bradford (Bradford 1976) using the BioRad dye reagent, with bovine serum albumin (BSA) as a standard.

29. Store the protein at –80 °C.

4.2.5.2 DsRNA Digestion by Recombinant Dicer

PROTOCOL 17

1. Prepare dsRNA (800–1000 nt) by T7-based in-vitro transcription (see Chapters 2 and 3 for several protocols).

2. Dilute up to 60 µg of dsRNA in 1–150 µl Dicer digestion buffer (Table 4.10).

Tab. 4.10

Dicer digestion buffer

HEPES pH 8.0	30 mM
NaCl	250 mM
EDTA	0.05 mM
$MgCl_2$	2.5 mM

▶ **Note:** Make sure that the volume of dsRNA added does not exceed half the volume of the reaction.

3. Add the appropriate amount of purified Dicer and fill up the with Dicer digestion buffer to 300 µl.

▶ **Note:** Before using Dicer, check the activity of the protein preparation in dicing assays.

4. Mix the reaction gently and incubate for 15 min to 1 h at 37 °C.

▶ **Note:** If the activity is not as high, the incubation can be performed for 14–18 h at 37 °C. The digestion of about 60 µg of dsRNA usually produces about 20–30 µg of siRNAs after purification, which is sufficient for at least 120–150 24-well transfections.

5. Add 10 µl of 1 mM EDTA pH 8.0 to stop the reaction.

6. Check the integrity of the siRNAs in either a 2–4% TAE agarose gel or in a 15–20% polyacrylamide gel using a 10-bp DNA size standard.

7. Purify the siRNAs from Dicer by Ni-NTA-agarose chromatography to separate and recover Dicer and collect the flowthrough.

8. Purify and desalt the flowthrough by gel filtration on G25 spin columns (Amersham Biosciences) or common gel filtration methods.

9. Dilute the siRNA in RNA annealing buffer prior to storage at −80 °C (Table 4.11).

Tab. 4.11

Annealing buffer 1 (1x)		Annealing buffer 2 (10x)	
Sodium acetate	100 mM	Tris-HCl, pH 8.0	100 mM
HEPES-KOH pH 7.4	30 mM	EDTA, pH 8.0	10 mM
Magnesium acetate	2 mM	NaCl	1 mM

▶ **Note:** The siRNAs can also purified by digestion of the Dicer mixture with proteinase K (0.2 mg/ml) and chloroform:isoamyl alcohol (24:1) extraction (Sambrook and Russell 2001). Another method for purification is the separation of the mixture on a 2–4% agarose gel and electroelution of the siRNAs.

Since Dicer is too large (~220 kDa) to be easily expressed in prokaryotic systems, which usually produce much higher protein yields and offers an easier handling than eukaryotic expression systems, some groups and companies have attempted to establish the use of RNaseIII from *E. coli* to cleave dsRNA into siRNAs (Yang et al. 2002 a). *E. coli* RNaseIII is much smaller in size (ca. 30 kDa), and can be expressed in *E. coli* as a glutathione-S-transferase (GST) fusion protein that supports the purification of the recombinant enzyme. The digestion of the dsRNA can be carried out without removal of the GST-fusion tag. Such endoribonuclease-prepared *si*RNA (esiRNA) usually has a size of between 12 and 25 bp. Optimization of the digestion conditions led to an enrichment of 21-mers even with RNaseIII. In contrast to long dsRNA, esiRNA mediates effective RNA interference without apparent nonspecific effects in cultured mammalian cells. Many companies now offer modified recombinant *E. coli* RNaseIII that specifically produces fragments of only 21–23 nt (New England Biolabs "ShortCut™ RNAi Kit" or Stratagene).

In the first experiments on knock-down gene expression during the development of mammalian embryos, esiRNAs were injected into the lumen of the neural tube at specific regions and delivered into neuro-epithelial cells by directed electroporation (Calegari et al. 2002). esiRNAs develop the silencing with the same or even higher efficiency than single siRNAs, since it is more likely that the siRNA pool contains one or more siRNAs with highest efficacy.

4.2.5.3 Production of Recombinant RNaseIII from *E. coli*

The following protocol is adapted from Buchholz and coworkers (Yang et al. 2002 a) and Nicholson and coworkers (Li et al. 1993).

PROTOCOL 18

1. Amplify the ORF of RNaseIII (NCBI Acc. no. X02946) from *E. coli* DH5α by adding 2 μl of bacterial overnight culture to the PCR mixture. Use the primers depicted in Figure 4.15.

   RNaseIII Primers

   Forward
   5'-CGC***GGATCC***AACCCCATCGTAATTAATCGGCTTCA-3'
        BamHI

   Reverse
   5'-GACGTCCGACGATGGCAA-3'

   Fig. 4.15   RNaseIII primers.

2. Digest the PCR product and the prokaryotic expression vector pGEX-2T or pGEX-4T (Amersham Biosciences) with *BamHI* and *SmaI*.

   ▶ **Note:** pGEX vectors allow the production of GST fusion proteins.

3. Ligate the RNaseIII ORF into the *BamHI* and *SmaI* site of pGEX-2T or 4T.

4. Transform the recombinant vector into the *E. coli* expression strain BL21(DE3) to produce GST-RNaseIII protein.

5. Let the bacteria grow in TB (terrific broth) medium at 37 °C to an $OD_{600} = 0.5$.

6. Induce with 100 μM isopropyl-β-D-thiogalactoside for 4 h.

7. Centrifuge the cell suspension at 10 000 g for 20 min at room temperature or at 4 °C.

8. Resuspend the pellet in pGEX lysis buffer (25 ml/l of culture) (Table 4.12).

   Tab. 4.12

pGEX lysis buffer		Pre-elution buffer		Elution buffer*	
HEPES pH 7.6	20 mM	HEPES-NaOH pH 7.0	25 mM	Glutathione	50 mM
KCl	100 mM	NaCl	100 mM	(reduced)	
EDTA pH 8.0	0.2 mM	EGTA (or EDTA)	0.1 mM	Pre-elution buffer	
Triton X-100	1%				
Glycerol	20%				

   * Must be freshly prepared; re-adjust to pH 8.0

9. Add 1 ml of lysozyme (10 mg/ml) and a small amount of lyophilized DNase and incubate for 15 min at room temperature.

   ▶ **Note:** Place a stir bar into the suspension to facilitate dislodging of the pellet. Lay it on an angle in a plastic container on top of a stirring plate; leave it stirring until everything is suspended.

10. Transfer the suspension into a Dounce homogenizer and homogenize with 10 up-and-down strokes.

11. Spin the homogenate at 100 000 g for 1 h at 4 °C.

12. Pellet 1 ml of 50% slurry glutathione-Sepharose beads (Amersham Biosciences), and resuspend them in a small volume (~0.5–1 ml) of pGEX lysis buffer.

13. Rotate the supernatant of the ultracentrifugation end-over-end for 2 h at 4 °C.

▶ **Note:** Generally, 1 ml of 50% agarose slurry is incubated with 10–25 ml of GST-RNaseIII supernatant.

14. Pour the suspension into a polypropylene Econo-Column (Bio-Rad) and collect the flowthrough in the same tube.

15. Pour the flowthrough over the column two to three more times in order to collect as many beads as possible. Freeze the flowthrough at –80 °C.

16. Wash several times (20–30 bed volumes) with pGEX lysis buffer.

17. Wash several times (20–30 bed volumes) with Pre-Elution Buffer (Table 4.12).

18. Prepare five to six numbered Eppendorf tubes. Put the column into the first tube and add 200 µl of Elution Buffer (Table 4.12). Once the beads are "dry", transfer the column to the next tube and add 500 µl Elution Buffer. Continue eluting with 500 µl Elution Buffer.

19. Set up 0.5 ml aliquots of the BioRad protein assay in glass tubes. Add 10 µl of the eluates and mix well. Identify the fractions of high protein content by their intense blue color. Pool the peak fractions.

▶ **Note:** Use the Bradford protein assay here. Glutathione interferes massively with the BCA assay!

20. Dialyze the purified protein against the RNaseIII storage buffer (Table 4.13A) and store at –20 °C.

Tab. 4.13

A) RNaseIII storage buffer		B) RNaseIII digestion buffer	
Tris-HCl pH 7.9	20 mM	Tris-HCl pH 7.9	20 mM
EDTA	0.5 mM	EDTA	0.5 mM
NaCl	140 mM	NaCl	140 mM
KCl	2.7 mM	KCl	2.7 mM
DTT	1 mM	DTT	1 mM
$MgCl_2$	5 mM	$MgCl_2$	5 mM
Glycerol	30%	Glycerol	5%

21. To prepare esiRNAs, digest 100 µg of dsRNAs with 1 µg of recombinant RNaseIII in 200 µl RNaseIII digestion buffer (Table 4.13B) for 15 min at 37 °C.

▶ **Note:** Do not extend the incubation time, as this will lead to smaller fragments such as 12–15- nt siRNAs.

22. Terminate the reaction by adding EDTA pH 8.0 to a final concentration of 20 nM.

23. Separate the siRNAs in a 12% PAGE using 1x TBE buffer (Sambrook and Russell 2001), or in a 2–3% TAE-agarose gel as described for DNA (for both protocols, see the next section).

24. Use a 10-bp DNA marker to estimate the migration of RNA duplexes.

25. To purify the esiRNAs, either extract the appropriate agarose gel slice using gel extraction reagents such as the QIAquick gel extraction kit from Qiagen, or elute them from the polyacrylamide gel by soaking the gel slice in 1 M ammonium acetate at 37 °C overnight.

26. Recover the esiRNAs by ethanol precipitation (Sambrook and Russell 2001).

27. Dissolve the esiRNAs in TE buffer.

28. Alternatively, the digestion products can be purified using the QIAquick nucleotide removal kit (Qiagen) to remove the enzyme.

For laboratories that do not want to embark on protein expression and purification, there is an option to purchase Dicer and RNaseIII from various companies. Besides Stratagene and New England Biolabs (both provide RNaseIII), Invitrogen is offering a complete dsRNA dicing system. The BLOCK-iT™ Dicer RNAi Transfection kit provides the necessary reagents to generate enough diced product to perform up

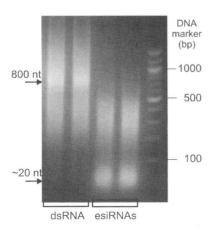

Fig. 4.16 Digestion of in-vitro-transcribed dsRNA of about 800 nt by recombinant RNaseIII from *E. coli*. The 2% TAE-agarose gel shows the separation of dsRNA prior and after digestion that results in the production of enzymatically produced siRNAs (esiRNAs).

to 150 transfection experiments in 24-well plates with up to five genes. The kit includes recombinant Dicer and purification reagents for diced siRNAs, as well as the Lipofectamine™ 2000 transfection reagent for the highly efficient delivery of siRNAs to cells.

4.3
Delivery of siRNAs into Cells

Although RNA has demonstrated great potential as a therapeutic tool, its efficient delivery to targets remains a major challenge. Besides the regular methods, novel approaches such as the use of bacteria as DNA carriers might facilitate delivery by using adjuvants or photochemical internalization, although these systems are in their very early stages of development.

One way to induce RNAi in cells is transient introduction of synthetic or enzymatically transcribed siRNAs by transfection, electroporation, or microinjection for example. Several companies offer specialized siRNA-delivery reagents especially for the transfection method. Delivering siRNAs directly to whole vertebrate animals is more problematic than it is for invertebrates or cell lines, because the animals cannot absorb the siRNAs or any kind of naked nucleic acids through the skin, and for many other purposes simple injection into the bloodstream has proved ineffective. In 2002, Hannon and coworkers and Lewis and colleagues independently employed a "hydrodynamic transfection method" to deliver naked siRNAs to mice via tail-vein injection (Lewis et al. 2002; McCaffrey et al. 2002). These authors observed downregulation of a reporter gene by 80–90% in the liver, kidney, spleen, lung, and pancreas, but the effect was relatively short-lived and lasted for only a few days. This procedure will be discussed later in the chapter.

The following section will deal with short outlines of delivery methods, as they can vary widely according to the type and supplier of the chemical reagents.

4.3.1
Transfection

To date, the most common delivery method for siRNAs is regular transfection, as described for DNA. Although recently, electroporation has attracted more attention, transfection remains the most reliable technique. Nevertheless, it should be noted that transfection using calcium phosphate is less efficient for siRNA than it is for DNA. Following the use of calcium phosphate and polybrene, various cationic liposomal formulations were developed during the early 1990s. As yet, transfection reagents based on cationic lipids, dendrimers, and polyethyleneimine (PEI) in general result in better transfection efficiency of up to 90%, depending on the cell type. The most widely used transfection reagent is the Lipofectamine™2000 (Invitrogen), but other companies offer similarly efficient products; examples include RNAiFect (Qiagen) or siPORT™Lipid and siPORT™Amine transfection reagents (Ambion). For more information, the reader is referred to the manuals of the transfection reagents.

The following protocol is a short outline of the Lipofectamine™2000 transfection protocol. A very detailed protocol can be found on the Invitrogen web site, or in the reports by Gitlin et al. (2002) and Yu et al. (2002).

PROTOCOL 19

1. Refer to Table 4.14 for the appropriate reagent amounts and volumes to add for different culture dishes.

Tab. 4.14   Relative amounts of siRNAs, transfection reagent and cells. Modified from the Lipofectamine™2000 application manual (Invitrogen).

Culture dish	Relative surface area	Growth medium	Amount of synthetic siRNAs or T7 SiRNAs		Amount of Dicer or RNaseIII digested siRNAs		Lipofect- amine™2000 dilution volume	
96x well	0.2	100 µl	~4 pmol in	25 µl	20 ng in	25 µl	0.6 µl in	25 µl
24x well	1.0	500 µl	~20 pmol in	50 µl	50 ng in	50 µl	1.0 µl in	50 µl
6x well	5.0	2 ml	<100 pmol in	250 µl	250 ng in	250 µl	5.0 µl in	250 µl
10 cm ⌀	5.0	10 ml	>500 pmol in 1.25 ml		1.25 µg in 1.25 ml		25.0 µl in 1.25 ml	

▶ **Note:** Use the recommended amounts of transfection reagent and siRNA as a starting point to optimize conditions for your cell line.

2. One day before transfection, plate cells in the appropriate amount of growth medium without antibiotics to 30–50% confluency at the time of transfection.

3. Dilute siRNAs in the appropriate amount of Opti-MEM® I Reduced Serum Medium (Invitrogen) or in any other serum-free medium, and mix gently.

4. Mix Lipofectamine™2000 gently before use and dilute the appropriate amount in Opti-MEM® I Reduced Serum Medium or any other serum-free medium.

5. Incubate for 5 min at room temperature.

6. Combine Lipofectamine™2000 with the siRNAs, and mix gently.

▶ **Note:** Combine the diluted Lipofectamine™2000 with the diluted d-siRNA within 30 min to avoid a decrease in activity.

7. Mix gently, and incubate for 20 min at room temperature.

▶ **Note:** The solution may appear cloudy due to complex formation, but this has no influence on the transfection efficiency.

8. Add the transfection mixture to the appropriate culture dish without removing the growth medium.

9. Mix gently by rocking the plate back and forth.

10. Incubate the cells at 37 °C for 24–96 h.

▶ **Note:** It is not necessary to replace the medium for the incubation time required for the knock-down.

11. Analyze the cells for the knock-down by either RT-PCR, Western blot, cell imaging, or other assays.

4.3.2
Electroporation

Although siRNA delivery by chemical transfection works well for many adherent cell lines, suspension cell lines or even many primary and nondividing cells are difficult to transfect with calcium phosphate or cationic lipids as well as other transfection reagents. Hence, other methods must be applied to deliver siRNAs to the cells. An alternative to chemical delivery is physical delivery, such as microinjection or electroporation. While microinjection is very laborious and is not suitable for large cell cultures, electroporation has been used to efficiently deliver siRNAs to cells (Dunne et al. 2003; Heidenreich et al. 2003; Scherr et al. 2003; Walters and Jelinek 2002).

While cationic lipid-complexed siRNAs are incorporated via endocytosis and endosomal escape, a homogenous cytosolic distribution is observed after electroporation, indicating a direct entry of siRNAs into the cytoplasm.

The estimation of uptake efficiency by each cell line is usually performed by electroporation of fluorescently labeled siRNAs (Walters and Jelinek 2002). However, it should be noted that the use of negatively charged dye molecules such as fluorescein can give false-negative results. It was speculated that this might be due to their interaction with the RISC complex, leading to a quenching of the fluorescence (Dunne et al. 2003). Positively charged dyes such as Cy3 or Cy5 do not show those quenching effects, and are therefore recommended for monitoring the silencing with siRNAs (Dunne et al. 2003).

PROTOCOL 20

The following protocol is devised for electroporation of MDCK cells modified from (Ge et al. 2003).

1. Grow MDCK cells in DMEM supplemented with 10% FCS, 2 mM L-glutamine, 100 units/ml penicillin, and 100 µg/ml streptomycin at 37 °C/5% CO_2.

2. Trypsinize logarithmic-phase MDCK cells and resuspend in serum-free RPMI medium.

3. Check the electroporation efficiency with Cy3-labeled siRNAs using different cell concentrations and different settings for voltage and capacity.

▶ **Note:** Refer to the manual of your electroporation device for the optimized electroporation conditions for each cell line. Cell number and capacities can vary.

4. Resuspend cells at 2×10^7 cells/ml in RPMI serum-free medium.

5. Place 500 μl of the cell suspension in electrode gap cuvettes with a 4-mm gap (Invitrogen). Keep on ice.

6. Add siRNA to the cell suspension to a 2 μM final concentration.

7. Gently mix the suspension by flicking the cuvette.

8. Incubate on ice for 5 min.

9. Place the electroporation cuvettes in the electroporator and pulse once at 400 V and 975 μF.

▶ **Note:** The voltage and the capacities can vary for different electroporation devices and cell types. These value are for the use of the Gene Pulser apparatus (Bio-Rad). For example, for mouse bone marrow cells the settings are 1×10^6 cells/ml, 320 mV, 1600 μF (Oliveira and Goodell 2003).

10. As quickly as possible, add 500 μl of DMEM/10% FCS medium to the cuvette

▶ **Note:** To check the viability of the cells after electroporation, take an aliquot and count the cells with trypan blue. Cell viability can also be performed by FACS analysis using 2 μg/ml propidium iodide (PI) for dead cell discrimination.

11. Gently transfer the cell suspension into a 15-ml sterile tube containing 8 ml DMEM/10% FCS.

12. Mix gently, and divide the cell solution into three wells of a 6-well plate and culture in DMEM/10% FCS for 48 h at 37 °C/5% CO_2.

▶ **Note:** The incubation time depends on the turnover of the targeted protein.

14. Analyze the cells for the RNAi phenotype.

One of the most interesting innovations in the field of physical delivery techniques is the Nucleofector™System by Amaxa (http://amaxa.com). The Nucleofector™ Technology is a major improvement of the well-known electroporation technology. It consists of a unique combination of electroporation and optimized solutions for delivery of nucleic acids directly into the cell's nucleus or cytoplasm for each cell type. Direct delivery into the nucleus is a prerequisite for most nondividing cells when treated with DNA. Besides, specialized solutions for the electroporation of siRNAs

Fig. 4.17 The Nucleofector™ electroporation device (Amaxa, http://www.amaxa.com). Until now, most of the protocols from the company have been designed for human cell lines and primary cells but some data are available, e.g. for mouse and rat, that have been published by several users (Krauss et al. 2003; Verrecchia et al. 2003).

allow very efficient internalization of the siRNAs. Although it is not inexpensive, the siRNA internalization results from this technology are very convincing (Chun et al. 2002). Depending on the cell lines, transfection efficiencies can reach 99% with fluorescently labeled siRNAs, whereas transfection efficiencies are normally 70% on average. Currently, Amaxa is marketing an electroporation device (Nucleofector™) (Figure 4.17) that is equipped with a variety of programs, each of which represents different electrical parameters that are unlike any other commercially available electroporation parameters. The programs have been adapted to the requirements of a particular cell type.

4.3.3
Cell Penetration Peptides (CPPs) for siRNA Delivery

Although considered to be established methods, transfection and electroporation still suffer from many drawbacks. As mammalian cells do not possess an amplification mechanism for siRNAs, the concentration of siRNAs is diluted with every cycle of cell division, leading to a recovery of the mRNA levels after 5–9 days. Therefore, the treatment with siRNA must be repeated if the analysis of the phenotype requires longer knock-down periods.

Moreover, in many cell lines the common transfection techniques are not 100% efficient, and most primary cells and non-dividing cells are not susceptible to transfection altogether.

However, the transient approach bears some positive features, which makes this approach a valuable tool. In some experimental set-ups it may be desirable to knock-out genes for a short period of time to be able to observe the recovery of the loss-of-function phenotype. To study the function of genes essential for development in fully grown organisms, a transient approach permits the generation of the desired loss-of function phenotype once the decisive stages of development have been completed. Finally, clinical applications often demand only a temporary change in expression profile without causing permanent changes. This is important to suppress viral infections or treating inflammatory processes (Arenz and Schepers 2003; Boden et al. 2003; Das et al. 2004; Giladi et al. 2003; Hamasaki et al. 2003; Hilleman 2003;

Jacque et al. 2002; Lawrence 2002; Lindenbach and Rice 2002; Pooggin et al. 2003; Pusch et al. 2003; Song et al. 2003; Yamamoto et al. 2002).

Especially for clinical applications, the question of in-vivo delivery needs to be resolved. A highly efficient delivery is required to reduce costs for the production of siRNAs. More importantly, lower doses ensure a more specific action of the drug with fewer side effects. And finally, none of the common methods used to introduce siRNAs into cells is suitable for the treatment of patients. While most chemical transfection reagents work well in cell culture, they usually show moderate effects in humans or other mammalian organisms. Likewise, mechanical procedures such as electroporation, particle bombardment, or hydrodynamic injection are not applicable in human beings.

4.3.4
Cell-Penetrating Peptides (CPPs) or Protein Transduction Domains (PTDs)

Protein transduction domains offer an alternative to the traditional methods of siRNA delivery. These short amino acid sequences are able to interact with the plasma membrane in a way that leads to a highly efficient uptake into the cytoplasm. This feature was first observed for the third helix of the *Drosophila* homeoprotein Antennapedia (AntP), and for the HIV-1 Tat protein (TAT) (Derossi et al. 1994, 1998; Prochiantz 1996; Spencer et al. 1993). Closer studies revealed that only short stretches, mainly consisting of positively charged amino acids as arginine and lysine, were responsible for the translocation of the PTDs through the plasma membrane – hence the name *cell-penetrating peptides* (CPPs). Stretches of these amino acids help to maintain the helical structure of the PTD, which is the structural prerequisite for membrane penetration. Many natural and synthetic CPPs are known that contain those structural properties; a selection is listed in Table 4.15.

The mechanism by which protein transduction takes place remains to be elucidated in detail, though several reports on this topic were published recently (Richard

Tab. 4.15 CPPs currently in use for many delivery processes, as modified from Fischer et al (2001). (Brugidou et al. 1995; Derossi et al. 1994; Elliott and Ohare 1997; Fischer et al. 2001; Nagahara et al. 1998; Pooga et al. 2001; Vives et al. 1997; Wender et al. 2001; Williams et al. 1997).

CPP name	Peptide sequence	References
Antennapedia peptide, penetratin	RQIKIWFQNRRMKWKK	Derossi et al. 1994
Retro-inverso penetratin	KKWKMRRNQFWVKVQR	Brugidou et al. 1995
W/R penetratin	RRWRRWWRRWWRRWRR	Williams et al. 1997
HIV TAT	GRKKRRQRRRPPQ	Vives et al. 1997
HIV TAT	YGRKKRRQRRR	Nagahara et al. 1998
Rz (or R8 and R9)	RRRRRRR	Wender et al. 2001
Transportan	GWTLNSAGYLLGKINLKALAALAKKIL	Pooga et al. 2001
HSP Vp 22	DAATATRGRSAASRPTERPRAPARSASRPRRPVE	Elliott and Ohare 1997

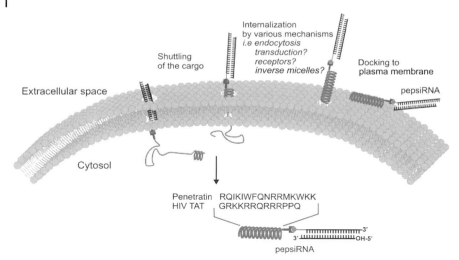

Fig. 4.18 Cartoon of the penetration process supported by CPPs, as it was suggested in the early days of CPP research. With recent studies on the mechanism of entry, it has become clear that endocytosis is involved, but might not be the only means of entry. After uptake, the disulfide bond that links the two moieties (CPP and siRNA) is reduced under the conditions of the cytosol to release the short interfering RNA (siRNA).

et al. 2003; Silhol et al. 2002; Vives et al. 2003). It has been shown that cellular uptake of the CPPs takes place in a receptor-independent fashion. CPPs are mostly amphipathic molecules that interact with the negatively charged head groups of the plasma membrane via their positive amino acid residues. In the following step, the peptide enters the cell by either forming an inverse micelle, disrupting the membrane structure, or by some specific form of endocytosis. The experimental findings are contradictory however, and it appears that several mechanisms may be used simultaneously, which vary in preference with the peptide sequence, cell type, and cargo molecule.

Those CPPs can be produced by recombinant expression and solid-phase synthesis and coupled to a variety of reporter molecules. It turned out that conjugates of TAT with the 120 kDa protein β-galactosidase entered almost all cell types when injected into the bloodstream of mice (Schwarze et al. 1999). The conjugates were even found beyond the blood–brain barrier and likewise, CPPs where found to cross the water-permeability layer of the skin. Experiments in cell culture revealed that CPPs entered almost all types of cells with very high efficiency and rapid uptake kinetics (Schwarze and Dowdy 2000). CPPs are still functional when coupled to large proteins of up to 100 amino acids, antisense DNA of up to 80 nt, fluorescent probes of different structures, and even paramagnetic nanoparticles, liposomes and recombinant viruses (Christiaens et al. 2002; Gratton et al. 2003; Schwarze and Dowdy 2000; Villa et al. 2000; Wunderbaldinger et al. 2002). Their great versatility with respect to cargo and cell type – as well as their high efficiency in delivering cargo molecules into cells – makes the CPPs a valuable tool for the introduction of siRNAs into mammalian cells and even fully grown organisms. The following protocol describes the generation of CPP-coupled siRNAs, which in my laboratory are referred to as pepsiRNAs (*pep*tide-coupled *si*RNAs) (Schmitz et al. submitted; Schmitz and Schepers 2004). The protocol is based on the covalent coupling of the two building blocks, which allows the use of modified synthetic siRNAs as well as the use of modified enzymatically synthesized siRNAs. The covalent coupling of the peptides with oligonucleotides – in contrast to proteins or peptides – includes some difficulties that require a degree of chemical expertise, but this can be overcome by custom synthesis of the two building blocks.

4.3.4.1 Coupling of siRNA to CPPs

A variety of CPPs has been coupled to many classes of cargo molecules, ranging from fluorescent probes and magnetic beads to antibodies, liposomes, and even viruses (Christiaens et al. 2002; Gratton et al. 2003; Schwarze and Dowdy 2000; Villa et al. 2000; Wunderbaldinger et al. 2002). Depending on the purpose and the reactivity of the molecular units to be linked, many coupling strategies have been developed and refined for the individual systems. Many cargo molecules bear functionalities that permit coupling via amide bonds, esters, and thioethers leading to stable linkages for the mostly artificial probe molecules (Table 4.16).

Active functionalities occurring at the N-terminus of the CPPs or in cargo molecules such as the 2'-OH or adenine/guanine-NH_2 of nucleic acids may be used for multiple labeling, which can be desirable for the detection of low quantities of DNA/RNA. In other cases, a defined number of attached probe molecules is required for quantification purposes.

Finally, some cargo molecules may only be coupled at one defined site, where the chemical modification does not alter their functionality. This can be achieved by making use of single functional groups that either naturally occur in the cargo molecule or may be introduced by the utilization of chemically modified building block in the synthesis of the CPPs. This system can be exploited for efficient delivery of siRNAs into cells and into animals. For the coupling of the siRNA both, the CPP and the siRNA must be modified by functional groups. The most common and versatile approach for the synthesis of CPPs and siRNAs is solid-phase synthesis, where a broad

Tab. 4.16 A selection of cargo molecules and coupling linkers

Cargo	Peptide (up to 100 AS)	
	Fluorescent probes	
	ssDNA (up to 80 nt)	
	Antisense RNA	
	PNAs	
Coupling groups	Amide bonds	CPP-**CO-NH**-Cargo
	Thioethers	CPP-**CH₂-S-CH₂**-Cargo
	Disulfide bonds	CPP-**S-S**-Cargo

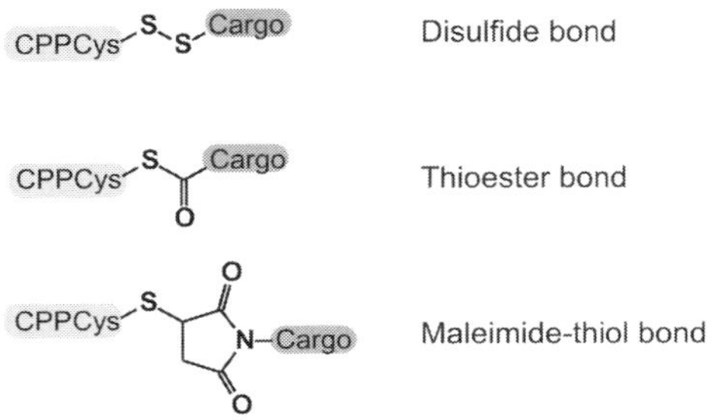

CPPCys–S–S–Cargo   Disulfide bond

CPPCys–S–C(=O)–Cargo   Thioester bond

CPPCys–S–(maleimide)–N–Cargo   Maleimide-thiol bond

Fig. 4.19 Selection of different CPP-cargo linkers that are suitable for ribonucleotide coupling. Modified from Fischer et al. (2001).

spectrum of modified building blocks can be incorporated at any given site of the molecule. This method has the advantage that the building blocks are custom-made, fully analyzed (HPLC, mass spectrometry, PAGE, concentration), and ready to be coupled. The main disadvantages are the high costs and the long delay between placing the order and the shipment (around 4–6 weeks). In contrast, in-vitro transcription by T7 RNA polymerase permits the selective incorporation of modified ribonucleotides at the 5′-end of the selected siRNA strand, if they are active as initiation nucleotides (Zhang et al. 2001). Most synthetic probe molecules are commercially available with coupling functionalities such as hydroxyl, amine, boron, carboxyl or thiol groups, and can be easily linked to the modified cargo molecules. Among the available coupling functionalities the disulfide bond is the most suitable for the purpose of delivering functional siRNAs into cells. Others are depicted in Figure 4.19.

Due to the reducing environment of the cytosol, the disulfide bond is cleaved upon its entry into the cell, thereby releasing the siRNA from the peptide moiety with only a minimal residue on the siRNA that is unlikely to interfere with the enzymes of the RNAi mechanism. Moreover, the cleavage of the CPP prevents translocation of the siRNA into the nucleus, where many protein transduction domains tend to accumulate. In vivo, this system enables local siRNA application, which may be interesting for the study of essential genes in individual tissues, where a systemic loss would be lethal.

Even though the antisense strand of the siRNA can tolerate some modification without any effect on the silencing efficacy of the siRNA, it is recommended to modify the 5′-end of the sense strand. Modification of the sense strand 5′-end can either be obtained by solid-phase synthesis (custom-made siRNAs available at Dharmacon (http://www.dharmacon.com)) (Figure 4.20A), or by chemical synthesis of a linker that can be added to the 5′-end during T7-based transcription (Schmitz et al. submitted; Schmitz and Schepers 2004) (Figure 4.20B). In-vitro transcription with excess of 5′-thiophosphate-functionalized guanosine (GSMP, 5′-desoxy-5′-thioguanosid-monophosphorothioate) (Schmitz et al. submitted; Zhang et al. 2001) as an initiator nucleotide and subsequent dephosphorylation yields in a 5′-thiol-modified sense-RNA (Figure 4.20B). The antisense strand is transcribed without modification, and the two purified strands of RNA are hybridized to form the appropriate siRNA.

Besides by chemical synthesis, CPPs for disulfide coupling can be obtained by recombinant expression of fusion peptides (i.e. with GST) adding an extra C-terminal cysteine residue for thiol-functionalization or by chemical synthesis. This can be per-

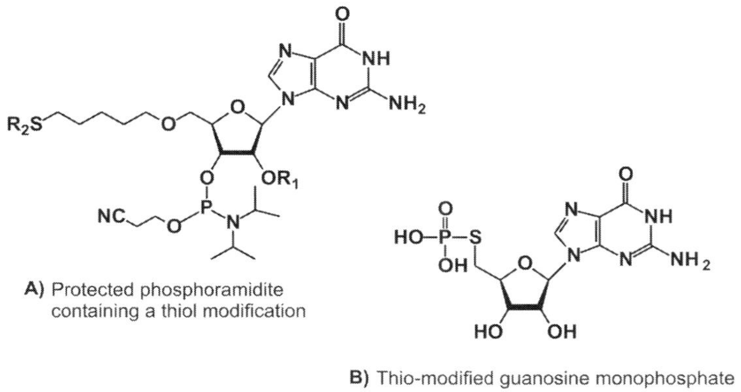

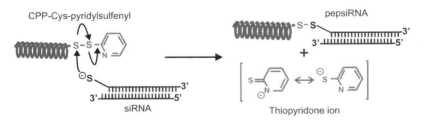

A) Protected phosphoramidite containing a thiol modification

B) Thio-modified guanosine monophosphate

Fig. 4.20 Selection of linkers that can be used for: (A) solid-phase synthesis (R1 and R2 = protection groups); (B) T7-based enzymatic synthesis of 5′-sense-GSMP-modified siRNAs (GSMP, 5′-desoxy-5′-thioguanosid-monophosphorothioate; Schmitz et al. submitted; Schmitz and Schepers 2004; Zhang et al. 2001).

formed by altering the sequence of the peptide's ORF. Additionally, a biotin or fluorescence label can be introduced at the N-terminus to monitor uptake of the conjugates inside the cell. Thus, thiolated oligonucleotides can be conjugated with CPP-Cys-building block via disulfide bond formation. To favor formation of the heterodimeric conjugate and to prevent the formation of homodimers, the thiol group of either peptide *or* siRNA, is activated with a pyridylsulfenyl-group (Figure 4.21). A disulfide-exchange between the free thiol and the pyridylsulfenyl-activated thiol group is usually employed (Bernatowicz et al. 1986; Houk et al. 1987). The activation group is released in the form of a thiopyridone ion, which is stabilized by mesomeric effects, thus preventing a back-reaction (Figure 4.21). Over 90% of heterodimers are formed in this reaction at room temperature in less than 1 h (Bernatowicz et al. 1986; Houk et al. 1987; Vives and Lebleu 1997).

Procedure of the coupling reaction can be monitored by the release of thiopyridone ion, which exhibits an absorption maximum at 314 nm, and this can be de-

Fig. 4.21 Coupling mechanism of the modified 3-nitro-2-pyridylsulfenyl derivative of cysteine; this was prepared and used to facilitate the formation of an unsymmetrical disulfide bond (Bernatowicz et al. 1986; Vives and Lebleu 1997).

tected either photometrically or with the naked eye, by its brown color. The siRNA-peptide conjugates can be additionally tested in an SDS-PAGE using a 20% SDS-polyacrylamide gel obtained from a 19:1 acrylamide-bisacrylamide solution. The conjugate in nonreducing Laemmli buffer should run slightly higher than the conjugate in reducing Laemmli buffer (Table 4.19 B), which is equal in size to the uncoupled peptide sample.

4.3.4.2 Chemical Synthesis of 5′-Thiol-Modified siRNAs

To obtain the thiol-modified siRNAs made by solid-phase chemistry, proceed to your siRNA provider and order custom-made siRNAs containing a 5′-C$_6$-SH spacer. Not all siRNA providers offer custom-made synthesis and such modifications. Exceptions are Dharmacon (http://www.dharmacon.com) and Proligo (http://www.proligo.com), but there may be others. Notably, thioate modifications are quite common but should not be mistaken with the 5′-thiol modification.

4.3.4.3 Enzymatic Synthesis of 5′-Thiol-Modified siRNAs

Before starting with the enzymatic synthesis of the 5′-thiol-modified siRNA, it is recommended that the reader consults Sections 4.2.3 and 4.2.4. These sections describe the steps from the design of the siRNA to the final enzymatic synthesis. The synthesis of 5′-thiol-modified siRNAs is based on a slightly modified version of this protocol, and the steps are presented in Figure 4.22.

For the design, it should be noted that according to latest findings the efficiency of the siRNA can be enhanced up to a 100-fold if the internal stability of the 5′-end of the antisense strand is much lower than the stability of the sense strand. This set-up facilitates the action of the RISC helicase, and favors incorporation of the antisense strand over the sense strand (Schwartz et al. 2003). Since it appears that only one strand of the siRNA is required for RISC action, while the other one is dismissed and subsequently degraded, it is possible to alter the 3′-nucleotides of the sense strand to form mismatches or wobble base pairs with the corresponding antisense sequence, thus decreasing the stability of the antisense 5′-end significantly. If there are no data available on the efficacy of a single siRNA, it is recommended to use three to five different siRNAs that cover different sites along the target mRNA. It has been shown that mixtures of siRNAs exhibit a cooperative effect, leading to a silencing efficiency which exceeds that of the most efficient siRNA of the set (Ji et al. 2003; Zhang et al. 2002). For coupling to CPPs, mixtures of three to five siRNAs are used. To ensure specific silencing, an extensive homology search should be conducted with the selected siRNA sequences to ensure that no genes with homologous sequences are accidentally targeted.

As the following enzymatic synthesis is based on the protocol described in Section 4.2.4, only the critical steps will be repeated. For the whole protocol, refer to the respective section. As a 5′-sense linker, GSMP, 5′-desoxy-5′-thioguanoside-monophosphorothioate (Schmitz et al. submitted; Zhang et al. 2001), is used as depicted in Figure 4.20. GSMP is not available commercially, and so must be obtained either by chemical synthesis or by custom synthesis from a company usually producing nucleotides. The synthesis requires either chemistry laboratory equipment or a chemis-

A

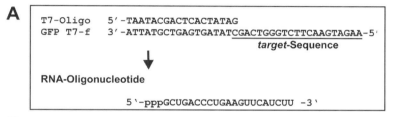

```
T7-Oligo    5'-TAATACGACTCACTATAG
GFP T7-f    3'-ATTATGCTGAGTGATATCGACTGGGTCTTCAAGTAGAA-5'
                              target-Sequence
```

RNA-Oligonucleotide

5'-pppGCUGACCCUGAAGUUCAUCUU -3'

B

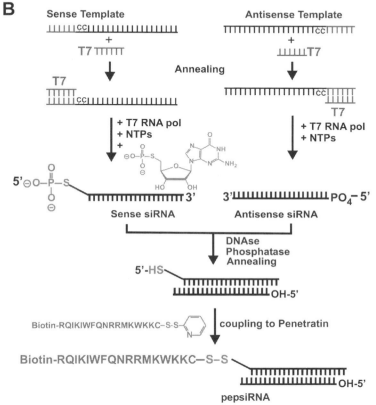

Fig. 4.22 Production of 5'-sense thiol-modified siRNAs by T7 in-vitro transcription. (A) In-vitro transcription of DNA-oligonucleotides containing a double-stranded T7 RNA polymerase recognition site leads to RNA-transcripts complementary to the single-stranded DNA-sequence. The antisense strand will be synthesized in 5'→3' direction, beginning with the last G of the 3'-end of the double stranded T7 sequence adjacent to the recognition site. The transcription preferentially starts with a GG. (B) For the generation of 5'-sense thiol-modified siRNAs, an eightfold excess of synthetic thiol-modified gua-nosine (GSMP = 5'-deoxy-5'-thioguanosine-5'-monophosphorothioate) is added to the reaction mixture that forms the sense strand. GSMP is preferentially incorporated at the 5'-end of the transcript. Following the T7 transcription, sense and antisense strands were treated with DNase and calf intestinal phosphatase to degrade residual DNA template and to remove the 5'-phosphate. The sense and antisense strands are hybridized to form siRNAs. The free SH-group can then be coupled to a Cys-pyridylsulfenyl-activated Penetratin™ (Schmitz et al. submitted; Schmitz and Schepers 2004).

try facility. The synthesis protocol comprises three steps, and is based on a protocol by Zhang and coworkers, which was recently modified by Schmitz et al. due to many difficulties that occurred during the synthesis and purification (Schmitz et al. submitted; Schmitz and Schepers 2004; Zhang et al. 2001).

PROTOCOL 21

1. Design at least three to five T7-linked DNA template oligonucleotides for the sense and the antisense strand according to the design rules described in Sections 4.2.3 and 4.2.4.

▶ **Note:** Make sure that the resulting RNA starts with a G to allow the incorporation of the coupling linker.

2. Resuspend the lyophilized, custom-made DNA oligonucleotides in twice-distilled water or annealing buffer (Table 4.18) at a concentration of 100 µM.

3. Anneal the T7-linked sense DNA template oligonucleotide with equal amounts of T7 sense primer as described in Section 4.2.4, and add RNase-free water to a final volume of 15 µl.

4. Repeat step 3 with the T7-linked sense DNA template oligonucleotide in a separate tube.

▶ **Note:** Do not mix both reactions!

5. Incubate the solutions for 5 min at 80 °C in a heating block.

6. Briefly centrifuge to collect the evaporated sample.

▶ **Note:** Spin down briefly! Do not let the sample cool down to room temperature.

7. Anneal the oligonucleotides by returning the samples into the heater, switching off the heating block, and allowing the solution to slowly cool down until the heating block reaches a temperature below 30 °C.

8. Spin down briefly.

9. To introduce the 5'-thiol modification at the sense strand 5'-end, apply the following modification of the protocol in Section 4.2.4.

10. Add the following components of the RiboMAX™T7 in-vitro transcription kit (Promega) to a final volume of 100 µl (Table 4.17). Use separate tubes for the sense and antisense strands.

Tab. 4.17

Sense strand		Antisense strand	
T7 Reaction components	**Reaction**	**T7 Reaction components**	**Reaction**
RiboMAX™ T7 transcription buffer (5x)	20 µl	RiboMAX™ T7 transcription buffer (5x)	20 µl
rNTPs (25 mM)	6 µl	rNTPs (25 mM)	6 µl
GSMP (25 mM)	**49 µl**		
DNA oligonucleotide (100 µM)	15 µl	DNA oligonucleotide (1–10 µg)	15 µl
Enzyme Mix T7 Express	10 µl	Enzyme Mix T7 Express	10 µl
Nuclease-free water	none	Nuclease-free water	49 µl
Total volume	100 µl	Total volume	100 µl

▶ **Note:** Add some RNase inhibitor (RNasin, Promega), when you are not sure if your GSMP is RNase-free.

11. Carefully mix the components.

12. Incubate the reaction mixtures overnight at 37 °C.

13. Add 1 µl of DNase and 4 µl of CIP to the sense and antisense transcript according to the manufacturer's instructions to digest the DNA template and to remove the phosphates.

▶ **Note:** Removal of the phosphate releases the 5′-thiol function of the sense strand and removes the ppp-residue from the 5′-end of the antisense strand that has been recently shown to induce interferon response (Kim et al. 2004).

14. Incubate both solutions for 15 min at 37 °C.

15. Add 0.1 vol of 3 M sodium acetate buffer and 2.5 vol of ethanol to precipitate the RNA.

16. Store at –20 °C for at least 30 min.

17. Centrifuge for 20 min at 4 °C and 14 000 r.p.m.

18. Carefully remove supernatant.

▶ **Note:** The supernatant may be transferred into fresh RNase-free tubes. Addition of more ethanol can enhance the yield, and avoids the risk of accidentally losing the pellet.

19. Dry the pellets.

▶ **Note:** Do not let the pellets dry for too long, as fully dried RNA samples are very difficult to dissolve.

20. Dissolve the pellets either in 50 µl of annealing buffer containing 100 mM DTT (Table 4.18) or in 100 mM DTT solution.

Tab. 4.18

Annealing buffer (1x)	
Sodium acetate	100 mM
HEPES-KOH PH 7.4	30 mM
Magnesium acetate	2 mM
DTT	100 mM

▶ **Note:** DTT is required to prevent homodimerization of the sense strand by disulfide bonds.

21. Determine the RNA concentration photometrically.

22. Combine equimolar amounts of the sense and antisense solutions into one microfuge tube.

23. Incubate the solutions for 5 min at 80 °C in a heating block.

24. Briefly centrifuge to collect the evaporated sample.

▶ **Note:** Only brief centrifugation! Do not let the sample cool down to room temperature.

25. Anneal the siRNA duplex by returning the mixture back into the heater, switching off the heating block, and allowing the solution to slowly cool down until the heating block reaches a temperature below 30 °C.

4.3.4.4 Coupling of Cys-modified CPPs to siRNAs

As mentioned before, the cysteine-modified CPPs can be obtained either by expression of recombinant GST-CPP-Cys fusion proteins and cleavage of the peptides after purification (refer to common protein expression protocols), or by standard chemical synthesis. Some of the peptides are commercially available in the cys-pyridylsulfenyl activated form, such as Penetratin™ from Q-biogene (http://www.qbiogene.com). The peptides can be activated with the pyridylsulfenyl-group as described elsewhere (Bernatowicz et al. 1986; Houk et al. 1987; Zhang et al. 2001).

PROTOCOL 22

1. Add 30 μl of 100 mM DTT solution to 100 μl of the thiol-modified siRNAs to cleave the homodimers.

2. Incubate overnight at 37 °C.

3. Dissolve Cys-pyridylsulfenyl-activated Penetratin™ in an RNase-free, degassed 400 mM NaCl solution to obtain a 1 μg/μl solution.

▶ **Note:** Peptides are extremely sensitive to oxidation. Never vortex the peptide samples, and avoid getting bubbles into the solution by mixing with the pipette. If possible, saturate the RNase-free NaCl solution with argon or nitrogen gas and regularly flood microfuge tubes with argon/nitrogen while handling the peptide.

4. Split the peptide into aliquots of 50 µl and freeze overlaid with argon/nitrogen at –80 °C.

5. Gently vortex the resin of the gel filtration columns (MicroSpin™ G-25, Amersham Biosciences) to resuspend the matrix.

▶ **Note:** This step is required to remove the DTT (that would prevent coupling) by reducing the activated peptide and all the other disulfide bonds.

6. Open the cap of the column by a half-turn and snap open the bottom closure.

7. Place the column in an RNase-free 1.5-ml microcentrifuge tube without a cap.

8. Centrifuge at 3000 r.p.m. (= 735 g) in microcentrifuge for 1 min at room temperature.

▶ **Note:** Use a timer!

9. Place the column in a fresh RNase-free microfuge tube without a cap.

10. Open the column and carefully apply 100–150 µl of siRNA solution to the resin.

▶ **Note:** Add the solution dropwise in the center of the resin surface, without disturbing the resin.

11. Centrifuge for 2 min at 3000 r.p.m.

12. Determine the concentration of the flowthrough by measuring the absorption in a photometer at 260 nm.

▶ **Note:** Remember to switch the photometer to dsDNA mode. It is absolutely essential to measure the concentration, because the recovery after gel filtration is usually reduced (in some cases up to 40%). Another possibility is to perform an ethanol or isopropanol precipitation (Sambrook and Russell 2001).

13. Pipet an equimolar amount (see siRNA concentration) of peptide into an RNase-free microcentrifuge tube flooded with argon.

14. Slowly add the siRNA solution straight into the peptide solution for coupling to pepsiRNAs without causing bubbles.

▶ **Note:** A yellow to brownish stain indicates procedure of the coupling reaction.

15. Incubate the coupling reaction for at least 1 h at 37 °C.

▶ **Note:** Higher temperatures may favor the formation of insoluble aggregates.

16. Split the conjugate solution into aliquots and store at –80 °C.

▶ Optional: Run a 20% SDS-PAGE (acrylamide:bisacrylamide 19:1) with one aliquot of the conjugate in non-reducing Laemmli buffer (Table 4.19A), and another aliquot and a peptide sample in reducing Laemmli buffer (Table 4.19B) to ensure that the coupling was successful. The peptide and the pepsiRNAs can be visualized by silver staining or colloidal Coomassie staining.

Tab. 4.19

A) Non-reducing Laemmli buffer		B) Reducing Laemmli buffer	
Tris/HCl pH 6.8	0.2 M	Tris/HCl pH 6.8	0.2 M
SDS	5%	SDS	5%
Glycerol	25%	Glycerol	25%
Bromophenol blue	0.075 mg/ml	Bromophenol blue	0.075 mg/ml
		β-Mercaptoethanol	12.5%

4.3.4.5 Treatment of Cells with pepsiRNAs

PROTOCOL 23

1. Calculate the concentration of the pepsiRNA solution.

2. Prepare a mix of three to five pepsiRNAs by diluting the conjugate with the corresponding amount of serum-free medium resulting in a 2x concentrated solution (i.e. 1 ml per well of a 6-well culture dish; 0.25 ml per well of a 24-well culture dish) (Figure 4.23).

▶ **Note:** Avoid serum in the medium. The peptide moiety of the conjugates may form insoluble complexes with the free DNA contained in serum.

▶ **Note:** To prevent the pepsiRNAs sticking unspecifically to the culture dishes and microcentrifuge tubes, the plastic ware may be siliconized (Sambrook and Russell 2001).

3. Grow the cells (depending on the cell type and the culture conditions) to confluence in 6-well plates or 24-well plates.

▶ **Note:** Avoid too much free plastic in the culture dish to prevent the pepsiRNA sticking to the dish.

▶ **Note:** Some cells do not allow a confluent layer without differentiation. For those cells, apply higher concentrations of pepsiRNAs.

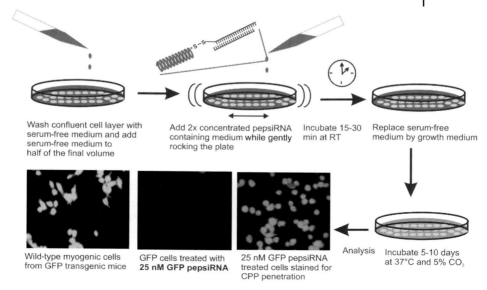

Fig. 4.23 Illustration of cell treatment with pepsiRNAs and subsequent analysis by cell imaging (Schmitz et al., submitted).

4. Remove the medium and wash the cell layer two or three times with serum-free medium to remove free DNA and serum.

▶ **Note:** Make sure that cells don't dry out between washing steps.

5. Add 1 ml of serum-free medium per 6-well (250 μl per 24-well).

▶ **Note:** Some cells should not be cultured in serum-free medium (i.e. myogenic cells). Shorten the treatment with pepsiRNAs, or add a very small amount of serum.

6. Mix three to five pepsiRNAs in a microcentrifuge tube.

7. Slowly add 1 ml (250 μl) of a 2x concentrated pepsiRNA-containing medium to each well by gently swirling the plate.

▶ **Note:** Gently swirl the culture dish, while dropping the medium with pipette from above, and vary the position of the pipette to allow an even distribution of the pepsiRNAs. The conjugate enters the cells very rapidly, so that individual cells might take up different amounts of conjugate if the solution is added onto one spot.

8. Incubate for 30 min at 37 °C.

9. Replace serum-free medium with regular serum-containing growth medium.

10. For cell imaging and immunofluorescent staining, trypsinize the treated cells and replate them on coverslips at the appropriate density.

11. Analyze the rest of the cells for silencing using the appropriate assay system (i.e. Western blot, Northern blot, RT-PCR, protein activity, etc.).

Target-RNA levels should be significantly decreased after incubation overnight. Protein levels should be decreased according to the respective turnover. In dividing cells, the RNA silencing effect lasts out about 5 days, whereas in slowly growing cells, it may last longer. In order to down-regulate protein levels with a longer half life-time, the procedure must be repeated at 5-day intervals.

We usually see an efficient silencing with pepsiRNA concentrations between 25 and 50 nM. However, each pepsiRNA or each pepsiRNA mixture must be evaluated in a titration assay for each cell type. It is highly recommended to perform a serial dilution of a 100 nM pepsiRNA solution.

4.4
Analysis of the siRNAs

4.4.1
PAGE of siRNAs

An analysis of the appearance of siRNAs in the cytosol is often required for the publication of information. There are many ways to separate siRNAs from the remaining mRNA or other RNA species. For example, one can separate the total RNA content of the cytosol in a formaldehyde-containing agarose gel, which is sufficient to separate mRNA from smaller RNA species, such as the siRNAs, but does not allow for discrimination between different siRNA sizes. The separated RNA can be eventually transferred to a nylon membrane to probe for the loss of mRNA and the appearance siRNA species at the same time. The other method that can be applied to analyze the size of the resulting siRNAs in the cell is based on the separation of RNA in an 8 M urea:12–15% polyacrylamide gel electrophoresis (urea-PAGE). The gel can be either denaturing or non-denaturing, depending on the type of analysis. Although this method is straightforward, it requires some special laboratory equipment. For a good separation of siRNAs, the polyacrylamide gel must be longer (at least 50–55 cm) than a regular gel used to separate proteins. If available, a gel apparatus can be used that is usually dedicated to DNA sequencing. The sequencing gels are perfect for this application, both in size and thickness. If such an apparatus is not available, then large vertical protein gel separation units of at least 30 cm height can be used. These can be rebuilt from the scheme presented in Figure 4.24A. Although the gel is quite large, there is no need to use a special gel-casting stand. Taping the sides (as is required for sequencing gels) to avoid leakage of the gel solution is also unnecessary when casting the gel in a horizontal position (Figure 4.24B). When using this method, two different glass plates are required. The bottom plate should be longer than the top plate, which should have two appendages at each side called "ears". Both plates should be pretreated as described in the following protocol, and assembled as shown in Figure 4.24B.

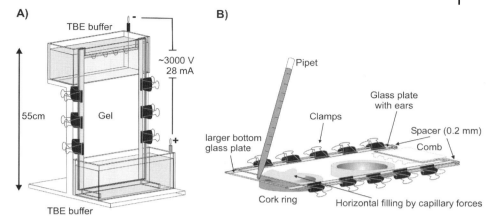

Fig. 4.24 (A) Schematic view of the minimum electrophoresis equipment necessary to run the urea PAGE for siRNA separation. The gel length should be up to 55 cm for good separation. The reservoirs are filled with 1 × TBE buffer. Better results are obtained if the gel temperature is controlled using a thermoplate. (B) For casting the gel, place the prepared glass plates on the bench surface, using two cork rings. One of the glass plates should be shorter than the other, and should contain 'ears' to facilitate the loading. Pipet the gel solution onto bottom glass plate and allow it to be drawn between the two plates by capillary forces.

PROTOCOL 24

1. Thoroughly wash the glass plates with tap water and soap to remove dust, fat, and residual gel debris.

▶ **Note:** Wear gloves to avoid contaminations!

2. Rinse the plates with tap water and dry with paper towels.

3. Evenly place the glass plates on cork rings with the gel side facing upwards.

4. Wipe both plates with ethanol.

5. Spread 200 µl of bind-silane (Amersham Biosciences) on the top glass plate with appendices (ears), and evenly distribute silane with a paper towel.

▶ **Note:** Don't inhale the vapor!

▶ **Note:** If the gel is to be subjected to Northern blot analysis or the siRNAs have to be transferred to nylon membranes, don't use bind-silane for the top plate. This would prevent the gel from coming off the glass. This method is only suitable for radioactively labeled siRNAs. Instead, one can use the Re-pel-silane (Amersham Biosciences) for both plates, as this facilitates removal of the gel from the glass. This method is used prior to Northern blot detection.

6. Spread 1 ml of Repel-silane (Amersham Biosciences) onto the bottom glass plate and distribute it evenly with a paper towel.

▶ **Note:** Don't inhale the vapor, and change gloves to avoid cross-contamination of bind- and Repel-silane!

7. Wait about 1 min until the silane has dried; then polish each glass plate with a paper towel soaked in ethanol.

8. Cover the glass plate with paper towel until use.

9. Place spacers (0.2–0.4 mm) at the rim of the bottom glass plate.

10. Place the top glass plate on the bottom glass plate, with the coated side facing downwards, and line up the ears with the upper end of the bottom plate.

▶ **Note:** This will leave a free space at the opposite end of the bottom plate to facilitate the gel casting.

11. Fix the sandwich with large paper clamps (Staples) (Figure 4.24B).

▶ **Note:** Make sure to avoid bulges in the spacers.

12. Prepare the gel solution using the recipe in Table 4.20 by dissolving urea in water and 1x TBE at 55 °C. Add the 30% acrylamide/bisacrylamide (19:1) mixture to the cooled solution and filter it through a 0.45-mμ filter to remove any particles.

Tab. 4.20

12% Gel solution (100 ml)	
Urea	42 g
Acrylamide/bisacrylamide (19:1) (40%)	30 ml
TBE (10x)	10 ml
Water	25 ml
TEMED*	40 µl
Ammonium persulfate (APS) (10 mg/ml)[a]	800 µl

[a] Add immediately before casting

13. Add 40 µl TEMED and 0.8 ml APS immediately before use.

14. Pipet the gel solution onto the bottom glass plate as shown in Figure 4.24B, and allow the solution to enter the space between the two glass plates.

▶ **Note:** The space will be filled without any bubbles and leaks by capillary forces until the solution reaches the upper end of the glass plates.

15. Insert the comb (0.2–0.4 mm) between the ears of the top plate.

16. Put two heavy flasks on top to exert pressure on the gel during polymerization.

17. Clean the glass plates free of any extending pieces of gel and residues of urea buffer, and dry the surface of the glass plates with a paper towel.

18. Place the gel into the gel apparatus according to the manufacturer's instruction, or as shown in Figure 4.24A.

▶ **Note:** Apply lubricant/silicon oil to seal the chamber.

19. Fill the cathode and anode reservoirs with 1x TBE buffer (Table 4.21).

Tab. 4.21

10 x TBE buffer (1 l)	
Tris	108 g
EDTA pH 8.0 (0.5 M)	40 ml
Boric acid	55 g

20. Pre-run the gel for 30 min at ~3000 V and 20–27 mA.

21. Mix equal amounts of siRNAs or the RNA samples and 2x RNA loading buffer (Table 4.22).

Tab. 4.22

2 x RNA loading buffer	
Formamide	95 %
EDTA pH 8.0	20 mM
Bromophenol blue	0.05 %

22. Heat samples to 99 °C for 5 min, and then cool down on ice immediately.

23. Briefly spin down condensed liquid.

24. Apply ~5 µl of sample to the gel.

25. Use a 10-bp DNA marker to estimate the migration of RNA duplexes.

▶ **Note:** Use a very fine Hamilton syringe. Carefully guide the syringe with the fingers to avoid bending the needle. Move the piston slowly. Carefully wash the syringe with water after each sample. Wash the syringe thoroughly after use.

26. Run the gel at ~3000 V and 20–27 mA for at least 1–2 h.

27. Use Table 4.23 to achieve correct separation of the siRNAs.

Tab. 4.23

Polyacrylamide	Range of separation	Bromophenol blue	Xylene cyanol
3.5%	1000–2000 bp	100 bp	460 bp
5.0%	80–500 bp	65 bp	260 bp
8.0%	60–400 bp	45 bp	160 bp
12.0%	40–200 bp	20 bp	70 bp
15.0%	25–150 bp	15 bp	60 bp
20.0%	6–100 bp	12 bp	45 bp

28. Open the clamps and remove the gel plate from the electrophoresis chamber.

29. Place the gel sandwich on the cork rings and remove the clamps.

30. Separate the two glass plates using a spatula.

▶ **Note:** The gel should stick to the top plate in the case of pretreatment with bind-silane. If no bind-silane is used, carefully remove the top glass plate and avoid detaching the gel from the bottom plate. In this case, the gel can be transferred onto a 3 MM Whatman paper for further processing (see sequencing gel protocols). Be careful, because the gel is very thin and delicate!

31. Wash the gel bound to the top plate in 10% acetic acid for 30 min to allow fixation.

32. Wash the gel plate in distilled water for 5–10 min to remove the urea.

33. Dry the gel on the top plate for 1 h at 80 °C in an oven.

▶ **Note:** Binding of the gel to the top glass plate prevents deformation and destruction of the gel, as it is usually very delicate.

34. Excise the region containing the siRNAs and transfer them onto a positively charged nylon membrane using electrotransfer for polyacrylamide gels (Sambrook and Russell 2001).

35. For recovery of siRNAs: elute short RNAs of appropriate sizes from gel slices by soaking in 1 M ammonium acetate at 37 °C overnight.

36. Recover the siRNAs by ethanol precipitation.

4.4.2
Determination of dsRNA and siRNAs by Agarose Gel Electrophoresis

dsRNAs and siRNAs may be may be crudely separated and visualized by agarose gel electrophoresis. They can be quantified by comparison to known amounts of dsDNA. This technique allows an easy detection of specific siRNAs or a simultaneous detection of mRNA and siRNAs either by Southern or Northern blot. Due to the thickness of the gels, the procedure is less delicate than polyacrylamide gel electrophoresis. However, it is not sensitive enough to distinguish siRNAs of different length.

4.4.2.1 Detection of dsRNA and siRNA Duplexes

PROTOCOL 25

1. Prepare a 2–3% agarose gel in 1x TAE buffer, as described for DNA electrophoresis.

2. Dilute the siRNA- or dsRNA-containing sample in 1x gel loading buffer (Table 4.24).

   Tab. 4.24

Gel loading buffer (10x)	
Ficoll	25%
EDTA, pH 8.0	1 mM
Bromophenol blue	0.25%

3. As a marker, add DNA ladders from 10 to 100 bp (100 to 9000 bp for dsRNA).

4. Visualize the separated dsRNA or siRNA by staining the gel in 0.5 mg/ml ethidium bromide.

   ▶ **Note:** Avoid incorporation of the stain into the gel, as it can alter the migration rate of the dsRNA and make accurate size determination difficult.

5. To probe the siRNAs for their specificity, transfer them to a positively charged nylon membrane and perform a Southern blot with a gene-specific probe, as described for DNA (Sambrook and Russell 2001).

4.4.2.2 Simultaneous Detection of siRNA and mRNA

To assess single-stranded RNA integrity, either denaturing gel electrophoresis can be performed using formaldehyde agarose gels as described for RNA separation (Sambrook and Russell 2001), or non-denaturing gels can be loaded with denatured RNA. While denaturing gels provide the greatest resolution of the denatured RNA, those non-denaturing gels loaded with denatured RNA still provide acceptable results.

Denaturing Gels

RNA is usually separated by denaturing agarose gels containing 2.2 M formaldehyde, and by using RNA sample buffer that contains formamide and formaldehyde. This method is well established, and can be found in most laboratory manuals for molecular biology (Sambrook and Russell 2001). Formaldehyde forms Schiff bases with the imino group of guanines, and therefore prevents intermolecular base-pairing. After separation in a denaturing gel, the RNA can be eventually transferred to a nylon membrane and probed for specificity by Northern blotting. Since the protocol for denaturing RNA agarose gels and Northern blotting is well established, it will not be discussed in detail here, and the reader is referred to a typical molecular biology laboratory manual, for example that published by Sambrook and Russell (2001).

Non-denaturing Gels

PROTOCOL 26

1. Add 5–20 µg of RNA (2 µl) to 18 µl of RNA denaturing buffer (Table 4.25).

Tab. 4.25

RNA denaturation buffer (1x)		RNA gel loading buffer (10x)	
TAE (10x)	10%	Glycerol	50%
Formaldehyde	20%	EDTA, pH 8.0	10 mM
Formamide	50%	Bromophenol blue	0.25%
		DEPC water	

2. Add 2 µl of RNA gel loading buffer, and heat the sample for 5–10 min at 65–70 °C prior to loading onto the gel.

3. Prepare a 2–3% 1x TAE-agarose gel to perform electrophoresis under standard conditions used for the analysis of DNA samples (Sambrook and Russell 2001).

4. Include the appropriate RNA marker on the gel to determine the size and integrity of the RNA sample with denatured RNA.

5. The denatured dsRNA may be gel-quantified by comparison to ssRNA or denatured DNA

▶ **Note:** The linear control DNA will appear rather diffuse on a native gel.

6. Following electrophoresis, visualize the RNA with ethidium bromide stain (0.5 mg/ml).

▶ **Note:** Avoid incorporation of the stain into the gel, as it can alter the migration rate of the dsRNA and make accurate size determination difficult.

7. Transfer the RNA to a positively charged nylon membrane by Northern blot and probe for specificity as described by common protocols (Sambrook and Russell 2001).

4.5
RNAi with Short Hairpin RNAs (shRNA)

The application of synthetic siRNAs is severely restricted by low to moderate transfection efficiency and short-term persistence of transient gene expression. To overcome this limitation, expression vectors are currently in use employing siRNAs or short hairpin RNA (shRNA) expression cassettes that resemble pre-miRNAs and undergo processing by Dicer (Brummelkamp et al. 2002; Lee et al. 2002; Miyagishi and Taira 2002b; Paddison et al. 2002a; Paul et al. 2002; Sui et al. 2002). Like synthetic siRNAs, they are designed to pair perfectly with the target mRNA to induce RNAi. To induce robust silencing, the shRNA transcript must be efficiently transported from the nucleus to the cytoplasm, where it has to be recognized as a Dicer and RISC substrate. These siRNA hairpin expressing plasmids are designed for either transient or persistent suppression of specific gene expression, allowing the analysis of loss-of-function phenotypes that develop over longer periods of time. Compared to the synthetic siRNA duplexes, they show similar potency to trigger RNAi in mammalian cells.

Promoters
Most RNAi expression vectors contain strong RNA polymerase III promoters that control expression of the shRNAs (Brummelkamp et al. 2002; Lee et al. 2002; Miyagishi and Taira 2002b; Paddison et al. 2002a; Paul et al. 2002; Sui et al. 2002). Those promoters include the human H1 and the mouse U6. RNA polymerase III usually transcribes a limited number of genes including 5S RNA, tRNA, 7SL RNA, U6 snRNA, and a number of other small stable RNAs that are involved in RNA processing (Paule and White 2000). As shown for the expression of antisense oligonucleotides, ribozymes, and RNA decoys, RNA polymerase III expression systems reveal a great potency in stable expression of short inhibitory RNAs in vivo and in vitro (Ilves et al. 1996; Jennings and Molloy 1987; Ojwang et al. 1992; Sullenger et al. 1990). Although the U6 snRNA and H1 RNA promoters appear functionally similar, the efficiency of U6 transcription is more sensitive to the initiating (+1) nucleotide, which for optimal expression should be guanosine. Usually, U6 cassettes express shRNAs as chimeras with the first 27 nucleotides of endogenous U6 snRNA (U6 + 27). Notably, all products of RNA Pol III-driven gene expression are initiated at a specific nucleotide lacking the 5'-cap and 3'-poly(A) tail that distinguish them from RNA Pol II transcripts.

Instead, RNA polymerase III (Pol III) transcription generally terminates at a run of four to six thymidine (T) residues, but some Pol III genes contain runs of T residues that are not recognized as termination signals (Gunnery et al. 1999). Remarkably, the human small nuclear RNA U6 gene (*U6*) even reveals higher expression levels as the human tRNA gene promoters (tRNA[Met], tRNA[Val]) (Ilves et al. 1996). However, modifications of common tRNA[Met]-derived (MTD) promoters can exhibit up to 60% more expression (Boden et al. 2003). When compared to other RNA Pol III cassettes, tRNA[Met] and tRNA[Val] are found to enhance the nuclear export ability of the shRNA transcript (Kawasaki and Taira 2003), while there is evidence for nuclear retention of shRNAs expressed from the U6 promoter (Paul et al. 2002) (Table 4.26).

Tab. 4.26 Steady-state expression levels of RNA Pol III expression cassettes

$tRNA^{Met}$ (modified MTD) > U6 (6 + 1; 6 + 27) > H1 > $tRNA^{Met}$ > $tRNA^{Val}$

Recently, it has been shown that Pol III promoters can work in concert with Pol II promoters driving the expression of a reporter plasmid. Besides the established RNA Pol III expression systems, RNA Pol II-based siRNA cassettes have also been generated using promoters like CMV or EF1α (Diallo et al. 2003a; Xia et al. 2003) (see also Section 4.6). RNA Pol II-derived shRNAs are 3'-polyadenylated and possess a 5'-cap. The stem-loop structure is supposed to be released from the RNA Pol II transcript in nuclear processing events. Instead of driving ubiquitous shRNA expression, they can be used for tissue-specific expression in whole organisms. Moreover, these promoters can potentially be used to drive expression of a reporter gene and the shRNA if the shRNA sequence is inserted into the 3'-untranslated region (UTR) of the reporter gene transcript (Zeng and Cullen 2003) (Figure 4.25).

RNA pol III expression

RNA pol II expression

Fig. 4.25 Comparison of RNA Pol III and Pol II transcription products. Pol III promoters reveal shRNA structures with 5'-G and 3'-uridine overhangs, whereas Pol II promoters express stem loop structures that are 5' methylguanosine-capped and extended by a 3'-poly A tail.

4.5.1.
RNAi with shRNA-Expressing Vectors

Currently, many shRNA expression vectors are commercially available from different companies (Ambion, OligoEngine) designed for many purposes (Figure 4.26). However, the most favored vector system is still pSUPER (OligoEngine), the first generation vector designed by Agami and coworkers (Brummelkamp et al. 2002). This is based on a pBluescript backbone (Stratagene) modified with a human H1 RNA Pol III cassette. The vector can be readily constructed by inserting two complementary 64-mer DNA oligonucleotides, which encode a 19-nt sense and antisense strand of the desired shRNA separated by a short spacer region of 9 nt, which has no homology to either one of the 19-nt sequences (Figures 4.26 and 4.27).

When expressed under the control of a human H1 RNA Pol III promoter, the RNA transcript is supposed to fold back into a short hairpin RNA structure with a 19-bp dou-

A)

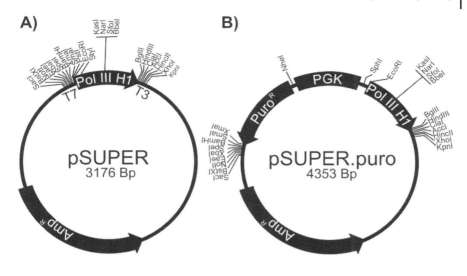

B)

C)

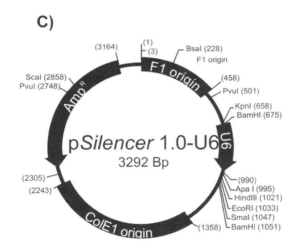

Fig. 4.26 DNA map of selected shRNA expressing vectors. The pSUPER vector (OligoEngine) is based on the pBlueScript KS-phagemid vector, and features the H1-RNA Pol III promoter. The pSUPER sequence file and its variant pSUPER-puro containing a puromycin resistance cassette for selection are available by downloading from the pSUPER RNAi System section of the OligoEngine Web site (http://www.oligoengine.com). Another example is the p*Silencer* vector from Ambion (http://www.ambion.com), which features U6 Pol III controlled shRNA expression (www.ambion.com).

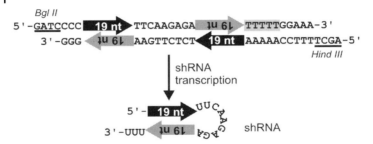

Fig. 4.27 Design of DNA oligonucleotides for cloning of shRNAs in pSUPER. The black arrows correspond to the coding sequence of the siRNA; the gray arrows correspond to the complementary sequence. The direction of the arrows show the 5′ → 3′ orientation

ble-stranded region. The insert is designed such that the first two 5′-bases of the loop region are expressed as a double uridine mimicking the two bases overhang generated by RNaseIII cleavage of dsRNA. It also contains a T5-terminator sequence that additionally incorporates two uridines at the 3′-end of the hairpin transcript (Figure 4.27). The shRNA has been designed to be an optimal substrate for Dicer, which recognizes the hairpin structure and subsequently processes the 3′-ends with 3′-uridine overhangs.

Suitable 19-nt sequences for the shRNA design can be found using siRNA target finder programs and incorporated into the cassette sequence (for a detailed review on the design, see Section 4.2.3). Both, complementary 64-mer DNA oligonucleotides can be ordered either unmodified or 5′-phosphorylated, which enhances ligation efficiency of the double strand insert. As shown in Figure 4.27, the shRNA expression cassette comprises sticky ends encoding core sequences of endonuclease restriction sites (here *BglII* and *HindIII*) hybridizing to pre-digested pSUPER with the corresponding enzymes. Prior to the ligation of the DNA-insert, both oligonucleotides will be annealed to result in a double-stranded DNA.

The cloning of the insert has some drawbacks. Because of the complementary sequences, recombination events within the shRNA expression cassette are very likely. Bacterial strains such as *E. coli* DH5α containing recombinase activity can process the insert by either excision or recombination of insert sequences. Therefore, cloning efficiency can be slightly enhanced using recombinase-deficient (recA-) *E. coli* strains like SURE II (Stratagene). A more detailed discussion on the bacterial amplification can be found in Section 4.6. Nevertheless, it needs screening of several bacterial colonies to isolate intact shRNA expression cassettes containing vectors. Although the cloning efficacy for shRNAs is low and selection of the mammalian cell clones is very time-consuming (see Section 4.6.1), this technique is valuable for long-term studies, which demand down-regulation rates of gene expression to higher extents than are achievable by transient applications.

Other vectors are available that allow the tetracycline/doxycycline-inducible expression controlled by the Pol III promoter (Miyagishi and Taira 2002a; Vigna et al. 2002). Likewise, there are first reports on the combination of the Pol III-driven expression of shRNAs and the Cre/lox system.

Vector-mediated systems for specific siRNA expression in mammalian cells using Pol III promoters allowing high levels of transcription activity have been developed in the past few years, widening the usage of RNA interference. The "Cre-On" system developed by Taira and coworkers could switch on the expression of shRNAs in the presence of Cre recombinase, which can be either accomplished by co-transfection with a Cre recombinase expressing vector or by direct delivery of Cre recombinase protein from the medium into the cells (Kasim et al. 2003). As shown by Edenhofer and colleagues (Peitz et al. 2002), Cre recombinase can be efficiently delivered from the medium to the nucleus of the cell when expressed as a TAT-NLS fusion protein. This fusion tag comprises the TAT peptide (an arginine-rich peptide derived from HIV-1) and nuclear localizing signal (NLS) (Figure 4.28). Upon addition of TAT-NLS-Cre, complete and functional siRNAs are generated, and reporter activity is suppressed.

Fig. 4.28 Design of recombinant Cre recombinase (from Peitz et al. 2002). Fusion of histidine-tagged Cre recombinase with the poly cationic HIV1-TAT peptide and a nuclear localization sequence (NLS) significantly improves its cellular uptake and nuclear localization.

4.5.1.1 Loop Structures

Several sizes of loops have been tested for silencing activity of the shRNA transcript. It has been shown for pSUPER that the length of the loop has a dramatic impact of the shRNA silencing rates. Testing loops of 5, 7 and 9 nt length revealed that a loop length of 9 nt produces the greatest silencing activity (Brummelkamp et al. 2002). Other loops from 3 nt (Jacque et al. 2002; Yu et al. 2002) to 23 nt (Paddison et al. 2002 a; Paddison et al. 2002 b) have been employed. However, it should be noted that many loops that have been described in the literature contain at least some bases capable of pairing (Arendt et al. 2003). In terms of the nucleotide sequences of the loop, they are as different as the length encompassing restriction sites, palindromic sequences, and even loops from a variety of miRNAs like the *mir-23* (Kawasaki and Taira 2003), which greatly enhanced the cytoplasmic localization of an shRNA controlled by the U6 promoter.

4.5.1.2 Interferon Response

Despite the observation of Tuschl and coworkers (Elbashir et al. 2001 a) that siRNAs shorter than 30 nt do not induce an interferon response, attention has been focused on the nonspecific effects induced by endogenously expressed shRNAs. Recently, it was found that either transfection of siRNAs or transcription of shRNAs result in interferon-mediated activation of the Jak-Stat pathway and global up-regulation interferon-stimulated genes within the nucleus (Bridge et al. 2003; Sledz et al. 2003; Stark et al. 1998). This effect is mediated by the dsRNA-dependent protein kinase, PKR, which is activated by intracellular presence of 21-base-pair (bp) siRNAs and required for up-regulation of IFN-α and possibly other, cellular signaling molecules. These are still controversial results, as comparative studies have been conducted on interferon induction by siRNAs and their shRNA counterparts which showed a significantly stronger interferon response after application of shRNA (Bridge et al. 2003). Notably, almost all plasmid vectors can induce interferon upon transfection, independently of the type of insert they are bearing (Akusjarvi et al. 1987).

Therefore, caution must be exerted in the interpretation of data from experiments using RNAi technology for suppression of specific gene expression (Sledz et al. 2003). Induction of interferon response by siRNAs is not only of concern for the use of RNAi technology in basic research, but also for possible therapeutic applications (Gitlin et al. 2002; Jacque et al. 2002). One should test for interferon response before attributing any effect to siRNA or shRNA application (see Section 4.6). A simple precaution to limit the risk of interferon induction is to use the lowest effective dose of siRNAs or shRNA vectors (Bridge et al. 2003).

4.5.1.3 Advantages

Since the knock-down effect of siRNAs is typically limited to seven to ten days post transfection, it makes them unsuitable for use in long-term knock-down studies. The main advantage of shRNA expression vectors is therefore the ability to overcome this limitation. Vectors with different antibiotic resistance cassettes have been shown to mediate permanent suppression of specific gene expression for several weeks and longer, allowing the analysis of loss-of-function phenotypes that develop over longer periods of time. Selection of cells by FACS sorting or by transient application of selection markers can permit the enrichment of cells positive for the vector, which can compensate for low transfection efficiencies. Furthermore, new-generation vectors even allow the integration of shRNA expression cassettes into the genome of the mammalian host cell (Rubinson et al. 2003).

4.5.1.4 Limitations

Because cloning is involved, the procedure is not suitable for a quick validation of siRNAs. It is time-consuming, with no guarantee that the chosen shRNA will sufficiently reduce gene expression. However, this limitation will be balanced by the ability to produce large quantities of a silencing construct. To avoid getting into cloning of inefficient siRNAs, it is more feasible to begin the search for highly effective siRNAs in transient RNAi experiments with synthetic duplex siRNAs.

Similar to the transient RNAi approach, the correct design of the shRNA sequence seems to be crucial for highly efficient gene silencing. Although various new algorithms based on thermodynamic stability and the asymmetric nature of siRNAs recently revealed better design rules (Khvorova et al. 2003; Reynolds et al. 2004; Schwarz et al. 2003), the selection process remains empirical (for more details, see Section 4.2.3). Since the knock-down efficacy depends on good target recognition, validation of the siRNA or shRNA sequence is still very important before embarking on these cloning approaches. However, even after the enrichment of shRNA-expressing cell clones, none of these attempts usually revealed a complete silencing of gene expression. Likewise, Tuschl and coworkers only realized a knock-down of 80–90 % by using their synthetic siRNAs (Elbashir et al. 2001a). As a single siRNA inhibits expression of the target gene rather incompletely, it is still important to find ways to increase the inhibition. Experiments with more than one siRNA, or cotransfection of more than one shRNA expressing vector, are showing cooperative silencing effects (Ji et al. 2003; Yang et al. 2002a).

4.5.1.5 Protocols

Cloning of pSUPER-shRNA Expression Vectors

The most efficient siRNA sequence can be obtained by synthetic siRNA transfection and validation or by using the novel siRNA design programs that are freely available on the internet (see Section 4.2.3). Particular criteria must be considered in order to optimize the silencing effects. A key criterion is to avoid regions within the first 50–100 nt of the target mRNA as the usually high GC content decreases cloning efficiency and mRNA targeting. Sequences which are highly enriched in GC often display internal loop structures and highly complex secondary structures. Those structures are competing with the stabilized shRNA stem and significantly reduce its efficiency. The uniqueness of the sequence is essential for the specificity. To avoid unspecific knock-down of related genes showing a certain homology to the shRNA of interest (off target effect), a systematic homology search of the sequence should be performed (http://www.paralign.org). Insert the chosen 19-nt sequence into the insert masks of the respective programs as described below (Figure 4.29), and order the DNA oligonucleotides. DNA oligonucleotides can be ordered as depicted in Figure 4.29 or 5'-phosphorylated, which enhances ligation efficiency.

```
            sense strand                complementary strand
5'-GATCCCCAACTTGCTACAAGTATCTCTTCAAGAGAGAGATACTTGTAGCAAGTTTTTTTGGAAA-3'
    3'-GGGTTGAACGATGTTCATAGAGAAGTTCTCTCTCTATGAACATCGTTCAAAAAAACCTTTTCGA-5'
          complementary strand              sense strand
```

Fig. 4.29 Design of shRNA inserts for pSUPER. Shown is a 19-nt sequence chosen from squalene synthase mRNA

PROTOCOL 27

1. Resuspend the lyophilized custom-made DNA oligonucleotides in twice-distilled water or annealing buffer at a concentration of 50 μM.

▶ **Note:** To enhance the annealing efficiency of DNA oligonucleotides, use the ligation buffer or annealing buffer to dilute the oligos.

Tab. 4.27

Annealing buffer 1 (1x)		Annealing buffer 2 (10x)	
Sodium acetate	100 mM	Tris-HCl, pH 8.0	100 mM
HEPES-KOH pH 7.4	30 mM	EDTA, pH 8.0	10 mM
Magnesium acetate	2 mM	NaCl	1 mM

2. Pipette a 1:1 solution of both DNA oligonucleotides (10 µl each) into a new microcentrifuge tube and mix the solution (the final concentration will be 25 µM).

3. Incubate the solution for 10 min at 95 °C in a heating block.

4. Briefly centrifuge to collect the evaporated sample.

▶ **Note:** Only brief centrifugation! Don't let the sample cool down to room temperature.

Switch off the heater and let the solution slowly cool down until the temperature of the heating block is below 30 °C.

5. Centrifuge the annealed insert and proceed with the method; alternatively, store it at 4 °C for several days or at –20 °C for several months.

6. Verify the integrity of your annealed dsOligo on a 2–4% TAE agarose gel, if desired. A suitable amount would be 5 µl of a 500 nM stock and comparing it to an aliquot of each starting single-stranded oligo 5 µl of a 500 nM solution. As a standard, use a 10-bp DNA Ladder (Invitrogen, Cat. #10821–015).

▶ **Note:** When analyzing an aliquot of the annealed dsOligo reaction by agarose gel electrophoresis, one will see a detectable higher molecular weight band representing annealed dsOligo, and a lower molecular weight band representing unannealed single-stranded oligos.

7. In the meantime: digest 1 µg of pSUPER with 5 u *BglII* and *HindIII* each according to the enzyme manufacturer's instructions.

8. Purify the digested vector either by phenol extraction (Sambrook and Russell 2001) or on a 1.0% agarose gel, followed by gel extraction.

9. In a new microcentrifuge tube, mix the insert and the purified digested vector at a molar ratio of 8:1. Do not exceed the final concentration of 50 ng.

▶ **Note:** If an equimolar ratio of the ligation components is used, one should not exceed 50–60 fmol of each component in a final concentration of 10 µl.

10. Add ligase buffer according to the manufacturer's instructions, but do not exceed a volume of 20 µl.

11. Add a high-concentration T4-ligase and incubate for 20 min at room temperature.

▶ **Note:** Room temperature ligations require specially concentrated T4 ligases and ligation buffers that contain polyethylene glycol to enhance the viscosity of the solution and to prevent Brownian motion of the DNA.

12. For the transformation, incubate 200 µl of competent *E. coli* SURE II cells (recA-) with at least 5 µl of the ligation reaction for 20 min on ice.

▶ **Note:** Using smaller volumes of competent cells will decrease transformation efficiency.

13. Briefly heat-shock the bacteria for 1 min at 42 °C and place them back on ice for 2 min.

▶ **Note:** For transformations, both temperatures and incubation times are crucial.

14. Resuspend the bacteria in 1 ml S.O.C medium and incubate them for 1 h at 37 °C, while constantly agitating.

15. Plate the bacteria on LB-agar plates and incubate at 37 °C for 20 h.

16. Pick the colonies and incubate them in 5 ml LB medium overnight at 37 °C.

17. Insertion of the stem-loop sequence can be verified by colony PCR or by restriction digest. Insertion causes a band shift of ~60 bp when compared to the parental vector. For PCR, use the following primers (Table 4.28).

Tab. 4.28

PCR primers for pSUPER

Name	Sequence	Position (bp)
T7 primer binding site	5′-AATACGACTCACTATAG-3′	627-643
T3 primer binding site	5′-CTTTAGTGAGGGTTAAT-3′	989-1005
M13(-20) primer binding site	5′-GTAAAACGACGGCCAGT-3′	600-616
M13 reverse primer binding site	5′-CATGGTCATAGCTGTT-3′	1023-1038

18. Analyze the restriction fragments in a 2% TAE-agarose gel electrophoresis.

19. Sequencing can be performed using the forward and reverse primers previously used for colony PCR.

4.5.2
Lentiviral RNAi Approach

4.5.2.1 Delivery Issues

Despite the effectiveness of the RNAi technology, there remain many drawbacks in its application within tissue culture and whole mammalian organisms. One way to induce RNAi in cells is to introduce synthetic siRNAs using physical techniques such as microinjection or electroporation of synthetic RNAs, or to use chemical delivery reagents such as liposomes or polycationic macromolecules, which are commercially available. Likewise, a variety of endogenous siRNA expression systems are employed. Some use an RNA polymerase III (Pol III) promoter to drive separate expression of the sense and antisense strands, which then hybridize in vivo to produce the siRNA. Other systems use Pol III promoters to drive the expression of short hairpin RNAs (shRNA), individual transcripts that result in RNA stem-loop structures, which are processed into siRNAs by Dicer. The introduction of those expression vectors into cells also requires electroporation, microinjection, or liposomal transfer of the DNA precursor vector into the cells or tissues. Most cell lines are easily transfectable, and recombinant cell clones can be selected that are permanently expressing RNAi phenotypes. Exceptions are primary cells, stem cells, and most nondividing cells. Without cell division, the shRNA (DNA) construct cannot be introduced into the nucleus where the DNA will be transcribed, and so it resides passively in the cytosol. There are also important cell types which have been resistant to the introduction of both siRNAs and shRNAs, mainly because of their resistance to chemical delivery systems.

Delivering siRNAs directly to whole vertebrate animals is more problematic than it is for invertebrates or cell lines, however. Whole-body applications of siRNAs using liposomal or chemical approaches have not been proven to be successful, whereas physical techniques such as the "hydrodynamic transfection method" to deliver naked siRNAs to mice via tail-vein injection has been shown to knock-down a reporter gene by 80–90% in the liver, kidney, spleen, lung, and pancreas (Lewis et al. 2002; McCaffrey et al. 2002). The effect is relatively short-lived, lasts only a few days, and not all organs and cell types can be reached.

Yet another method has been reported that can circumvent many of those difficulties. This method makes use of viral vectors to infect cells with the dsRNA-expression construct. The viral delivery systems comprise retroviral vectors (Hemann et al. 2003) such as adenoviral vectors, and the so far predominantly applied lentiviral vectors (Rubinson et al. 2003; Tiscornia et al. 2003). This retroviral approach is capable of delivering shRNAs into almost every cell or tissue, including stem cells and neurons.

The use of lentiviruses allows a systematic test of gene function in the context of the entire organism. It further permits a quick generation of animal models to determine which genes are important to the function of different tissues and organs, and which might be effective therapeutic targets in diseases.

Beside their capability of infecting noncycling and post-mitotic cells (Naldini 1998; Naldini et al. 1996), lentiviruses have additional advantages over other DNA delivery systems. They can be used to generate transgenic animals through infection of embryonic stem cells or embryos as the transgenes are not silenced during development of the organism (Lois et al. 2002; Pfeifer et al. 2002; Scherr et al. 2002; Scherr and Eder 2002). Major disadvantages of the lentiviral system are the biosafety issues that restrict the use of retroviruses in most laboratories.

The modified lentiviral systems currently in use include a significant number of safety features designed to minimize its relation to the wild-type, human HIV-1 virus. Since it has long been known that vectors derived from human immunodeficiency virus (HIV) are highly efficient vehicles for in-vivo gene delivery, the complexity of the HIV genome can be exploited to provide lentiviral vectors with novel features. Significant progress was achieved in the biosafety of HIV-derived vectors by eliminating all of the viral sequences that were nonessential for transduction. In addition to the structural genes, HIV contains two regulatory genes, *tat* and *rev*, that are essential for HIV replication, and four accessory genes that encode critical virulence factors.

This third-generation lentivirus vector uses only a fractional set of HIV genes: gag, pol, rev, and HIV-1 chimeric long terminal repeats (LTR). Moreover, genes encoding the structural and other components required for packaging the viral genome are separated onto four plasmids and are missing any regions of homology to prevent undesirable recombination events leading to the generation of a replication-competent virus. They are dependent on upstream elements and trans complementation for expression to produce viral progeny (e. g. gal, pol, rev, env) in the 293FT producer cell line. This split-genome, conditional packaging system is based on existing viral sequences and acts as a built-in device against the generation of productive recombinants (Dull et al. 1998; Naldini 1999; Zufferey et al. 1998).

Since none of the plasmids contains LTRs or the Ψ packaging sequence, the HIV-1 structural genes are excluded in the packaged viral genome, and their expression is prevented in infected cells. The G glycoprotein gene (VSV-G) gene from vesicular stomatitis virus (VSV) is used instead of the HIV-1 envelope (Griffiths et al. 1993; Yee et al. 1994 a). This overcomes the limitation of the tropism and low titers of most retroviral vectors, thus allowing the production of a high-titer lentivirus with a significantly broadened host cell range (Griffiths et al. 1993; Yee et al. 1994 a, b).

A) Envelope vector pMD2G

B) Packaging vectors

pMDL-g/p RRE

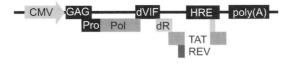

pRSV-Rev

Fig. 4.30 Maps of the four vector split lentiviral genome.
(A) The envelope protein encoding vector, pMD2G, encoding for VSV-G instead of its lentiviral counterpart. (B) The packaging vectors, pMDL-g/p RRE and pRSV.

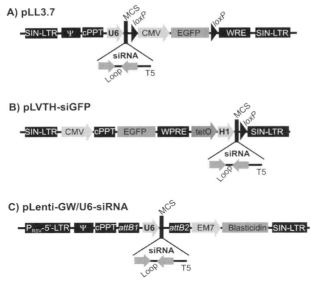

Fig. 4.31 Maps of different lentiviral vectors. (A) pLentilox3.7 or pLL3.7 designed by the van Parijs laboratory, (B) pLVTH-siGFP, which allows the doxycycline-regulated inducible expression of the shRNA. (C) The commercially available pLenti-GW/U6-siRNA vector from Invitrogen allows cloning based on the Gateway technology.

Van Parijs and colleagues created a system based on a disarmed lentivirus, and even induced RNAi in transgenic animals by infecting zygotes. In those mice, RNAi-directed gene down-regulation occurs throughout the animal (Rubinson et al. 2003). These authors designed the lentiviral vector pLentiLox 3.7 or pLL3.7 (Figure 4.31A), which allows the generation of efficient high-titer lentiviral stocks providing long-term stable RNAi effects. Besides, a variety of lentiviral vectors are available that even allow the inducible expression of the siRNA (Wiznerowicz and Trono) (Figure 4.31B).

4.5.2.2 Limitations

As the lentivirus is pseudotyped with VSV-G, it is capable of infecting human cells. Therefore, the shRNA sequence must be carefully designed to avoid any damage to humans. The use of this system requires certain biosafety working permissions, depending on the transgenes and promoters used in recombinant viruses, but the biosafety level must be at least BL2 (USA) or S2 (Germany). For more information, see the biosafety guidelines provided by the Centers for Disease Control (CDC, USA) (http://www.cdc.gov/od/ohs/biosfty/bmbl4/bmbl4toc.htm) or the Robert Koch Institut (http://www.rki.de/GENTEC/GENTEC.HTM). For other countries, look for the biosafety requirements at the respective governmental institutions. Furthermore, be aware of the risks when creating lentivirus that express shRNAs targeting human genes involved in controlling cell division (tumor suppressor genes), immune system, and apoptosis. A homology search of the shRNA sequence with the human genome database (BLAST, or Smith–Waterman; see Section 4.2.3) can verify whether there are potential human targets. If lentiviruses are used to infect human primary cells or stem cells, caution is recommended and a serious and careful analysis of the biohazardous potential of the experiment must be carried out (see Robert Koch Institute, Germany, http://www.rki.de/GENTEC/GENTEC.HTM or the Centers for Disease Control, CDC, USA, http://www.cdc.gov/od/ohs/biosfty/bmbl4/bmbl4toc.htm). The use of lentiviral expression systems of the new generation provides the advantage that they generate a replication-incompetent lentivirus. Those systems are based on studies of Naldini and coworkers (Dull et al. 1998).

In the recombinant virus genome the U3 region of the 3′-LTR is deleted, which does not prevent production in the infectious virus in the virus-producing cell line, but facilitates self-inactivation of the lentivirus after transduction of the target cells to enhance the biosafety of the vector (Dull et al. 1998; Zufferey et al. 1998). The complete lentivirus can only be produced when the lentivirus genome is co-transfected with other plasmids that contain the genes encoding structural proteins and other factors required for packaging of the virus (e. g. gal, pol, rev, env). The final virus is replication-incompetent, and no longer capable of producing fully packaged lysogenic viruses after transduction of the target cells.

Nevertheless, it is highly recommended that users should be well trained in working with viruses in general, and especially with retroviruses. For more information refer to the following references: (i) Retrovirus biology and the retroviral replication cycle (Buchschacher and Wong-Staal 2000); and (ii) Retroviral and lentiviral vectors (Naldini 1998, 1999; Naldini and Verma 2000).

4.5.2.3 Cloning of RNAi Cassette

Pol III H1-based Vectors (pLVTH-siGFP)
The lentiviral vector pLVTH-siGFP generated by the Trono laboratory allows the simultaneous expression of a reporter gene (EGFP) under the control of a Pol II promoter, and the shRNA under the control of the H1-Pol III promoter. The expression of shRNA can be controlled with tetracycline/doxycycline by a tetO module upstream of the Pol III promoter. A great advantage of this vector is the easy cloning procedure for shRNA cassettes. If the shRNA cassette is already available in pSUPER (see Figure 4.26), for constitutive expression it can be easily transferred into the lentiviral vector pLVTH (Figure 4.31B). Together with the H1 Pol III promoter, the shRNA cassette is excised from pSUPER using *EcoRI* and *ClaI*. The H1 promoter in pLVTH is eventually replaced with the H1-shRNA cassette from pSUPER. Further, a new version of pLVTH (pLVTHM) has been developed that permits the direct cloning of annealed shRNA into the lentiviral vector. In that case, you would need to design your shRNA as depicted in Figure 4.32.

Pol III U6-based Vectors (pLentilox3.7, pLenti GW/U6 siRNA)
Other systems are already commercially available. One of the easy-to-use systems is the Gateway technology-based lentiviral system BLOCK-iT™ and ViraPower from In-

Fig. 4.32 Design of shRNA inserts for pLVTH-siGFP. Shown is the 19-nt sequence, the loop, and the T5 termination signal.

vitrogen (Figure 4.31C). This provides a simple method to clone multiple shRNA target sequences for transient transfections. The shRNA cassette-containing vector can be used for lentiviral delivery. This RNAi cassette contains the sense and antisense DNA oligonucleotide separated by a 4-nt loop and flanked by a human U6 Pol III promoter and a Pol III termination signal. Once the cloning of the vectors pLL3.7, pLVTH, or the U6 RNAi Entry Vector is complete, all constructs are ready for use in initial transient screening experiments. While pLL3.7 and pLVTH already contains the long terminal repeats (LTR) that are required for the recombination with viral genome, the RNAi cassette of the BLOCK-iT™ U6 RNAi Entry Vector must be transferred into the final lentiviral (pLenti6/BLOCK-iT™-DEST) or other retroviral vectors. The transfer follows a recombination process using the Gateway technology (Invitrogen). The final destination vector has all the required components for efficient packaging of the U6 RNAi cassette into lentivirions. To generate those virions, the final lentiviral vector must be transfected into 293T cells together with the respective

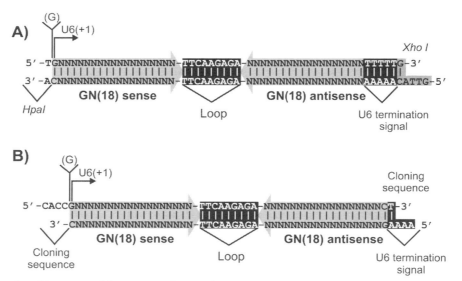

Fig. 4.33 Design of shRNA inserts for pLentilox3.7 (A) and pLenti GW/U6 siRNA (B). Shown are the 18-nt sequence plus 1 G at the 5′-end, the loop, the T5 termination signal, and the cloning sites as sticky ends.

viral packaging vectors (for example the ViraPower™ Lentiviral Support Kit; Invitrogen). The same procedure must be followed to generate virions from the original pLentiLox3.7 vector (pLL3.7) and the doxycycline-inducible vector pLVTH-siGFP.

In order to find appropriate targeting sequences for the shRNA sequence, one must use the currently available algorithms for the design of synthetic siRNAs (see Section 4.2.3). Likewise, for the cloning procedure of shRNA expression vectors controlled by a U6 promoter (Section 4.5) and the algorithms in Section 4.2.3, one must choose a G as a +1 site of transcribed sequence (Figure 4.33).

Choose sequences that start with a G in the coding oligo, and add a complementary C to the 3′-end of the top strand oligo, since G is the preferential starting base of the U6 promoter. If G is not the first base of the target sequence, Invitrogen recommends adding a G to the 5′-end of the top strand oligo following the CACC overhang but skipping the addition of the complementary C at the 3′-end coding strand, which has been shown to reduce the activity of the shRNA

pLentilox3.7 (pLL3.7) Vector

This vector is available from the van Parijs Laboratory at MIT, and information on the system can be obtained from the laboratory's website (http://web.mit.edu/ccrhq/vanparijs/). The vector contains a multiple cloning site (MCS) following the U6 promoter. Digestion of the MCS with *Hpa*I and *Xho*I allows insertion of the shRNA encoding sequence. *Hpa*I leaves a blunt end prior to the –1 position in the promoter, subsequently removing a T from the U6 promoter sequence. Therefore, the shRNA oligo must comprise an additional T at the 5′-end in order to reconstitute the –1 nucleotide of U6 promoter. Beside the core of the sense-loop-antisense sequence, one must add the U6 termination sequence of five T bases and the respective overhang of the *Xho*I site (Figure 4.33A). Since the loop sequence and length is essential for efficient expression of the shRNA, it corresponds to the established 9-nt TTCAAGAGA sequence of the shRNA expression vector pSUPER (Brummelkamp et al. 2002). Figure 4.34 shows the map of the pLentilox3.7 vector.

BLOCK-iT™ U6 RNAi Entry Vector

The design of the BLOCK-iT™ U6 RNAi Entry Vector RNAi cassette differs only at the end sequences from one described for pLentilox3.7 (Figure 4.33). The BLOCK-iT™ U6 RNAi Entry Vector can directly be applied in transient RNAi experiments, and is comparable to the pSUPER system (OligoEngine) (see Section 4.5). For lentivirus generation, it must be recombined with the pLenti6/BLOCK-iT™- DEST vector, which contains the 3′- and 5′-LTRs of the lentivirus. Recombination occurs via *att-L* and *att-R* sites, as described for the Gateway cloning system with LR-Clonase™ enzyme (Invitrogen) (Figure 4.36). The Invitrogen manual does not especially recommend a specific sequence requirement for the loop structure. Since the 9-nt loop described above (Brummelkamp et al. 2002) has been shown to be very effective for shRNA expression and function, it can be also recommended for this lentiviral cloning vector.

A)

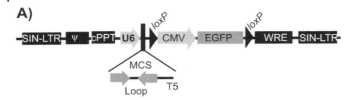

B)

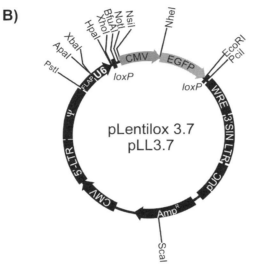

Fig. 4.34 Map of the pLentilox3.7 vector. (A) Linear map; (B) circular map. The lentilox vector allows the simultaneous Pol III expression of an siRNA and the Pol II controlled expression of a reporter gene, which can be excised in Cre mice since it is flanked by loxP sites.

4.5.2.4 Cloning Procedure

Both, the coding and the complementary DNA-oligonucleotide, should be ordered with a 5′-phosphate modification to enhance the ligation efficiency. Due to the fact that each oligonucleotide has a length of 60–70 nt and displays strong secondary structures, it must be extensively denatured before any annealing with the complementary strand can be performed. The annealing procedure depends on the ligation procedure performed afterwards. Therefore, several annealing buffers are suitable for the procedure. One can use the regular ligation buffer (without ligase enzyme mix), or another buffer that contains a high salt concentration, such as that recommended by the van Parijs laboratory (Table 4.29). Further cloning steps are similar to the procedure described in Section 4.5 for shRNA cloning.

Cloning of pLentilox 3.7

The entire procedure is modified after the procedure from the van Parijs laboratory (http://web.mit.edu/ccrhq/vanparijs/).

Tab. 4.29

Annealing buffer 1 (1x) (van Parijs lab)		Annealing buffer 2 (10x) (Invitrogen)	
Sodium acetate	100 mM	Tris-HCl, pH 8.0	100 mM
HEPES-KOH pH 7.4	30 mM	EDTA, pH 8.0	10 mM
Magnesium acetate	2 mM	NaCl	1 mM

PROTOCOL 28

1. Digest 1–2 µg pLL3.7 with 5 u of *XhoI* and *HpaI* each.

2. Dephosphorylate digested pLL3.7 with calf intestinal phosphatase (CIP) (New England Biolabs).

3. Resuspend the lyophilized custom-made DNA oligonucleotides in annealing buffer (Table 4.29) at a concentration of 50 µM.

4. Pipette a 1:1 solution of both DNA oligonucleotides (10 µl each) into a new microcentrifuge tube and mix the solution (the final concentration will be 25 µM).

5. Incubate the solution for 10 min at 95 °C in a heating block.

6. Briefly centrifuge to collect the evaporated sample.

▶ **Note:** Only brief centrifugation! Don't let the sample cool down to room temperature. Switch off the heater and let the solution slowly cool down until the temperature of the heating block is below 30 °C.

7. Centrifuge the annealed insert and proceed or store it at 4 °C for several days or at –20 °C for several months.

8. Ligate oligonucleotides into the *XhoI* and *HpaI* site of the pLL3.7 vector and transform into the recombinase deficient bacteria.

▶ **Note:** Use the endA-*E. coli* strain STBL2 or 3 (Invitrogen) to transform the ligation reaction. This strain is particularly well-suited for use in cloning unstable DNA such as lentiviral DNA containing direct repeats.

9. Analyze the insertion of insert on a 2% agarose gel. The insert causes a shift of ~60 bp in an *XbaI/NotI* fragment, when compared to parental vector.

10. Sequence the clone using the following primer corresponding to the FLAP cassette (Table 4.30):

Tab. 4.30

FLAP primer
5′-CAGTGCAGGGGAAAGAATAGTAGAC-3′

▶ **Note:** Using this sequencing primer allows the sequencing into the U6 promoter and stem loop.

11. Purify the recombinant vector prior to transfection of 293T cells by using the Qiagen Endo-Free prep Kits to avoid any contaminations of the DNA.

Cloning of the pLenti6/GW/U6 Vector

To generate the pENTR™/U6 precursor vector, anneal the oligonucleotides as depicted in Figure 4.33 as described in the previous protocol for pLentilox3.7.

▶ **Note:** pENTR™/U6 vector is not a circular plasmid. It is provided as a pre-digested linear vector containing ends that are complementary to the dsDNA oligonucleotide ends shown in Figure 4.35.

The lentiviral destination vector contains the following elements: RSV, Rous Sarcoma Virus enhancer/promoter for TAT-independent production of viral mRNA in the producer cell line (Dull et al. 1998); LTRs modified from HIV-1 5′- and 3′-LTRs for viral packaging and reverse transcription of the viral mRNA (Dull et al. 1998); Ψ, HIV-1 psi packaging sequence for viral packaging (Corbeau et al. 1996); RRE, HIV Rev response element for Rev-dependent nuclear export of unspliced viral mRNA (Kjems et al. 1991; Malim et al. 1989a–c), CmR, chloramphenicol resistance gene; *ccd*B, blasticidin resistance gene.

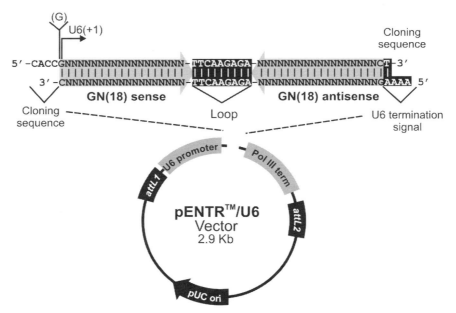

Fig. 4.35 Map of linearized pENTR™U6.

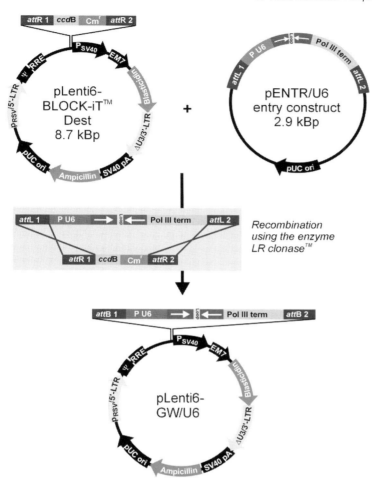

Fig. 4.36 The lentiviral destination vector contains the following elements: RSV, Rous Sarcoma Virus enhancer/promoter for TAT-independent production of viral mRNA in the producer cell line (Dull et al. 1998); LTR, Long Terminal Repeats modified from HIV-1 5′ and 3′ LTRs for viral packaging and reverse transcription of the viral mRNA (Dull et al. 1998); ø, HIV-1 psi packaging sequence for viral packaging (Corbeau et al. 1996); RRE, HIV Rev response element for Rev-dependent nuclear export of unspliced viral mRNA (Kjems et al. 1991; Malim et al. 1989a; Malim et al. 1989b; Malim et al. 1989c), CmR, chloramphenicol resistance gene; *ccd*B, blasticidin resistance gene.

PROTOCOL 29

1. Anneal the dsDNA oligonucleotides with the purified and linearized pENTR™/U6 using the following ligation protocol. Set up a 20 µl ligation reaction at room temperature using the following reagents.

Tab. 4.31

5X Ligation buffer	4 µl
pENTR™/U6 (0.5 ng/µl)	2 µl
dsoligo (5 nM; 1 : 10,000 dilution)	1 µl
DEPC water	12 µl
T4 DNA ligase (1 u/µl)	1 µl
Total volume	20 µl

2. Incubate for 5 min at room temperature.

▶ **Note:** The incubation can be extended to up to 2 h to yield higher numbers of colonies.

3. Transform the mixture into One Shot® TOP10 Competent *E. coli* according to Invitrogen's instructions.

▶ **Note:** The use of S.O.C. medium will enhance transformation efficiency.

4. Plate the bacteria on LB-agar plates supplemented with 50 µg/ml Kanamycin.

5. Pick colonies and prepare overnight cultures in 5 ml LB/50 µg/ml Kanamycin each.

6. Isolate the DNA using the Qiagen spin prep kit according to the manufacturer's instructions.

7. For the recombination of pENTR™/U6 with pLenti6/pBLOCK-iT™-Dest, follow the protocol of the manufacturer (Invitrogen).

▶ **Note:** The pENTR™/U6 contains two recombination sites, *att*L1 and *att*L2, flanking the U6 promoter, the insert, and the Pol III termination signal. It can recombine with the respective recombination sites, *att*R1 and *att*R2, in the destination vector pBLOCK-iT™-Dest (see: http://www.invitrogen.com).

8. Transformation requires endA-strain STBL-2 or 3 (Invitrogen) (Rubinson et al. 2003).

Lentivirus Production

Recombinant lentivirus can be produced by cotransfecting 293T cells with helper vectors that contain the packaging genes. The virus will be eventually obtained by collecting the supernatant. This supernatant can be used to infect cells or concentrated for use in embryo infections.

Beside the regular human 293T cell line, one can use the derivative 293FT to facilitate optimal lentivirus production (Naldini et al. 1996). This stably and constitutively expresses the SV40 large T antigen from pCMVSPORT6TAg.neo, and must be maintained in medium containing G418.

PROTOCOL 30

1. Culture 293FT cells in the appropriate medium (DMEM supplemented with 10% fetal calf serum (FCS), 2 mM L-glutamine, 0.1 mM MEM Non-essential amino acids, and 1%, penicillin/streptomycin).

2. The day before transfection, plate 12×10^6 293T cells or the modified 293FT cells in a tissue culture dish (10 cm diameter) and cover with 15–20 ml medium.

▶ **Note:** It is essential to use a low passage number of cells.

3. Mix 0.5 µg/µl of the following endotoxin-free DNAs diluted in TE buffer (Table 4.32).

Tab. 4.32

3-Plasmid-system		4-Plasmid-system[a]		ViraPower™-system	
Vector	10 µg	Vector	20 µg	Vector	10 µg
VSVG	20 µg	VSVG	10 µg	ViraPower™ [b]	30 µg
Δ 8.9	15 µg	RSV-REV	10 µg	Packaging mix	
		pMDL g/p RRE	10 µg	(1 µg/µl)	

[a] Recommended
[b] Contains pLP1, pLP2, and pLP/VSVG

▶ **Note:** These concentrations are also used for the Lipofectamine™2000 transfection method (Invitrogen).

4. Perform calcium phosphate or Lipofectamine™2000 transfection.

Calcium Phosphate Transfection

1. Prepare a 1.25 M $CaCl_2$ solution and filter-sterilize.

2. Prepare $2 \times$ HBS buffer (Table 4.33).

3. Add 400 µl of 1.25 M $CaCl_2$ and 1.5 ml H_2O and mix gently.

4. Add 2 ml of 2x HBS buffer dropwise to DNA mixture while bubbling with a 200-µl pipette.

▶ **Note:** Tap the tube with a finger after adding each drop of 2x HBS. People with long fingers are usually capable of permanently tapping the tube while adding the 2x HBS.

Tab. 4.33

2x HBS buffer			
Solution 1		Solution 2	
NaCl (280 mM)	1.64 g	Na$_2$HPO$_4$ (anhydrous)	0.213 g
HEPES (50 mM)	1.19 g	Water	10 ml
Total	80 ml		

Add 1 ml of solution 2 to solution 1
Adjust the pH exactly to 7.1!!
Fill up to 100 ml and filter sterilize

5. After addition of 2× HBS, bubble the solution for another 15–30 s by "injecting" air with a 200-µl pipettor.

6. Vortex for 5 s.

7. When finished, continue to bubble for 12–15 s.

8. Incubate for additional 30 min.

9. Take plate of 293T (293FT) cells out of the incubator prior to pipetting of the DNA solution.

▶ **Note:** Because of the pH change, leave the plate in incubator for as long as possible.

10. While gently rocking the plate back and forth, add the transfection mixture dropwise all over the plate and return it immediately to the incubator.

▶ **Note:** Thorough mixing of the medium with the transfection reagent will avoid high local concentrations and premature precipitation of the phosphate.

11. Remove media after 4 h.

▶ **Note:** The solution should be slightly cloudy or milky from the phosphate precipitate and one should see crystals of phosphate located at the boundaries of the cells.

12. Wash twice with 10 ml warm phosphate-buffered saline (PBS).

13. Add 20 ml warm 293FT growth media and incubate the cells for 48 h at 37 °C (Table 4.34).

Tab. 4.34

293FT Growth medium	
DMEM	
MEM non-essential amino acids	0.1 mM
Sodium pyruvate	1 mM
L-glutamine	2 mM
Penicillin/streptomycin	1%
FCS	10%

14. Harvest the supernatant, which contains recombinant lentivirus and remove the cells and debris by centrifugation for 10 min at 2000 r.p.m. and 4 °C in a 50-ml tube.

15. Filter the virus-containing media through a 0.45-μm filter to remove smaller particles.

16. Add the filtered virus to a sterile ultracentrifuge tube (use sterile Parafilm-sealed tubes for the Beckmann SW-28 rotor or polycarbonate tubes with caps for the Beckman Ti-45 rotor).

▶ **Note:** It is useful to leave some of the supernatant aside to determine the titer during concentration.

17. Pellet the virus by ultracentrifugation (SW-28 rotor, 25 000 r.p.m., 90 min, 4 °C or Ti-45, 40 000 r.p.m., 90 min, 4 °C).

▶ **Note:** A swinging-bucket rotor is preferred as the pellet is concentrated at the bottom of the tube. If a swinging-bucket rotor is not available, a fixed-angle rotor can be used, but this may require more elution buffer to recover the virus.

18. Decant and aspirate the supernatant, but avoid touching the delicate pellet.

19. Add 15–100 μl cold PBS and leave the tube at 4 °C for 12 h. Do not shake!

20. To resuspend the pellet, flush the PBS several times over the pellet, without touching it.

▶ **Note:** As the pellet does not only contain viral particles, there will be a residual pellet after the flushing procedure, but this can be discarded.

21. Collect the solution and either aliquot (for storage) or use the virus.

22. Keep some of the virus for titer determination.

23. Freeze the aliquots in liquid in liquid nitrogen and store at –80 °C.

▶ **Note:** Avoid multiple freeze–thaw cycles.

Lipofectamine™ Transfection

Although the calcium phosphate transfection method is the most inexpensive, it usually requires a high degree of skill to transfect a high number of cells. Critical factors are the pH of the buffers, the dropwise application and even distribution of the DNA solution that is essential for the formation of the fine precipitate. Therefore, many laboratories prefer to use transfection by cationic lipids or polyethylenimines. Many companies provide lipid-based transfection reagents such as the Lipofectamine™2000. Follow the procedure below to co-transfect 293FT cells. You will need 6×10^6 293FT cells for each sample. This protocol is adapted from the original BLOCK-iT™ Lentiviral RNAi expression system from Invitrogen (http://www.invitrogen.com).

PROTOCOL 31

1. In a sterile 5-ml tube, dilute the vector and the packaging vectors as depicted in Table 4.35 in 1.5 ml of serum-free Opti-MEM® I medium (Invitrogen). Mix gently.

Tab. 4.35

3-Plasmid-system		4-Plasmid-system[a]		ViraPower™-system	
Vector	3 µg	Vector	3 µg	Vector	3 µg
VSVG	6 µg	VSVG	3 µg	ViraPower™ [b]	9 µg
Δ 8.9	4.5 µg	RSV-REV	3 µg	Packaging mix	
		pMDL g/p RRE	3 µg	(1 µg/µl)	

[a] Recommended
[b] Contains pLP1, pLP2, and pLP/VSVG

2. In a separate sterile 5-ml tube, add 36 µl Lipofectamine™2000 to 1.5 ml of serum-free Opti-MEM® I medium. Mix gently immediately before use.

3. Incubate for 5 minutes at room temperature.

4. Combine the diluted DNA with the diluted Lipofectamine™ 2000 and mix gently.

5. Incubate for 20 min at room temperature.

▶ **Note:** The solution may become disperse and milky, but this does not affect the transfection efficiency.

6. In the meantime, trypsinize and count the 293FT cells.

7. Resuspend the cells at a density of 1.2×10^6 cells/ml in Opti-MEM® I medium supplemented with 10% serum.

8. Add the DNA-Lipofectamine™2000 solution to a 10-cm tissue culture plate containing 5 ml of Opti-MEM® I medium supplemented with 10% serum.

▶ **Note:** Do not include antibiotics in the medium.

9. Add 5 ml (6×10^6 total cells) of the 293FT cell suspension to the plate and mix gently by rocking the plate back and forth.

10. Incubate the cells overnight at 37 °C in a CO_2 incubator.

11. Replace the media with DMEM supplemented with 10% FBS, 2 mM L-glutamine, 0.1 mM MEM Non-essential amino acids, 1% penicillin/streptomycin, and 1 mM MEM sodium pyruvate.

▶ **Note:** After transfection, fusion of the cells is observed that is due to the expression of the VSV glycoprotein. It does not affect production of the lentivirus.

12. After 48 h, harvest the virus-containing supernatant.

13. Centrifuge at 3000 r.p.m. for 5 min at 4 °C to pellet the cell debris. Perform filtration and further purification steps as described in the calcium phosphate transfection method.

Transduction of Virus

From this stage on, the recommendation is to work with great caution and according to the respective national biosafety guidelines. The transduction of lentivirus into mammalian cells is usually straightforward, and almost all cells can be infected when using the appropriate titer and a cell culture is showing a viability of at least 90%. However, the transduction efficiency can be enhanced by addition of hexadimethrine bromide (Polybrene®). To obtain optimal expression and silencing efficiency of your shRNA of interest, transduction of the lentiviral construct requires a suitable MOI (multiplicity of infection). MOI is defined as the number of virus particles per cell, and generally correlates with the number of integration events. Typically, shRNA expression levels increase as the MOI increases. To acquire the MOI, the titer of the virus must be determined and correlated with the cell number. Before transduction, one should test the cells for sensitivity to Polybrene®; some cells, such as primary neurons, are sensitive to this material.

Preparing Polybrene®

Polybrene solution can be prepared by diluting a 6 mg/ml stock solution of Polybrene® (Sigma) in deionized, sterile water and subsequent filter-sterilization as described in the BLOCK-iT™ Lentiviral RNAi Expression manual (Invitrogen). Aliquots (1 ml) may be stored at –20 °C for up to one year. More than three freeze–thaw cycles may result in a loss of activity; therefore the stock solution may be kept at +4 °C for up to 2 weeks. Complete the culture medium with Polybrene® (Table 4.36).

Tab. 4.36

Transduction medium

DMEM[a)]
Sodium pyruvate	1 mM
L-glutamine	2 mM
Penicillin/streptomycin	1%
FCS	10%

Polybrene®
6–8 µg/ml

[a)] Or any other medium (type of medium
depends on the type of cells used)

Titration of the Virus

PROTOCOL 32

1. Plate 4×10^5 of any adherent cell line of choice including 293FT cells per well in a 6-well plate and incubate for 12–24 h.

▶ **Note:** It is recommended to titrate the lentiviral stock using the mammalian cell line or type that will later be used for the transduction experiments, though any other adherent mammalian cell line will do the same job. Some cells are not suitable, such as nondividing cells and primary cells.

2. Prepare serial dilutions (10^{-3}, 10^{-4}, 10^{-5}, 10^{-6}, 10^{-7}, 10^{-8}) of the virus in 1.5 ml per well of transduction medium (Table 4.36).

3. Replace the culture medium of the cells with the respective virus-containing medium.

4. Incubate overnight at 37 °C/5% CO_2.

5. Replace the medium with regular growth medium (Table 4.37).

Tab. 4.37

Growth medium

DMEM[a)]
Sodium pyruvate	1 mM
L-Glutamine	2 mM
Penicillin/streptomycin	1%
FCS	10%

[a)] Or any other medium (type of medium
depends on the type of cells used)

▶ **Note:** The supernatant may still contain virus. Be cautious!

▶ **Note:** From here on, the protocol is divided into two parts, as titration of the recombinant virus derived from the pLentilox vector and the BLOCK-iT™ system is not based on the same assay!

PROTOCOL 33

A) Titration of the pLentilox-derived Virus

1. At 48 h after infection, trypsinize the cells and resuspend them in cold PBS for FACS analysis.

2. Analyze the infection rate by FACS and analyze for EGFP expression.

▶ **Note:** Beside the U6 expression cassette, the pLentilox vector contains an EGFP expression cassette under the control of a CMV promoter to allow monitoring of infection.

3. Take a dish from the serial dilution experiment, where you see an infection of 1–10 % of the cells.

4. Determine the titer by calculation using the equation below (Table 4.38).

Tab. 4.38

$$N \times 4 \times 10^5 = Y \longrightarrow Y \times A = viral\ particles/\mu l$$

N = percentage of cells that are EGFP positive
Y = positive cells/ml
A = dilution factor (µg virus/ml)

▶ **Note:** A sample calculation assuming 1% infection from the well with 0.1 µl of virus is shown in Table 4.39.

Tab. 4.39 For more information, go to the van Parijs laboratory web page http://www.web.mit.edu/ccrhq/vanParijs/

$$0.1 \times 4 \times 10^5 = 4 \times 10^3 \longrightarrow 4 \times 10^3 \times 10^3 = 4 \times 10^6\ viral\ particles/\mu l$$

▶ **Note:** In general, you should have at least 5×10^5 to 5×10^7 viral particles per µl (e.g. transducing units (TU)/ml) for cell and embryo infections. If the titer of your lentiviral stock is lower, it is recommended that you produce a new lentiviral stock.

B) Titration of the BLOCK-iT™ System-derived Virus

Since the recombinant virus does not lead to the expression of any fluorescent marker as described for the pLentilox-derived virus, transduction cannot be monitored by FACS analysis. However, this virus contains a Blasticidin resistance gene (Kimura and Ohyama 1994), which can help to select infected cells by application of the appropriate amount of Blasticidin in the medium (Takeuchi et al. 1958; Yamaguchi et al. 1965). All cells that are not infected will die during the relatively short selection period of 10–12 days. At 10–12 days after the first application of Blasticidin (at 14–16 days post infection), the culture is tested for the number of remaining cell colonies by staining with crystal violet (for a detailed protocol, see the user manual of the BLOCK-iT™ lentiviral expression system from Invitrogen). This assay can be performed with any other lentivirus that is expressing an antibiotic resistance gene. To determine the cell death rate per day, take uninfected cells and treat with different concentrations of antibiotic (e.g. 0, 2, 4, 6, 8, 10 µg/ml Blasticidin).

dsDNA oligonucleotides are annealed and incubated for 5 min at room temperature with the linearized pENTR™/U6 Vector. The mixture is transformed into One Shot® TOP10 Competent *E. coli*. The resulting plasmid DNA can be used immediately in a transient transfection. The shRNA expressed from the U6 promoter will form a hairpin that is processed into an siRNA molecule.

4.6
RNAi with Long Hairpin RNAs (lhRNAs)

Since the striking discovery that siRNAs mediate RNAi without induction of the interferon response (Caplen et al. 2001; Elbashir et al. 2001 a), RNAi has been extensively studied in mammalian organisms. The main aim has been to develop strategies for its application in medical therapy in order to combat all types of infectious diseases, and even cancer.

However, the first indications that mammals can also induce dsRNA-dependent RNA silencing came from the observation that after injection and transient application of long dsRNAs into early mouse embryos, mouse embryonic cells induce sequence-specific silencing (Billy et al. 2001; Paddison et al. 2002 b; Svoboda et al. 2000, 2001; Wianny and Zernicka-Goetz 2000). Long dsRNA were transfected or electroporated expressing the RNAi phenotype without any obvious nonspecific side effects. The genes that were silenced were either encoding reporter proteins or proteins with a rapid turnover in the cells, meaning that an interferon response might have been undetected. It is further known that in undifferentiated cells and during early development, the nonspecific responses to dsRNA are attenuated, whereas in somatic cells the interferon response machinery is fully active, thus preventing any exploitation of the RNAi pathway by exogenously applied long dsRNA. This observation and some preliminary experiments in somatic cells led to the conclusion that long dsRNA rather than siRNAs are not suitable for RNAi in mammals.

Nevertheless, the RNAi approaches based on either transient transfection of synthetic or in-vitro-transcribed siRNAs or the endogenous expression of shRNAs usually do not reveal a complete knockout of the respective gene due to the efficacy of mRNA targeting and binding (see Section 4.2.3). Likewise, siRNA and shRNA display potential of off-target effects due to their non-strand-specific incorporation into RISC.

This effect might be avoided, when long dsRNAs are cut by Dicer, which is assumed to discriminate between the sense and antisense strands. Despite their advantage of exhibiting persistent RNAi, shRNAs do not mediate complete silencing (Parrish et al. 2000; Tuschl et al. 1999; Yang et al. 2000), while regions of homology between the dsRNA and the target gene as short as 23 nt can mediate post-transcriptional gene silencing in tobacco, when the homology is contained within a longer dsRNA. The fact that purified siRNAs cleaved from long dsRNA (Nykänen et al. 2001) can efficiently mediate RNAi in vitro suggests that long dsRNAs are more effective because they are more efficiently processed into siRNAs. This can be due to a highly cooperative binding or cleavage by Dicer (Zamore 2001). An important aspect of this method is that the siRNAs processed from the dsRNAs by Dicer are available in their natural form, and therefore might exhibit stronger potency in RNAi than the synthetic ones.

Recent publications have now shown that despite some concerns, which are discussed within the field of RNAi, endogenously expressed long hairpin dsRNAs (lhRNAs) are capable of inducing an RNAi phenotype in mammalian somatic cells including human primary fibroblasts, melanocytes, HeLa cells (Diallo et al. 2003a, b), and even whole mice (Shinagawa and Ishii 2003). Several somatic cell lines were transfected with long inverted repeat DNA constructs as described for the application of RNAi in *C. elegans* and *Drosophila*. As for shRNA, expression of the inverted repeats results in RNA molecules that are supposed to fold back into hairpin-like structures by intra-molecular hybridization. Those long hairpin RNAs (lhRNAs) usually display about 500–800 nt in each sense and antisense direction, and the resulting RNA is supposed to be effectively double-stranded. It has been considered that these dsRNAs, will be processed by the mammalian Dicer into the 21- to 23-nt siRNAs, that will be able to induce the specific silencing of the corresponding gene in the same way as it occurs in *C. elegans* and *Drosophila*.

Although the cloning efficacy for long inverted repeats of a cDNA is very low, and the selection of cells is very time-consuming, this technique is valuable for long-term studies, which demand down-regulation rates of gene expression to higher extents than are achievable by expression of siRNAs or siRNA transgenes. Nevertheless, it should be noted that there might be a potential disadvantage of using lhRNAs as compared with shRNAs. If there are regions encompassing homology with other related or nonrelated genes, then nonspecific silencing or off-target effects might occur. However, this must be determined for each experiment.

To assess the idea that endogenously expressed dsRNA molecules encompassing several hundred nucleotides can be used to overcome problems related to the dsRNA-dependent nonspecific interferon response in cultured mammalian cells, different groups have designed different approaches. To drive the long hairpin RNA (lhRNA) expression, all of these investigators have used strong Pol II instead of Pol III promoters usually described for shRNA expression.

The following methods are based on the expression of lhRNAs from:
- Inverted repeat DNA.
- Direct repeat DNA.
- Inverted repeat DNA missing 5′-cap and poly A tail.

4.6.1
lhRNAs from Inverted Repeat DNA

The first method is based on the simple expression of long DNA inverted repeats comprising 200–800 bp in a consecutive sense and antisense orientation separated by a 6-bp restriction site as a loop (Figure 4.37).

To combine the sense- and antisense-oriented strands, both fragments were inserted into the respective restriction site of any suitable eukaryotic expression vector (e. g. pcDNA3.1; Invitrogen) generating the plasmid pHairpin. The expression of the corresponding lhRNA in eukaryotic cells may be controlled by a strong CMV promoter or an EF1α promoter. Following transfection, cell clones of the different cell types were selected for 30–40 days against the respective antibiotic (e. g. G418 (300–1200 µg/ml) (Diallo et al. 2003 a, b).

A restriction site as an inversion point between the sense and antisense orientation is supporting directional cloning. In contrast to the techniques for the persistent expression of shRNAs described above, the inverted repeat does not harbor more spacer

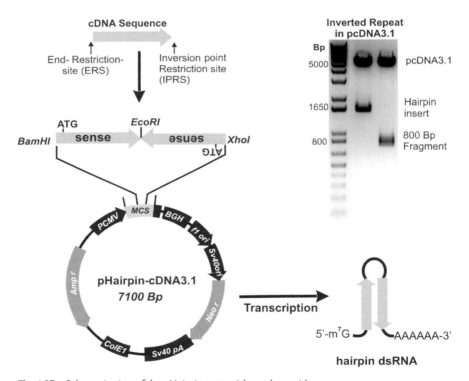

Fig. 4.37 Schematic view of the pHairpin vector (shown here with pCDNA3.1 as a backbone) containing the inverted repeat cassette flanked by different restriction sites to avoid the direct repeats that can facilitate cruciform structure formation. Right: control digestion of the recombinant vector.

than this restriction site separating the inverted fragments. Indeed, a long loop formed by a putative spacer is known to negatively influence the efficiency of lhRNAs to trigger RNAi (Smith 2000). Therefore, the fragments are directly linked, ensuring the formation of a loop with the minimum of 6 nt, which also is palindromic.

After an extensive period of selection with antibiotic (e.g. G418; geneticin), cells with permanent null phenotypes for various targeted genes are obtained. The reason why those cells do not undergo programmed cell death is unknown. Despite some very recent reports (Bridge et al. 2003; Sledz et al 2003), which showed that short hairpin RNAs (shRNAs) could indeed induce the interferon response, no induction of the interferon response has been detected in the lhRNA-expressing cells.

Possible explanations for this observation could be that the endogenous dsRNA is rapidly processed to smaller fragments, which are eventually processed into active siRNAs, avoiding an accumulation of longer dsRNA in the cytosol. It is also suggested that only those cells can escape selection, which produce long lhRNA in levels to low for interferon response induction, but sufficient for interference with endogenous mRNAs.

4.6.1.1 Inverted Repeat DNA

As described in Chapters 2 and 3 for *C. elegans* and *Drosophila*, the cloning of such inverted repeat constructs is very laborious and time-consuming due to the recombination events that take place during replication of the constructs. In vivo, large DNA palindromes are intrinsically unstable sequences (Leach 1994). Inverted repeats may initiate genetic rearrangements by the formation of hairpin secondary structures that block DNA polymerases or are processed by structure-specific endonucleases. The inverted repeat base-pairing results in cruciform structures, which have proved difficult to detect in bacteria, suggesting that they are destroyed. Besides, shorter fragments of the inverted repeats and the plasmid vector can often be isolated, thereby undermining this hypothesis. It has been shown that sbcCD, an exonuclease of *E. coli*, is responsible for the processing and cleavage of large palindromic DNA sequences in *E. coli* (Davison and Leach 1994; Leach 1994), thus preventing the replication of long palindromes. SbcCD is cleaving cruciform structures in duplex DNA followed by RecA-independent single-strand annealing at the flanking direct repeats, generating a deletion.

There are two possibilities to prevent the degradation. One is the insertion of a spacer between the inverted repeat, since it has been shown that inverted repeats with an interruption of the pairing at the center are less likely to form cruciform structures than perfect pairing inverted repeats (Bzymek and Lovett 2001; Sinden et al. 1991; Zheng et al. 1991). The other possibility is to use of nuclease- (sbcCD) or recombinase- (recA, B, J) deficient *E. coli* strains such as SURE II (Stratagene), JM105, JM103, or CES200 (ATCC, at http://www.atcc.org) (Table 4.40).

The SURE II competent cells were designed to facilitate cloning of inverted repeat DNA by removing genes involved in the rearrangement and deletion of these DNAs such as the UV repair system (*uvrC*) and the SOS repair pathway (*umuC*) genes. This results in a 10- to 20-fold increase in the stability of DNA containing long inverted repeats. Furthermore, mutations in the sbcC and RecA,B,J genes involved in

Tab. 4.40 *E. coli* strains that facilitate the inverted repeat cloning (Hanahan 1983; Wyman et al. 1985; Yanisch-Perron et al. 1985).

Strains	Genotype	References	Source
SURE II	e14– (McrA–) ? (mcrCB-hsdSMR-mrr)171 endA1 Sup E44 thi-1 gyrA96 relA1 lac recB recJ sbcC umuC::Tn5 (Kan^r) uvrC [F′ proAB lacI^{q}Z? M15 Tn10 (Tet^r) Amy Cam^r	http://www.stratagene. com/manuals/200238.pdf	Stratagene
JM103	F′ traD36 proA+ proB+ laclq delta(lacZ)M15 delta(pro-lac) supE hsdR endA1 sbcB15 sbcC thi-1 rpsL lambda-	Hanahan 1983	ATCC
JM105	F′ traD36 proA+ proB+ laclq delta(lacZ)M15 delta(pro-lac) hsdR4 sbcB15 sbC? rpsL thi endA1 lambda-	Yanisch-Perron et al. 1985	ATCC
CES200	F- delta(gpt-proA)62 thr-34::Tn10 lacY1 ara-14 galK2 xyl-5 mtl-1 leuB6 hisG4 argE3 hsdR mcrB rac- sbcB15 recB21 recC22 rpsL31 rfbD1 kdgK51 thi-1 tsx-33 lambda-	Wyman et al. 1985	ATCC

recombination events greatly increase the stability of inverted repeats. The combination of *recB* and *recJ* mutations confers a recombination deficient phenotype to the SURE cells that greatly reduces homologous recombination, similar to a mutation in the *recA* gene.

Even though those strains increase the potential to recover a clone, the recombination and cruciform structure formation is not completely abolished. For more information, see Bzymek and Lovett (2001).

The cloning of the inverted repeat is based on PCR amplification of the respective cDNA sequence and ligation of the sense-restriction site-antisense sequence prior to its insertion into the expression vector. Since the protocol comprises several agarose gel purification steps, one should make sure of starting with a reasonable amount of PCR product. Regular PCRs yield 1–4 µg of PCR product, but this can be increased with a few improved Taq-polymerases (Long Expand Taq-polymerase, Roche; Takara). The PCR comprises the use of different restriction sites at both ends of the inverted repeat sequence to perform a directed cloning of the inverted repeat in the respective vector (Figure 4.37). The PCR product from this reaction can be processed as described in the following method.

PROTOCOL 34

1. Generate the sense and antisense DNA fragments for the inverted repeat by PCR. The fragment size should be between 200 and 1000 bp. Add two different suitable restriction sites at the 5′-end of the forward primers (Figure 4.38), depending on the MCS of the eukaryotic expression vector.

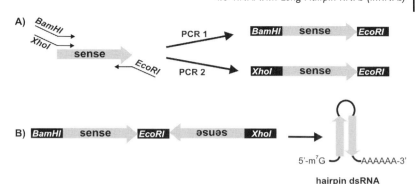

Fig. 4.38 (A) Schematic view of the primers used to generate the sense strand and the antisense strand of the inverted repeat. (B) The primers contain the respective restriction sites at their 5'-end that allow the directed cloning.

2. Digest the PCR fragments and the eukaryotic expression vector (i.e. pCDNA3.1) with the respective restriction enzymes to allow generation of the inverted repeat.

▶ **Note:** Most restriction enzymes are capable of directly digesting the DNA in the PCR mix (see Table 4.41). For an overview, check the booklet Lab FAQS on pp. 13–16 (chapter 1.1) from Roche, which can also be downloaded from their web site (http://www.roche-applied-science.com/LabFAQs/index.htm or http://www.roche-applied-science.com/frames/frame_support.htm). If the digest is not possible in the PCR buffer, one must purify the PCR product by agarose gel electrophoresis and gel extraction prior to digestion.

3. Purify the PCR fragments and the vector in a 1–1.5% agarose gel to remove the restriction enzymes. Extract the PCR products.

4. In a triple ligation, ligate your sense and antisense fragments into the respective restriction sites of the eukaryotic expression vector. The ligation protocol depends on the supplier of the T4-ligase.

▶ **Note:** The Rapid Ligation kit from Roche can be recommended, as it produces good results in the cloning of the inverted repeats.

▶ **Note:** If it is unavoidable to use only one restriction site to insert the inverted repeat into the vector, dephosphorylate the vector using calf intestinal phosphatase (1 µl/µg of DNA) in the digestion buffer for 15–60 min at 37 °C to prevent religation of the vector.

5. Incubate for 30 min at room temperature.

6. Transform 2–8 µl of ligated vector into *E. coli* SURE II cells (Stratagene) or others, as depicted in Table 4.40.

Tab. 4.41 Enzyme activity in the regular PCR buffer* (modified from LabFAQS, Roche)

Enzyme activity in PCR mix[a]			
Restriction enzyme	**Activity in %**	**Restriction enzyme**	**Activity in %**
Acc I	<5	Nco I	50
Alu I	100	Nde I	0
Apa I	100	Nhe I	100
Ava I	20	Not I	0
Ava II	<5	Nru I	75
Bam HI	100	Nsi I	100
Bcl I	0	Pst I	90
Bgl I	30	Pvu I	<5
Bgl II	0	Pvu II	100
Cla I	100	Rsa I	100
Dpn I	100	Sac I	100
Dra I	100	Sal I	0
Eco RI	50	Sau3A I	100
Eco RV	10	Sca I	<5
Hae III	100	Sfi I	10
Hind II	100	Sma I	100
Hind III	10	SnaB I	50
Hinf I	50	Spe I	0
Hpa I	100	Sph I	<5
Hpa II	40	Ssp I	0
Kpn I	50	Stu I	30
Ksp I	0	Xba I	60
Mlu I	<5	Xho I	<5
Nae I	0		

[a] 10 mM Tris HCl, pH 8.3, 50 mM KCl, 1.5 mM $MgCl_2$

7. Plate the bacteria on selection LB/agar plates supplemented with the appropriate antibiotic.

▶ **Note:** Due to the high rate of recombination events, the number of growing colonies is very low. Besides, the percentage of positive clones bearing the inverted repeat is often below 1%. To increase the amount of positive recombinant colonies, it is recommended to set up several ligation and transformation reactions for the same cloning procedure simultaneously. Likewise, the amount of DNA for ligation must be increased for the protocol to work well.

8. Pick the colonies and grow the bacteria in 1–3 ml LB medium + antibiotics.

▶ **Note:** For high-throughput DNA isolation, use either a pipetting robot such as the BioRobot system (Qiagen), or the 96-well turbo DNA isolation method as described in the following protocol.

4.6.1.2   High-throughput Inverted Repeat Isolation

This protocol is designed for small scale, high-throughput plasmid DNA isolation, and facilitates the analysis of more than a hundred colonies in a short period of time. Due to the high number of either recombination events or formation of cruciform DNA, the number of colonies that contain the full-length inverted repeat construct is usually low. The yield can range from 0.1–0.5% in regular *E. coli* strains to 1–50% in the strains listed in Table 4.40, depending on the cloning procedure. The analysis of up to 500 colonies is very laborious and time-consuming. PCR amplification of the inverted repeat would ease the analysis, but often this does not work well due to the highly stable secondary structure of the DNA. The procedure is facilitated by automated DNA isolation (BioRobot system; Qiagen), though for many laboratories such a robot is not affordable. An alternative for automation is a vacuum manifold (QIAvac 96; Qiagen), which supports DNA isolation and purification in a 96-well format. The DNA isolation system is accommodating filters (TurboFilter 96) and the well-known purification columns (QIAprep 96) in 96-well plates, and the DNA purification is performed by vacuum aspiration of the buffers (QIAvac 96) (Figure 4.39). The parallel preparations yield up to 20 µg of high-copy plasmid DNA from 1–5 ml of overnight cultures of *E. coli* grown in LB (Luria-Bertani) medium.

Fig. 4.39   QIAvac 96 vacuum device filled with QIAprep 96 purification columns.

E. coli cultures of 1.3–1.5 ml per colony are grown in a 96-well, flat-bottomed block culture dish, and all pipetting steps can be performed using multichannel pipettors. The following protocol is modified from the original protocol for the QIAprep 96 kit (Qiagen).

PROTOCOL 35

1. Pick the bacterial colonies and inoculate 1.3–1.5 ml LB medium per well of a 96-well, flat-bottomed block culture dish.

2. Grow the bacteria overnight or for 20–24 h in a tangential shaker, covering the culture block with either tape that is pierced with two to three

holes per well for aeration or with specialized AirPore™ microporous tape sheets.

▶ **Note:** The 96-well, flat-bottomed block culture dish and the corresponding porous sheets are available in the QIAprep 96 kit, or they can be purchased from several companies that distribute tissue culture plastic ware. The blocks are autoclavable.

3. Centrifuge the entire bacterial cultures at once for 20 min at 1500 r.p.m. and room temperature, using a rotor that permits the centrifugation of 96-well microtiter plates.

▶ **Note:** Cover the block with adhesive tape during centrifugation to avoid any spills.

4. Discard the medium and resuspend the pelleted bacterial cells in 250 µl Buffer P1 (a component of the QIAprep 96 kit), and transfer to the flat-bottomed block provided with the kit.

▶ **Note:** No cell clumps should be visible after resuspension of the pellet.

5. Add 250 µl Buffer P2 to each sample using an eight-channel pipettor. Seal the block with the tape provided, and gently invert it four to six times to mix. Incubate at room temperature for 5 min.

▶ **Note:** The procedure is mainly as used for the regular Qiagen DNA isolation kit, as it is based on the method described by Birnboim and Doly (1979). Avoid incubating for more than 5 min with buffer P2 as this will result in lower yields of the plasmids.

▶ **Note:** It is important to mix gently by inverting the block. Do not shake vigorously, as this may result in mixing the samples if the tape is leaky.

6. During incubation, prepare the QIAvac 96 by placing the filter plate (Turbo-Filter 96) in the QIAvac top plate.

7. Seal any unused wells of the filter plate with tape to increase the vacuum during the incubation procedure.

8. Place the plate holder inside the QIAvac base and place the column plate (QIAprep 96 Plate) into the plate holder.

9. Assemble the complete vacuum device by placing the QIAvac 96 top plate with the filters directly over the base containing the columns, and apply the vacuum.

10. Add 350 µl Buffer N3 to each sample, and repeat the sealing and mixing procedure from step 5.

▶ **Note:** Proceed with the purification as established for the regular DNA isolation kits.

11. Pipet the lysates from step 3 (850 µl per well) onto the filter plate.

▶ **Note:** Be careful to remember the orientation of application. It is easy to mix up the samples!

12. Apply the vacuum until all samples have passed through.

▶ **Note:** Regulate the vacuum by a three-way valve to an optimal flow rate of approximately 1–2 drops per second. This will increase the yield. It is recommended to open the valve slowly. If the sample is flushed through both of the plates, it cannot be recovered for re-application.

13. Switch off the vacuum and slowly ventilate the vacuum device.

14. Exchange the filter plate with the column plate containing the cleared lysates, and replace the plate holder in the base with the waste tray.

15. Apply the vacuum as described before.

▶ **Note:** The flowthrough is collected in the waste tray.

16. Switch off the vacuum, and wash the columns by adding 0.9 ml PB buffer to each well and re-applying the vacuum.

▶ **Note:** This step is absolutely necessary when using the *E. coli* JM105 or JM103 that express endonuclease activity (*endA* +). This step is not necessary with SURE II cells.

17. Repeat step 16 with 0.9 ml PE buffer.

18. Repeat step 17 and apply full vacuum for additional 10 min to dry the columns.

19. Remove the top plate from the vacuum device and vigorously tap the top plate on a paper towel to dry the nozzles of the columns.

▶ **Note:** Residual ethanol from the PE buffer will dilute the DNA and inhibit subsequent enzymatic reactions.

20. Replace the waste tray from the vacuum device base with the collection microtube plate.

▶ **Note:** Do not use a 96-well microtiter plate. For elution into a 96-well microtiter plate, replace the waste tray with an empty collection microtube rack holding the microtiter plate in place.

21. Incubate the DNA for 1 min with 50–100 µl of elution buffer EB (10 mM Tris-HCl, pH 8.5) or water.

22. Elute the DNA with maximum vacuum for 5 min.

23. Switch off the vacuum and ventilate the QIAvac 96 slowly. For later analysis, seal the plate and freeze at –20 °C.

24. After DNA isolation, verify insertion of the inverted repeat by double or triple digest with the appropriate restriction enzymes.

▶ **Note:** Use 96-well microtiter plates for pipetting the digestion reaction.

25. Tightly seal the plate and incubate in a 37 °C incubator to avoid evaporation of the samples. Carefully remove the sealing.

26. Separate the digestion products on a 1–1.5% TAE agarose gel.

▶ **Note:** Electrophoresis units and casting stands are commercially available that allow the separation of DNA/RNA in 96-well plate format. They facilitate the application of the samples by multichannel pipettors.

27. Precipitate the remaining DNA with 0.1 vol of 3 M sodium acetate and 2.5 vol of 100% ethanol.

▶ **Note:** Freeze at –80 °C for future use, or dilute in TE buffer.

28. Transfect the shRNA-expressing vector into the desired cell line and select cell clones expressing the lhRNA.

▶ **Note:** Not all clones necessarily show a complete knock-down of gene expression. There are still some that are expressing a knock-down phenotype.

4.6.2
Expression of lhRNA from Direct Repeat DNA

This method was originally developed by Greg Hannon and coworkers to avoid the laborious inverted repeat cloning and recombination events within the cells (Paddison 2002). They used a so-called Flip-cassette to produce the lhRNA, as depicted in Figure 4.40. In contrast to the other approaches, the lhRNA template DNA is not introduced as an inverted repeat (sense-antisense) but in a direct repeat (sense-sense) interrupted by a antibiotic resistance gene (e.g. Zeocin). The second sense fragment is flanked with two *loxP* sites ("locus of crossover P1") (Hamilton and Abremski 1984; Hoess et al. 1982) as a recognition site for the P1 bacteriophage Cre recombinase (for "causes recombination") (Figure 4.40).

The Cre recombinase is usually used to introduce different molecular alterations into the mouse genome (Lakso et al. 1992; Sadowski 1993). It mediates different effects on their DNA target sequences, depending exclusively on the orientation of the specific recognition sequence (*loxP* site) such as sequence excision, duplication, integration, inversion (Bockamp et al. 2002; Sadowski 1993).

The *loxP* sequence is a 34-bp sequence consisting of two 13-bp inverted repeats and an 8-bp asymmetrical core spacer region, which determines the orientation of the site (Figure 4.41).

To date, the most common application of the Cre recombinase is the reciprocal exchange of the regions that flank the *loxP* sites, which occurs in *trans*. It is mainly used for the generation of conditional knock-out mice (for a review, see Bockamp

A)

B)

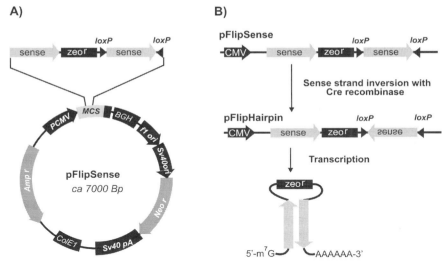

Fig. 4.40 (A) Schematic view of the direct repeat DNA vector (pFlipSense). The direct repeat is located in a so-called Flip-cassette. The repeat of the sense fragment is "floxed" (flanked by *loxP*). (B) After cloning of pFlipSense the vector is treated with Cre recombinase to convert the sense repeat into an antisense repeat to result in pFlipHairpin.

Fig. 4.41 Sequence of the *loxP* site comprising an inverted repeat domain and an asymmetrical core region (from Yu and Bradley 2001).

et al. 2002). A *cis* recombination event between two *loxP* sites in the same orientation will lead to the excision of the *loxP*-flanked DNA sequence as a circular molecule. If *loxP* sites are orientated in opposite directions, the *loxP*-flanking sequence will be inverted, meaning that flanking the second sense sequence with loxP sites in opposite directions will flip the sequence to result in antisense orientation as described for the pFlipSense vector (Figure 4.40) (Paddison 2002).

The pFlipSense vector (Figure 4.40) can be generated by several cloning steps, as described in the following protocol. To convert the pFlipSense into pFlipHairpin (Figure 4.44), the second sense sequence is flipped by Cre recombinase. This can either be obtained by in-vitro treatment of the vector using commercially available Cre recombinase protein (Stratagene, New England Biolabs), or by co-transfection of the pFlipSense with a Cre recombinase-expressing plasmid. Likewise, a very elegant conversion method is mediated by transient addition of a Cre recombinase to the pFlipSense transfected cells. Cre recombinase, which contains an N-terminal TAT-

NLS domain (TAT = cell penetrating peptide from HIV1 TAT protein; NLS = nuclear localization sequence), is able to transduce the cell membranes and translocate into the nucleus of the cells (Peitz et al. 2002).

PROTOCOL 36

1. Amplify 200–1000 bp of the target sequence in two PCRs using the following primer: PCR 1: primers 1 and 2, PCR 2: primers 5 and 6 (Figure 4.42).

First Sense Insert

1) 5′ – ▓Restriction site1▓ (N18) DNA forward –3′

2) 5′ – ▓Restriction site2▓ (N18) DNA reverse –3′

Zeocin Resistance Gene Loop

3) 5′ – ▓Restriction site2▓ (N18) Zeocin resistance gene forward –3′

4) 5′ – ▓Restriction site2▓ (N18) Zeocin resistance gene reverse –3′

Second "floxed" Sense Insert

5) 5′ – ▓Restriction site 2▓ ATAACTTCGTATAGCATACATTATACGAAGTTAT (N18) DNA forward
                              loxP site

6) 5′ – ▓Restriction site 3▓ ATAACTTCGTATAGCATACATTATACGAAGTTAT (N18) DNA reverse
                              loxP site

Fig. 4.42 Primers for the PCR or RT-PCR amplification of the different parts of the Flip-cassette. The restriction sites can be adapted from the eukaryotic expression vector of choice.

2. Amplify the antibiotic resistance gene Zeocin from the vector pCDNA3.1/ zeo (Invitrogen) using the primers 3 and 4 (Figure 4.43).

Zeocin cassette

Forward primer 5′-ATGGCCAAGTTGACCAGTGC-3′

Reverse primer 5′-TCAGTCCTGCTCCTCGGCCAC-3′

Fig. 4.43 Primer pair for the amplification of the zeocin cassette.

▶ **Note:** The Zeocin gene (Stratagene), later located between the repeats, maintains selection and stability of the Flip-cassette.

3. Digest the PCR products with the appropriate restriction sites according to the enzyme supplier.

4. Ligate all fragments into restriction site 1 and 3 of your eukaryotic expression vector (e.g. pCDNA3.1) using common ligation protocols (Sambrook and Russell 2001) (Figure 4.44).

5. To create the inverted repeat for lhRNA production, incubate the pFlipSense vector with the appropriate amount of Cre recombinase (New England Biolabs) *in vitro* according to the manufacturer's instructions.

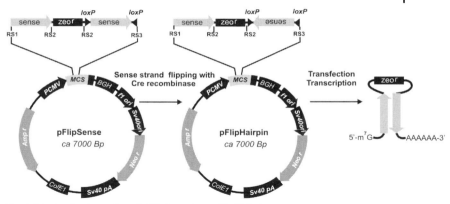

Fig. 4.44 In-vitro inversion of pFlipSense into pFlipHairpin by Cre recombinase. RS = restriction site.

6. Transform the resulting pFlipHairpin into one of the *E. coli* strains listed in Table 4.40; this permits the replication of DNA-containing cruciform structures, which tend to form from inverted repeats.

7. Isolate the DNA from colonies by the high-throughput method described in the last section, and screen for positive clones.

▶ **Note:** The yield of positive clones is much higher than described for the previous method.

8. Transfect cells with the pFlipHairpin vector according to the common transfection protocols.

9. Select stably transfected cell clones by Zeocin (Stratagene or Invitrogen) treatment as described in the last section or by the manufacturer's instructions.

4.6.3
Inverted Repeat DNA Missing 5'-Cap and Poly(A) Tail

One of the difficulties in using lhRNA transcribed from the Pol II promoter is the possible interferon response. Usually, Pol II transcripts are transferred to the cytosol immediately after transcription, resulting in an enrichment of dsRNA in the cytosol (Stark et al. 1998).

A powerful method for the use of lhRNA expression in mammals apart from any interferon response induction is based on a novel vector called pDECAP (*De*letion of *Cap* structure and *poly*(A)) (Shinagawa and Ishii 2003).

This prevents the export of the lhRNAs into the cytosol, since the lhRNA transcripts are lacking a 7-methylguanosine (m^7G) cap structure at the 5'-end and a poly(A)tail at the 3'-end. A prerequisite for efficient export of mRNA to the cytosol is the completion of those pre-mRNA processing steps, including 5'-capping, splicing,

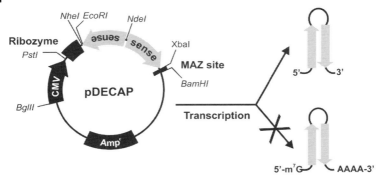

Fig. 4.45 Schematic view of pDECAP, which mediates the expression mRNAs lacking 5'-cap and poly(A) tail. It contains a ribozyme element to prevent capping, and a MAZ site to induce a premature stop and release of the RNA polymerase during transcription.

3'-polyadenylation, and binding to export proteins that specifically recognize mRNA species (Lewis and Tollervey 2000). It is assumed that type III ribonucleases localized in the nucleus, such as Drosha homologues or Dicer, are pre-cleaving lhRNAs into siRNAs. Those siRNAs may eventually exported into the cytosol to induce the degradation of target mRNA.

In eukaryotes, the majority of mRNAs contain a 5'-m^{7}G-cap that is added co-transcriptionally. The cap contributes to a variety of processes in the nucleus, including protection against 5'-3'-exonucleases, facilitating efficient pre-mRNA splicing, 3'-end formation, and especially mRNA nuclear export (Lewis and Izaurralde 1997; Lewis and Tollervey 2000). It has been shown that cytosolic factors, such as the eukaryotic translation initiation factor 4G (eIF4G), may be recruited to pre-mRNAs in the nucleus via its interaction with the cap-binding protein and accompanies the mRNA to the cytoplasm (McKendrick et al. 2001). To prevent 5'-capping a *cis*-acting hammerhead ribozyme (Huang and Carmichael 1996) is placed directly downstream of the RNA start site. A second prerequisite for a rapid nuclear export of the mRNA is polyadenylation of the 3'-end (Huang and Carmichael 1996). To prevent the premature export of the lhRNA, the poly(A) termination signal was replaced by a specific sequence that mediates Pol II transcriptional pausing. This sequence encodes the zinc finger protein MAZ (Yonaha and Proudfoot 2000), an element that is usually present between closely spaced human genes (Ashfield et al. 1994).

The nascent lhRNA is expected to be released from Pol II during transcriptional pausing and subsequently degraded into siRNAs by a nuclear member of the RNase III family. It is explicitly recommended not to add introns, because splicing sites may recruit proteins that enhance the rate of mRNA export into the cytosol (Luo and Reed 1999).

1) Pausing

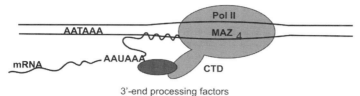

3'-end processing factors

2) Cleavage and polyadenylation

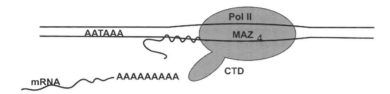

3) Degradation of 3'-product

4) Release of polymerase

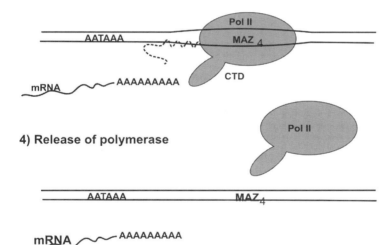

Fig. 4.46 Schematic view of the different steps during transcriptional pausing. During transcription, Pol II arrests at the MAZ$_4$ site and the processing of the 3'-end starts with cleavage and polyadenylation followed by residual mRNA degradation and the release of the polymerase (from Yonaha and Proudfoot 2000). CTD = C-terminal domain.

PROTOCOL 37

1. To generate your own pDECAP vector, ligate the hammerhead ribozyme (Huang and Carmichael 1996) encoding sequence into suitable restriction sites upstream of insertion site for the inverted repeat to prevent 5′-capping (Figure 4.47A).

A) Ribozyme sequence

Pstl   *Nhel*

```
5'-TGCAGCTGCAGCTCGAGATGCATGTCGCGGCCGCCTCCGCGGCCGCCTGATGAGTCCGTGAGGACGAAACATGCATAGGCTAGCG-3'
      |||||||||||||||||||||||||||||||||||||||||||||||||||||||||||||||||||||||||||||||||||
   3'-CGACGTCGAGCTCTACGTACAGCGCCGGCGGAGGCGCCGGCGGACTACTCAGGCACTCCTGCTTTGTACGTATCCGATCGCGATC-5'
```

B) MAZ sequence

Xbal   *BamHI*

```
5'-CTAGAGGCCCTTATCAGGGCCTCTGGCCTTGGGGGAGGGGGAGGCCAGAATGG-3'
      ||||||||||||||||||||||||||||||||||||||||||||||||||
   3'-TCCGGGAATAGTCCCGGAGACCGGAACCCCCTCCCCCTCCGGTCTTACCCTAG-5'
```

Fig. 4.47 Sequences of the ribozyme (A) and the MAZ site (B). The coding and complementary DNA strands can be ordered as oligo-nucleotides. Addition of the restriction site as sticky ends facilitates the insertion of the cassettes. The restriction sites depicted here are exchangeable with others.

2. Ligate the Pol II pausing site (MAZ) (Yonaha and Proudfoot 2000) encoding sequence into suitable restriction sites downstream of the insertion site for the inverted repeat to prevent addition of the poly(A) tail (Figure 4.47B).

▶ **Note:** This can be obtained by the replacement of the poly(A) signal of the respective vector by the MAZ cassette.

3. Generate the sense and antisense DNA fragments for the inverted repeat by PCR (see Section 4.6.1). The fragment size should be between 200 and 1000 bp. Add two different suitable restriction sites at the 5′-end of the forward primers (Figure 4.38), depending on the MCS of the your final pDECAP vector or any other eukaryotic expression vector.

▶ **Note:** The introduction of a 12-bp non-palindromic spacer (5′-GGTGCGCA-TATG-3′) as a loop may facilitate the cloning and the hairpin formation (Figure 4.48).

Spacer (Loop)

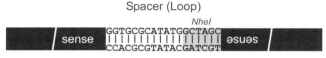

Fig. 4.48 Example for the construction of the 12 bp loop. One can also insert a restriction site at the 3′-end of the sense and antisense strand.

4. Digest the inserts and the vector with the appropriate restriction enzymes.

5. Ligate the two PCR products into the appropriate restriction sites of the prepared pDECAP vector.

6. Transform the vector into competent *E. coli* SURE II cells.

7. Plate the transformation reaction on the appropriate LB-Agar plates and incubate overnight at 37 °C.

8. Pick colonies and analyze by high-throughput DNA isolation and digestion (see Section 4.6.1).

9. Amplify positive colonies and store the DNA at –80 °C, or transfect cells according the respective transfection protocol.

4.6.4
RNAi versus dsRNA Deamination

Reports have been made on the antagonistic effect of RNAi and RNA editing by ADARs (adenosine deaminases that act on dsRNA). ADARs directly target dsRNA and inhibit the processing of siRNAs by this editing process (Scadden and Smith 2001 a, b).

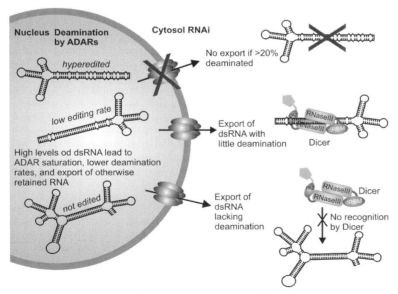

Fig. 4.49 Schematic view of the two contradictory events: the deamination process and RNAi. Long dsRNA (>50 nt) will be unspecifically deaminated by ADARs. If the editing rate is >20%, the dsRNA is prevented from being exported into the cytosol. However, export occurs when the ADARs are saturated by the presence of too much dsRNA. If hyper-edited dsRNA is translocated into the cytosol, it is often not recognized and cleaved by Dicer (Bass 2002; Knight and Bass 2002).

The ADARs are involved in an RNA-editing process that substitutes adenosines (A) by inosines (I) in cellular mRNA and viral dsRNA targets (Patterson and Samuel 1995). An A-to-I editing enzyme was discovered in *Xenopus* through its ability to unwind long dsRNAs by deaminating multiple "A"s to "I"s, which results in unstable I:U base pairs (Bass and Weintraub 1988). Although several well characterized RNA targets are edited in their coding region, it was recently found that A-to-I editing also occurs in 3'-UTR, intron, and noncoding RNA sequences in human and *C. elegans* RNA substrates (Morse et al. 2002). Possibly, hyperediting of viral or endogenously expressed long double-stranded RNAs by ADARs may provide a mechanism to remove dsRNA from cells, perhaps in conjunction with cytoplasmic or nuclear endonucleases now known specifically to cleave hyperedited dsRNA species (Scadden and Smith 2001 b). It has been shown for many early transcripts forming dsRNA stems that those stems are highly sensitive for hyperediting by ADARs. Probes that detect only modified RNAs revealed that these molecules are not highly unstable, but accumulate within the nucleus and are thus inactive with regard to gene expression due to a block of export into the cytosol (Kumar and Carmichael 1997). Substrates containing more than 50 nt that are perfectly paired can become nonspecifically deaminated (Nishikura et al. 1991; Polson and Bass 1994). Long hairpin RNA is therefore a perfect target for nonspecific deamination, which would disable their export to the cytosol to induce RNAi (Figure 4.49).

One could speculate that in lhRNA-expressing cell clones that survived the extensive selection procedure, the levels of siRNAs may be reduced by deamination of the overexpressed dsRNA. However, there are no reports of reduced activity of dsRNA or lhRNA processing, neither in the approach with the pHairpin vectors (Diallo et al. 2003 a, b), nor with the pDECAP vector (Shinagawa and Ishii 2003). This suggests that the activity of ADARs is not high enough to block the formation of siRNAs. It is speculated that ADARs can be saturated with dsRNA, thus preventing the deamination of further occurring dsRNA (i. e. from overexpression).

4.7
Test for Interferon Response

One of the limitations in the use of lhRNAs transcribed from the Pol II promoter is the interferon response. Due to a binding of the nascent Pol II transcripts by nuclear export factors those mRNAs are immediately transferred to the cytosol, where they induce an interferon response (Stark et al. 1998) (Figure 4.50). Interferons are cytokines that usually function as the host's firewall against viral infection.

The activation of interferons results in an induction of many complex signaling cascades mediated by a variety of proteins such as Jak1 and Tyk2 (Janus family tyrosine kinases), Stats1 and 2 (signal transducers and activators of transcription), and the IRF9 transcription factor terminated by the induction of interferon-stimulated genes (ISGs) within the nucleus (Haque and Williams 1998; Stark et al. 1998). Further, dsRNAs activates protein kinase R (PKR) and 2'-5'-oligoadenylate synthase leading to a block in translation and nonspecific mRNA degradation as part of the interferon response (Minks et al. 1979). Recently, it has been shown that beside long

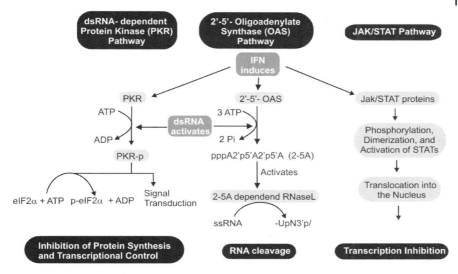

Fig. 4.50 Mechanism of dsRNA-induced interferon response (modified from Stark et al. 1998).

dsRNAs, even short hairpin RNAs (shRNAs) and enzymatically produced siRNAs can induce such an interferon response (Bridge et al. 2003; Kim et al. 2004; Sledz et al. 2003), making the application very complicated.

Although the nonspecific responses to dsRNA are attenuated in many nondifferentiated cells from early stages of development and in stem cells, they are present in most somatic cell types. There are some ways of circumventing those unspecific responses to exhibit RNAi in mammals, one of which is to use short dsRNA such as synthetic siRNAs. Another way is to use certain viruses for dsRNA delivery. Numerous mammalian viruses such as adenovirus or Vaccinia virus have evolved the ability to block PKR for efficient infection. Adenoviruses express VA RNAs, which mimic dsRNA with respect to binding but not to the activation of PKR (Clarke and Mathews 1995), while Vaccinia virus uses a different strategy to escape PKR activation. The Vaccinia virus encodes for a protein, E3L, which binds and covers dsRNAs (Kawagishi-Kobayashi et al. 2000) masking the dsRNA to prevent recognition. Further it encodes for a protein, K3L, which mimics eIF2α, the natural substrate of the PKR (Kawagishi-Kobayashi et al. 2000).

Co-transfection of the different lhRNA-expressing vectors with a vector that directs K3L, VA, or E3L expression attenuates the nonspecific response to dsRNA, but has no effect on the RNAi pathway. It was speculated that during the production of stable lhRNA-expressing cell lines, only those cells escape selection with antibiotics, which produce long lhRNA at levels to low to induce the interferon response, but sufficient to interfere with endogenous mRNAs (Diallo et al. 2003 a, b). The use of those viral factors now raises the possibility that blocking nonspecific responses to dsRNA will also facilitate the application of lhRNA in transient experiments.

Tab. 4.42

2'-5'-OAS	RT-PCR	Bridge et al. (2003)
	Northern blot	Diallo et al. (2003)
PKR	Western blot	Sledz et al. (2003)
	Imaging	Sledz et al. (2003)
	Kinase assay	Sledz et al. (2003)
elF2α	Phosphorylation assay	Srivastava et al. (1998)
	Western blot	Srivastava et al. (1998)
STATS1/2	Western blot	Sledz et al. (2003)
	Imaging	
Jak1/Tyk2	Western blot	Sledz et al. (2003)
	Imaging	
ISGs	ISG microarrays	Sledz et al. (2003)

Nevertheless, it is absolutely necessary to test for interferon response induction by some common assays. One assay is based on the measuring of the mRNA level of 2'-5'-oligoadenylate synthase (2'-5'-OAS) compared to wild-type cells by quantitative RT-PCR (real time PCR) or semi-quantitative PCR. Another method is to determine PKR activity. Further methods for detecting the interferon response are listed in Table 4.42.

Finally, it is recommended that cells be analyzed for apoptosis, for example by a DNA fragmentation assay (see the following protocol section).

4.7.1
Semi-Quantitative RT-PCR Analysis of 2'-5'-OAS Activation

PROTOCOL 38

1. Harvest wild-type and transfected cells and isolate RNA 48 h after transfection. (Use commercially available RNA isolation kits such as the RNEasy Kit from Qiagen.)

2. Quantify the RNA.

   ▶ **Note:** It is absolutely essential to determine the RNA concentration for validation of the RT-PCR product.

3. Pipette the RT-PCR reaction components for the amplification of 2'-5'-OAS and the housekeeping gene GAPDH as a control into two separate tubes.

   ▶ **Note:** This reaction can be performed as a multiplex PCR. The reverse primer should then be changed according to the sequence to avoid a size overlap of both products.

4. Amplify 2'-5'-OAS and GAPDH in an RT-PCR using the following primers (Figure 4.51).

▶ **Note:** Set up reactions in duplicates or triplicates from at least three transfection assay repeats.

▶ **Note:** For quantification of the RT-PCR product, use a PCR program with a
lower cycle number (ca. 20 cycles) to ensure that the amplification process is
still in the logarithmic phase.

OAS1 (human) (NM_0025349)

Forward primer : 5'-AGGTGGTAAAGGGTGGCTCC-3'
Reverse primer : 5'-ACAACCAGGTCAGCGTCAGAT-3'

GAPDH (human) (NM_002046)

Forward primer : 5'-ATGGGGAAGGTGAAGGTCG-3'
Reverse primer : 5'-CTTGAGGCTGTTGTCATACT-3'

GAPDH (murine) (NM_008084)

Forward primer : 5'-ATGGTGAAGGTCGGTGTGAA-3"
Reverse primer : 5'-TCGGCAGAAGGGGCGGAGAT-3'

Fig. 4.51 Selection of primers for semi-quantitative RT-PCR of 2'-5'-
oligoadenylate synthase 1 (OAS1) and the housekeeping gene glyce-
raldehyde-3-phosphate dehydrogenase (GAPDH). PCRs with those
primer pairs result in products of about 400–500 bp. NCBI accession
numbers are shown in brackets.

5. Analyze the 28S/18S rRNA content of each sample as a further internal control.

6. Analyze the PCR products on a 1.5% TAE agarose gel and stain with ethidium bromide (Sambrook and Russell 2001).

7. Quantify the amount of PCR product normalized to the endogenous reference (18S rRNA) and relative to the experimental control (GAPDH).

▶ **Note:** To obtain a more reliable quantification of the RT-PCR products, it is
recommended to perform real time RT-PCR (TaqMan, PerkinElmer).

4.7.2
DNA Fragmentation Analysis

PROTOCOL 39

1. Harvest cells in 0.5 ml of lysis buffer (Table 4.43).

2. Incubate lysates with 0.2 mg/ml proteinase K for 2 h at 50 °C.

3. Extract nucleic acid with equal volumes of phenol:chloroform:isoamyl alcohol (25 : 24:1) and re-extract with chloroform:isoamyl alcohol.

4. Precipitate the DNA overnight by addition of 1 volume of isopropanol at –70 °C.

Tab. 4.43

Lysis buffer	
Tris HCl pH 7.4	10 mM
EDTA	1 mM
NaCl	400 mM
SDS	1%

5. Collect precipitated DNA by centrifugation for 30 min at 14 000 g in a micro-fuge, and wash with ice-cold 70% ethanol.

6. Resuspend in 10 mM Tris-EDTA buffer (pH 8.0), and treat with RNase.

7. Separate DNA samples by electrophoresis on a 1.6% TAE-agarose gel and visualize by ethidium bromide staining.

▶ **Note:** In apoptotic cells a characteristic DNA fragment ladder should be visible (size ranging between 200 and 600 base pairs).

4.7.3
PKR Activity and eIF2α Phosphorylation

PKR activity can be measured by detection of eIF2α phosphorylation via [^{32}P] phosphate incorporation (Srivastava et al. 1998), or by Western blot analysis (Sledz et al. 2003).

eIF2α Phosphorylation by Western Blot Analysis

PROTOCOL 40

1. Determine the amount of cells in treated and nontreated cells, and normalize.

2. Lyze dsRNA-treated cells and wild-type cells with 1x Laemmli buffer (500 μl per 6-well or 100 μl per 24-well) (Table 4.44).

Tab. 4.44

5 x Laemmli buffer	
Tris HCl pH 6.8	0.2 M
Glycerol	2 ml
Bromophenol blue	0.0075%
β-Mercaptoethanol	1.25 ml
Total	10 ml

Dilute with PBS to 1x

3. Use a "rubber policeman" to scrape off the cells.

4. Collect the lysate in a microfuge tube and boil for 5 min at 95 °C.

5. Apply the appropriate amount of sample to 12 % SDS-PAGE under reducing conditions.

6. Subject the gel to Western blot analysis using PVDF membranes.

7. After blotting, incubate the membrane for 1 h at room temperature in 5 % Carnation low-fat milk (Nestlé) powder in TBS-T (Tris buffered saline-1 % Tween 20) (Sambrook and Russell 2001).

8. Incubate the membranes for 2 h at room temperature with a 1:1000 dilution of a monoclonal anti-phosphorylated eIF2α antibody (Cell Signaling, Beverly, MA, USA). Simultaneously probe the blot with an internal control antibody (such as for tubulin, or GAPDH, and total eIF2α) for later normalization of the results.

9. Wash three times for 5 min each with TBS-T.

10. Incubate the membrane with a 1:5000 dilution of an anti-mouse-HRP conjugated antibody (New England Biolabs).

11. After incubation with antibody, wash the membranes three times for 5 min each with TBS-T.

▶ **Note:** Avoid long washing times. HRP conjugates are sensitive to degradation.

12. Develop the blots using the ECL kit (Amersham Biosciences).

13. Chemiluminescence is evaluated in comparison to the endogenous control protein.

eIF2α Phosphorylation via [^{32}P]-Phosphate Incorporation

The phosphorylation status of eIF2α is determined in the shRNA/lhRNA-expressing cells or 60 h post-transfection with siRNAs. The labeling is performed by treating the cells with 400 μCi (per well of a 6-well plate) [^{32}P]-phosphate (orthophosphoric acid; DuPont NEN) for 4 h. Cell extracts are prepared and immunoprecipitated with a monoclonal anti-eIF2α antibody (Cell Signaling) using protein A-Sepharose as the immunoadsorbent. To evaluate the phosphorylation state of eIF2α, the immunoprecipitates are analyzed by SDS-PAGE and autoradiography (Srivastava et al. 1995).

Autophosphorylation Activity of PKR

PROTOCOL 41

1. For this experiment, treat the cells with dsRNA or use lhRNA- or shRNA-expressing cells. As negative controls, use wild-type cells treated with buffer alone or a 21-nt single-stranded RNA (ss). As a positive control, treat cells with Poly rI:rC (synthetic dsRNA).

2. For PKR immunoprecipitation, wash the cells with PBS and lyze 400 µg of total cell lysates from mock- and dsRNA-transfected cells in lysis buffer (Table 4.45) (Goh et al. 1999) about 90 min after transfection.

Tab. 4.45

Lysis buffer	
Tris/HCl pH 7.4	50 mM
NaCl	150 ml
NaF	50 mM
Glycerophosphate	10 mM
EDTA	0.1 mM
Glycerol	10%
Triton X-100	1%
PMSF	1 mM
NA-orthovanadate	1 mM
Protease inhibitors	2 µg/ml

3. Clarify the extracts by centrifugation (10 000 g) for 20 min at 4 °C.

4. Determine protein concentrations using the Bradford or bicinchoninic acid assays.

5. Incubate extracts containing normalized amounts of protein with a 1:5000 dilution of a monoclonal PKR antibody (source: Dr. Ara Hovanessian) according to (Goh et al. 1999) for 2 h at 4 °C.

6. Add 25 µl of equilibrated protein G-Sepharose beads to each sample and incubate overnight at 4 °C.

7. Collect the beads by centrifugation for 5 min at 1000 r.p.m. in a microfuge and wash thoroughly with lysis buffer (Table 4.45).

8. Add 5x Laemmli buffer to a final concentration of 1x to the mixture and boil the sample for 5 min.

9. Apply proteins to a 10% SDS-PAGE.

10. Transfer the proteins onto a PVDF membrane by Western blot and analyze PKR kinase activity by autoradiography.

11. As a control, perform immunoblotting of total PKR levels with a polyclonal PKR primary antibody followed by HRP-coupled anti-rabbit secondary antibody (Amersham Biosciences) and chemiluminescence.

12. For PKR activity assay, incubate purified PKR (100 ng) from immunoprecipitation for 30 min at 30 °C in 30 µl of kinase buffer containing [^{32}P]-ATP (Table 4.46) (Goh et al. 1999; Srivastava et al. 1998).

Tab. 4.46

Kinase buffer	
HEPES pH 7.5	25 mM
Mg-acetate	10 mM
ATP	50 μM
[^{32}P]-ATP	x μM

Immunodetection of Other Interferon-induced Proteins

Use Protocol A for the immunodetection of other proteins involved in the interferon pathway. Total protein (50 μg) is separated on 8% polyacrylamide gels and transferred to nitrocellulose membranes. Perform the immunostaining with monoclonal antibodies against STAT-1/2 (1:1000) (Santa Cruz Biotechnology) or PKR (1:5000).

Microarray Analysis

In addition to these methods, a semi-quantitative analysis of ISG can be performed using a microarray as described by Sledz et al. (2003). After isolation of total RNA from untransfected control, mock-transfected, and RNAi cells by normal methods, RNA is fluorescently labeled by direct incorporation in a cDNA synthesis reaction (Sambrook and Russell 2001). Wild-type control RNA is labeled with Cy3-dUTP (green), while RNA from RNAi cells is labeled with Cy5-dUTP (red). Samples are hybridized to the corresponding microarray overnight at 55 °C in slide hyb #3 (Ambion). Slides are washed with 2x SSC/0.1% SDS at 55 °C for 5 min and 0.2x SSC for 10 min at room temperature before scanning with the appropriate microarray reader. The subsequent data evaluation is performed according the bioinformatics software. For more information, see Sledz et al. (2003).

4.8
Inducible Persistent RNAi in Mammalian Cells

One of the major goals in the development of new techniques for the application of RNAi in mammals was the establishment of conditional vectors for shRNAs expression. Those vectors would allow the controlled expression of shRNA either in a temporal fashion or in defined tissues. An inducible system for RNAi allows a simultaneous analysis of loss-of-function phenotypes by comparing selected isogenic cell populations on the induced and noninduced levels. In addition, conditional RNAi permits the study of essential and multifunctional genes involved in complex biological processes by preventing inhibitory and compensatory effects caused by constitutive knock-down (Czauderna et al. 2003). Within the past year, several systems have been reported on this issue (Allikian et al. 2002; Calegari et al. 2002; Czauderna et al. 2003; Fritsch et al. 2004; Kasim et al. 2003; Matsukura et al. 2003; van de Wetering et al. 2003) (Figure 4.52).

Due to the nature of the Pol III promoter, which is not easily controllable in different tissues, the majority of these systems use the temporal inducible expression of

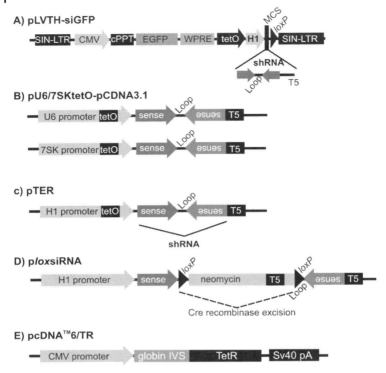

A) pLVTH-siGFP

SIN-LTR — CMV — cPPT — EGFP — WPRE — tetO — H1 — MCS / loxP — SIN-LTR

shRNA
Loop — T5

B) pU6/7SKtetO-pCDNA3.1

U6 promoter — tetO — sense — Loop — sense — T5

7SK promoter — tetO — sense — Loop — sense — T5

c) pTER

H1 promoter — tetO — sense — Loop — sense — T5

shRNA

D) ploxsiRNA

H1 promoter — sense — loxP — neomycin — T5 — loxP — sense — T5

Loop

Cre recombinase excision

E) pcDNA™6/TR

CMV promoter — globin IVS — TetR — Sv40 pA

Fig. 4.52 Maps of various vectors containing tetracycline operon (tetO)/ tetracycline repressor (tetR) system for the inducible Pol III-driven expression of shRNAs as: (A) the lentiviral vector pLVTH-siGFP (Wiznerowicz and Trono 2003); (B) U6-tetO and 7SK-tetO shRNA expression vectors (Czauderna et al. 2003; Matsukura et al. 2003); (C) pTER, a pSUPER-based H1-tetO shRNA expressing vector (van de Wetering et al. 2003); (D) the Cre/lox-dependent ploxsiRNA vector (Fritsch et al. 2004); (E) the tet repressor (tetR) expressing vector based on the pcDNA backbone (from Invitrogen).

shRNA. The system is based on the work of Ohkawa and Taira (2000) (Figure 4.54), who described the successful integration of the bacterial tetracycline operon (tetO) in the U6 promoter to allow the inducible expression of shRNAs. The tetO is regulated by the tetracycline repressor (tetR), which is expressed on a separate vector. In the absence of its substrate tetracycline, tetR is bound to the tetO element and represses the expression of the hairpin RNA (tet-off) (Matsukura et al. 2003). Doses of 1–10 μg/ml tetracycline or the more stable derivative doxycycline inhibit the tetR binding and allow transcription, depending on the strength of the promoter that drives the expression of tetR (Hillen and Berens 1994; Matsukura et al. 2003; Ohkawa and Taira 2000) (Figures 4.52 and 4.53A).

Likewise, the U6 promoter the H1 Pol III promoter has been shown to be functional with a tetO insertion directly upstream of the Pol III start site (van de Wetering

A)

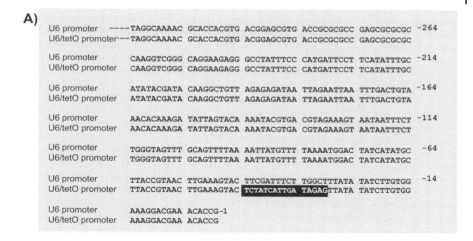

```
U6 promoter      ----TAGGCAAAAC GCACCACGTG ACGGAGCGTG ACCGCGCGCC GAGCGCGCGC -264
U6/tetO promoter---TAGGCAAAAC GCACCACGTG ACGGAGCGTG ACCGCGCGCC GAGCGCGCGC

U6 promoter         CAAGGTCGGG CAGGAAGAGG GCCTATTTCC CATGATTCCT TCATATTTGC -214
U6/tetO promoter    CAAGGTCGGG CAGGAAGAGG GCCTATTTCC CATGATTCCT TCATATTTGC

U6 promoter         ATATACGATA CAAGGCTGTT AGAGAGATAA TTAGAATTAA TTTGACTGTA -164
U6/tetO promoter    ATATACGATA CAAGGCTGTT AGAGAGATAA TTAGAATTAA TTTGACTGTA

U6 promoter         AACACAAAGA TATTAGTACA AAATACGTGA CGTAGAAAGT AATAATTTCT -114
U6/tetO promoter    AACACAAAGA TATTAGTACA AAATACGTGA CGTAGAAAGT AATAATTTCT

U6 promoter         TGGGTAGTTT GCAGTTTTAA AATTATGTTT TAAAATGGAC TATCATATGC  -64
U6/tetO promoter    TGGGTAGTTT GCAGTTTTAA AATTATGTTT TAAAATGGAC TATCATATGC

U6 promoter         TTACCGTAAC TTGAAAGTAC TTCGATTTCT TGGCTTTATA TATCTTGTGG  -14
U6/tetO promoter    TTACCGTAAC TTGAAAGTAC TCTATCATTGA TAGAGTTATA TATCTTGTGG

U6 promoter         AAAGGACGAA ACACCG-1
U6/tetO promoter    AAAGGACGAA ACACCG
```

B)

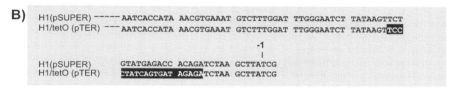

```
H1(pSUPER) ------AATCACCATA AACGTGAAAT GTCTTTGGAT TTGGGAATCT TATAAGTTCT
H1/tetO (pTER) ---AATCACCATA AACGTGAAAT GTCTTTGGAT TTGGGAATCT TATAAGTTCC

                                                           -1
                                                            |
H1(pSUPER)        GTATGAGACC ACAGATCTAA GCTTATCG
H1/tetO (pTER)    CTATCAGTGAT AGAGATCTAA GCTTATCG
```

Fig. 4.53 Comparison of the wild-type and the tetO-modified U6 and H1 promoters. The tetO sequence is depicted in the black squares. In both promoters the insertion was made almost directly upstream of the transcriptional start site.

et al. 2003). This 19-bp sequence from positions −23 to −5 does not generally affect promoter activity in vivo (Myslinski et al. 2001), and can be replaced by the tetO sequence. The modification can be performed in the pSUPER vector, resulting in pTER (van de Wetering et al. 2003) (Figures 4.52 and 4.53B).

Although those tetO-modified Pol III systems have been shown to work successfully, Pol III promoters usually are very sensitive to sequence modifications. Furthermore, many laboratories have reported a substantial leakiness of the tet-off system, leading to unwanted expression of the shRNA during selection of the stable cell line.

A very promising inducible system, which has been suggested for tissue-specific expression or temporal expression in mice, is based on a combination of Pol III-driven shRNA expression and the well established Cre/lox system (Fritsch et al. 2004; Rajewsky et al. 1996) (Figure 4.55). Expression of the shRNA is based on the presence of Cre recombinase. The major advantage of this method is the availability of a large variety of mice that express Cre recombinase in tissue-specific or tetracycline-controllable promoters. Further, many cell lines of those mice are commercially available. The most exciting feature of this system, however, is the transient administration of recombinant Cre recombinase in the medium.

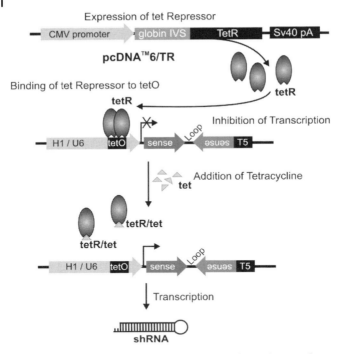

Fig. 4.54 Mechanism of the tetracycline or doxycycline induction of the tet operon. The tet repressor (tetR) is expressed from a second plasmid co-transfected with the tet operon (tetO) containing the shRNA expression vector. The tetR binds to the tetO sequence, thus blocking transcription. Binding of tetracycline/doxycycline releases the repressor and starts transcription of the hairpin.

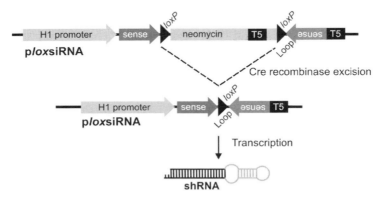

Fig. 4.55 Mechanism of transcription activation by excision of a floxed (flanked by *loxP* sites) neomycin cassette containing a termination sequence using Cre recombinase. Transcription of the full hairpin occurs after excision of the termination signal.

As shown by Edenhofer and colleagues (Peitz et al. 2002), Cre recombinase can be efficiently delivered from the medium to the nucleus of the cell when expressed as a TAT-NLS fusion protein. This fusion tag comprises the TAT peptide (an arginine-rich peptide derived from HIV-1) and NLS (as shown in Figure 4.28; see Section 4.5.1). Upon addition of TAT-NLS-Cre, complete and functional siRNAs are generated and reporter activity is suppressed. The shRNA-expressing vector (*ploxsiRNA*) is modified such as a "floxed" (flanked by *loxP* sites) neomycin cassette containing a Pol III termination site is placed between the sense and antisense oligonucleotide strands (Fritsch et al. 2004), thereby preventing synthesis of the antisense strand. In the presence of Cre recombinase, the neomycin cassette and the internal Pol III termination signal is excised and the transcription of sense and antisense strand occurs using the remaining *loxP* site as a loop (Figure 4.55).

4.9
Application of RNAi in Mice

4.9.1
Hydrodynamic Tail Vein Injection

The development of methods that allow an efficient expression of exogenous genes in animals would provide an exciting tool for the application of RNAi in functional genomics, and in the treatment of diseases. It has been reported recently that rapid tail vein injection of a large volume of plasmid DNA solution into a mouse results in high level of transgene expression in several organs, but especially in the liver. The effect of this hydrodynamics-based procedure is determined by the combined effect of a large volume and high-speed injection. It goes along with a transient irregularity of heart function, and a sharp increase in venous pressure resulting in an enlargement of liver and increased membrane permeability of the hepatocytes. It has been suggested by Wolff and coworkers that hydrodynamic injection generates membrane pores in hepatocytes to facilitate hepatic delivery (Zhang et al. 1999). Up to 40% of the liver cells can be reached after 8 h after injection, and repeated injections can enhance the phenotype (Liu et al. 1999). The uptake of radiolabeled DNA showed that the DNA is retained at the membrane surface for at least 1 h, and is rather slowly internalized into the liver cells. This explains the long-term effect of hydrodynamics-based transfection, and suggests that a pool of nucleic acids persists for slow- and long-term release.

It has recently been proven that RNA interference can be applied by the hydrodynamic transfection of siRNAs and shRNA expressing vectors to postnatal mice. Both, Kay (McCaffrey et al. 2002) and Herweijer (Lewis et al. 2002) and coworkers reported that the silencing effect persisted for three to four days after the injection of either shRNA expression vectors (McCaffrey et al. 2003) or synthetic siRNAs (Lewis et al. 2002; McCaffrey et al. 2002) (Figure 4.56).

The key step of this method is the high-speed injection (5–7 s) of a large volume of PBS buffer (1.8 ml) to transfer between 0.5 and 40 µg of synthetic siRNAs to-

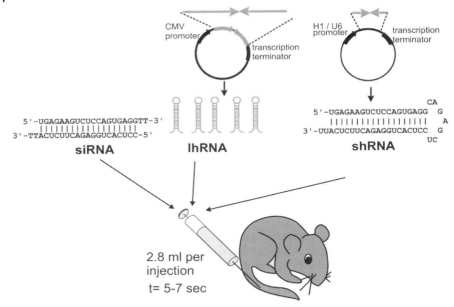

Fig. 4.56 Schematic view of the hydrodynamic tail vein injection. This method can be used to inject either siRNA (McCaffrey et al. 2002), lhRNA-expressing plasmids (Shinagawa and Ishii 2003), or shRNA-expressing plasmids (McCaffrey et al. 2003).

gether with 2–10 µg of luciferase expressing control plasmid pGL3 (Promega) and 800 units of RNasin (Promega) into the tail vein (or directly into the liver) of 6- to 9-week-old mice. The volume of saline buffer (PBS) injected must be equivalent to 8% of the body mass of the mouse (e.g. 1.6–1.8 ml for a mouse weighing 20 g) (Yang et al. 2002 b). After 72 h, the mice were anaesthetized and given 3 mg luciferin intraperitoneally 15 min before imaging to detect luciferase activity (McCaffrey et al. 2002). Kay and colleagues found an siRNA-mediated gene inhibition of up to 80%.

Even though in mice this method seems to be very suitable for the delivery of siRNAs, it will not be applicable in humans for therapeutic purposes such as treatment of a hepatitis B virus infection as the injection volume is too large (a person with a body mass of 60 kg would require an injection of 4.8 l to be given within seconds).

However, hydrodynamics-based methods have been developed for delivering DNA to the isolated rabbit liver using a minimally invasive balloon occlusion balloon catheter to occlude a selected hepatic vein (Eastman et al. 2002). A whole-organ technique was used wherein the entire hepatic venous system was isolated and the pDNA solution injected hydrodynamically into the *vena cava* between two balloons used to block hepatic venous outflow. Preliminary studies with this system have suggested a potential method for human gene therapy that is both therapeutically significant and clinically practicable (Eastman et al. 2002).

4.9.2
In-Vivo Electroporation

Besides the hydrodynamic tail vein injection, other methods are currently under investigation that are more applicable for future medical use in humans. There are successful reports on the in-vivo application of siRNA electroporation in the skeletal muscle of mice (Kishida et al. 2004). The electroporation procedure was shown to provide an effective means for transferring siRNA duplexes into the target tissue in vivo. Although this procedure has not shown the highest efficiency in gene silencing after the first application, the efficiency may be improved by repetitive treatments and/or several treatments into multiple portions of the target organ.

Previous experiments with DNA have shown the feasibility of electric pulses to target a variety of organs, including skeletal muscle (Aihara and Miyazaki 1998), the liver (Suzuki et al. 1998), skin (Vanbever et al. 1998), cornea (Oshima et al. 1998), blood vessels (Schwachtgen et al. 1994), and joints (Ohashi et al. 2002). The targeting of other tissues such as tumors, such as melanoma (Rols et al. 1998), hepatocellular carcinoma (Heller et al. 2000), colon tumors (Goto et al. 2000), and glioma (Yoshizato et al. 2000), is of special interest. Hence, electroporation may also promote the in-vivo transfer of siRNA into other organs of postnatal mammals (Kishida et al. 2004), similarly exhibiting the safety features required for clinical use.

A study by Kishida et al. showed a long-lasting RNAi effect after electroporation of siRNA into the skeletal muscle, even superseding the three to four days reported for the hydrodynamic tail vein injection (Lewis et al. 2002; McCaffrey et al. 2002). This effect can be ascribed to the different cell types into which the siRNA was introduced.

The protocol can even be applied to generate tissue-specific RNAi when administered to mouse embryos that have been re-implanted after electroporation (Calegari et al. 2002). For this procedure, the embryos must be dissected from the uterine walls and removed from the *deciduas capsularis* and Reichert's membrane. For electroporation, the embryos are embedded in agarose to prevent them from drying out. For further information, see Calegari et al. (2002).

PROTOCOL 42

The electroporation protocol for murine skeletal muscles is modified from previous reports using DNA (Ohashi et al. 2002).

1. Purchase female mice of the genetic background of interest (i. e. C57/BL6).

2. Anesthetize the mouse shortly prior to injection with 300 µl of 0.05 % xylazine-1.7 % ketamine in 0.9 % NaCl.

3. Dilute up to 8 µg of siRNA and, if desired, 2 µg of a reporter plasmid in 50 µl of PBS.

4. Prepare a sterile syringe with a 27-gauge needle to inject the siRNA solution into the tibial muscle.

5. Carefully inject the solution.

6. Immediately after injection, place a pair of stainless steel electrode plates (1.0 cm diameter) or the plugs of an electric cable onto the muscle, with the gap between the electrodes fixed at 0.33 cm (Figure 4.57).

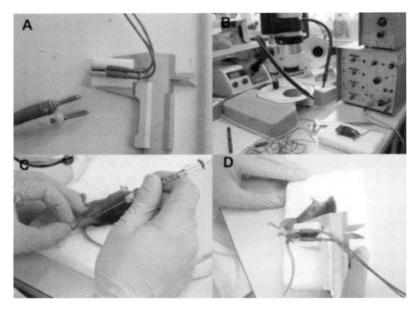

Fig. 4.57 Overview of the electroporation procedure. (A) A closer view of the condensor plates/cables. (B) A sedated mouse is placed on a piece of styrofoam next to the electroporator. (C) siRNA solution is injected into the muscle. (D) Cables are positioned and the electropulses applied. (Illustrations kindly provided by C. Arenz and R. Mundegar, University of Bonn.)

7. Apply the electric pulses with a square-wave electroporator (e.g. ECM 830; Q-biogen).

▶ **Note:** Deliver three square-wave pulses with a pulse length of 20–50 ms at 150 V/cm at a frequency of 1–2 Hz, followed by three other pulses with the opposite polarity.

8. At 2–4 days post electroporation, kill the mouse and analyze the phenotype.

4.9.3
Infection of Adult Mice with Adenovirus

Davidson and coworkers intravenously infected mice with an adenovirus vector carrying siRNA, which resulted in significant suppression of marker gene expression in

vivo (Xia et al. 2002). These authors established a highly suitable system for the efficient silencing of exogenous and endogenous genes in vitro and in vivo in brain and
liver, and successfully applied this strategy to a model system of a major class of neu-
rodegenerative disorders. As with the previous method, the adenoviral infection also
exhibited long-term persistence of the RNAi phenotype.

4.9.4
Generation of Transgenic Mice Using Lentiviral Infection

Lentiviral vectors are capable of stably expressing transgenes in a variety of stem cells
such as embryonic stem cells (ESC) and hematopoietic stem cells (HSC), without being
silenced during early development (Rubinson et al. 2003). Luk van Parijs and coworkers infected both stem cell types with recombinant lentiviral vectors and cultured the
cells according to common protocols. Lentivirus-infected cells were sorted by FACS
and injected into lethally irradiated congenic mice (Rubinson et al. 2003). Analysis of
the mice after 6–8 weeks revealed a contribution of the injected HSCs to all blood cell
lineages and to 20–40% of the lymphocytes. The same authors further infected mouse
ES cells to generate transgenic mice by injection of a single infected ESC into a mouse
blastocyst. The first-generation animals were found to be 60–90% chimeric.

However, the direct infection of single cell embryos was seen to be the most
powerful method for generating transgenic animals. While the silencing activity by
the shRNA expressing transgene is high during embryonic development, the expression of shRNA gene was maintained, albeit with reduced efficiency (5–60%). Likewise, ESC- or HSC-derived transgenic animals showed the same phenotype, though
this might be due to the low copy number of the transgene that is integrated into the
mouse genome. As shown from the studies of van Parijs and colleagues, this method
may be very useful for targeting organs other than the blood and the immune system, for example the liver, spleen, or brain, and may be broadly applied to the whole
organism, even in humans.

4.9.5
Generation of Transgenic Mice by Pronucleus Injection

Mice that express an RNAi phenotype can be generated either by viral infection of
ESC and transfer of the ESC into blastocysts, or by pronucleus injection of the transgene into fertilized mouse oocytes, as described for the generation of common transgenic mice. (For a detailed protocols and comments, see Hogan et al. 1994.) The latter procedure is straightforward, and allows a rapid production of founder animals.
However, one of the major drawbacks is the unspecific and uncontrollable insertion
of one to multiple copies of the transgene into the mouse genome, which makes
genotyping of the mice very difficult.

The method requires the use of an injection microscope device, as well as intensive
training in oocyte injection. Many companies or academic facilities now offer this injection service and genotyping after sending the DNA. The protocol below is a short description of the preparation of such DNA for pronuclear injection (Figure 4.58).

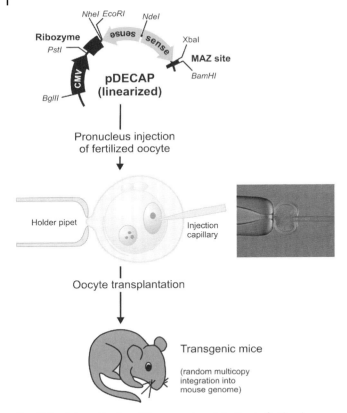

Fig. 4.58 Schematic view of the pronucleus injection in fertilized mouse oocytes. Linearized sh- or lhRNA-expressing plasmids are injected into the pronucleus, after which the fertilized eggs are reimplanted into the mouse uterus. The offspring (founder animals) are either positive or negative for the transgene and must be genotyped. The microscopic image was taken during a pronucleus injection in the laboratory of Hubert Schorle, University of Bonn.

PROTOCOL 43

1. Linearize 20 µg of the plasmid containing the shRNA/lhRNA-expressing insert and remove all unnecessary vector sequences by enzymatic digestion with the appropriate restriction enzymes.

▶ **Note:** The DNA should contain the promoter for the shRNA- or lhRNA-expressing insert and the termination signal.

▶ **Note:** The large amount of the DNA allows a stronger dilution at the end of the procedure.

2. Separate the DNA on a 1% TAE-agarose gel containing ethidium bromide using a large-slot comb.

▶ **Note:** Ethidium bromide does not interfere with the later steps.

3. Cut the DNA from the gel with a razor blade.

4. Elute the DNA by electroelution.

▶ **Note:** Do not use the glass bead extraction kit!

5. For electroelution, use either a device that is commercially available (such as the one by Schleicher and Schuell), or use dialysis tubing in a simple electrophoresis chamber (see protocol below).

Preparation of the dialysis tubing

6. Cut the dialysis tubing (flat width 25 mm) into pieces of about 7 cm length and boil for 5–10 min in 2% Na_2CO_3, 1 mM EDTA.

7. Wash the tubes in distilled water for some time and autoclave in 1 mM EDTA, pH 8.0. Store the tubing in solution at 4 °C, and wash with water before use.

Electroelution

8. Cut the DNA from the gel and place into the tubing.

9. Seal the tubing at one end with a plastic clamp and fill it with 350 µl of TE buffer.

10. Remove the air bubbles carefully and seal the other end of the tubing.

11. Place the tubing into an electrophoresis chamber filled with TAE buffer so that the DNA will move electrophoretically from the gel into the TE solution.

12. Apply power and allow elution for 1 h at ~100 V.

▶ **Note:** The elution can be controlled by exposing the tubing to UV light. The ethidium bromide stain must be visible in the solution. Avoid many exposures as this will damage the DNA.

13. When the elution is finished, reverse the polarization of the power supply for 1 min to allow detachment of the DNA from the tubing.

14. Pipette the TE solution from the tubing and resuspend carefully by pipetting up and down while rinsing the tubing walls.

15. Extract the DNA with 200 µl phenol:chloroform (1:1), and twice with the same volume of chloroform (Sambrock and Russel 2001).

16. Extract the DNA while vortexing for 5 min and 10 min centrifugation at high speed.

▶ **Note:** Carefully avoid any phenol contamination that would be toxic to the mice.

17. Precipitate the DNA with 0.1 volume 3 M Na acetate and 2.5 volumes ethanol for 30 min at –80 °C.

18. Spin down the precipitate for 20 min, 4 °C at 10 000 g.

19. Wash the pellet with ice-cold 70 % ethanol.

20. Dry the pellet for 5 min at room temperature.

21. Dissolve the pellet in 17 µl water.

22. Measure the concentration and dilute the DNA in injection buffer (Table 4.47) to a final concentration of 2 ng/µl.

Tab. 4.47

Injection buffer	
Tris HCl pH 7.5–8	10 mM
EDTA	0.1 mM

23. Check 10–20 µl on an agarose gel for purity.

24. Pass the remaining DNA through a 0.22 µm sterile filter and aliquot for injection purposes.

▶ **Note:** The filter step removes any particles and debris, which can clog the injection needle.

25. Store the DNA at –20 °C.

26. Inject DNA into a several (80–100) fertilized oocytes. For more information, see Hogan et al. (1994).

4.10
Alternative Methods: Ribozyme Cleavage of dsRNA

An alternative to the previously reported shRNA-expressing vectors is clearly the use of a self-cleaving, ribozyme-expressing vector to endogenously cleave an RNA precursor into siRNAs in mammalian cells (Figure 4.59). The studies of Taira and coworkers have been based on the observation that the hammerhead ribozyme, one of the smallest catalytic RNAs (Symons 1992; Uhlenbeck 1987; Zhou and Taira 1998), possesses a sequence motif with three duplex stems and a conserved catalytic core of nonhelical segments that is responsible for self-cleavage (*cis*-action). Its activity does not depend of a certain type of promoter, Pol III or Pol II, driving its expression.

These authors placed four ribozymes within transcripts that were expressed under the control of either a Pol II promoter such as CMV or a Pol III promoter such as

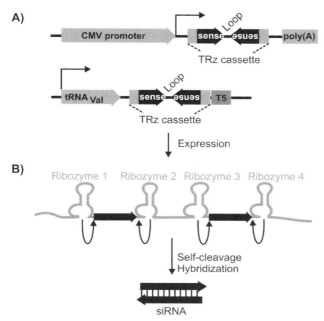

Fig. 4.59 Mechanism of hammerhead ribozyme-directed siRNA expression. (A) Map of the two types of expression systems containing the Pol III (tRNA$_{Val}$) and the Pol II (CMV) promoter. (B) Schematic representation of the TRz (trimming ribozyme) cassette. Sense and antisense siRNAs are produced during transcription by ribozyme-catalyzed self-cleavage of the transcript in *cis*. Arrowheads indicate the cleavage sites.

tRNA$_{Val}$ to excise specific sense and antisense sequences from primary transcripts in *cis*, which endogenously generate siRNAs in HeLa cells (Kato and Taira 2003). Few data are available on Pol II ribozyme expression, however. Like their hairpin RNA counterparts, those siRNAs generated by ribozymes were able to reduce the expression of a firefly gene for luciferase, which suggests that polymerase II (Pol II) systems – which are essential for the generation of many tissue-specific expression systems, as well as Pol III systems, in which siRNAs are generated by a trimming-ribozyme (TRz) system – can be used to suppress the expression of specific genes (Kato and Taira 2003).

The data from the comparison of both promoters revealed that the Pol III promoter is at least one order of magnitude more active than the Pol II promoter, which reflects data on shRNA expression using Pol II promoters. Nevertheless, the system could be used for tissue-specific expression using a variety of Pol II promoters with a decent silencing efficiency.

Other approaches are based on the use of allosteric ribozymes such as the one derived from the hepatitis delta virus (HDV) (Kertsburg and Soukup, unpublished) that performs catalysis upon recognition of specific effector molecules (Soukup and

Breaker 2000). It has been shown that allosteric ribozyme-encoded siRNAs are not recognized by Dicer, and are therefore protected from cleavage until induction of the ribozyme self-cleavage by 200 μM theophylline. Therefore, siRNA production is exclusively dependent upon allosteric ribozyme function.

4.11
High-Throughput Screens

As in *C. elegans* and *Drosophila*, the recent major breakthrough of RNAi in mammals was the establishment of several methods that allow large-scale and genome-wide screens in mammalian cells (Berns et al. 2004; Luo et al. 2004; Paddison et al. 2004; Shirane et al. 2004). This was made possible with the construction of libraries of shRNA-expressing vectors. The generation of such libraries were simultaneously performed by cloning of enzymatically retrieved shRNAs (Luo et al. 2004; Shirane et al. 2004) and by the specific design of shRNA encoding DNA oligos that are inserted into the Pol III promoter-containing vectors by the previously described methods (Berns et al. 2004; Paddison et al. 2004) (see Section 4.5). Other initiatives have already validated several thousand siRNAs for an siRNA genome-wide screen (Dharmacon, http://www.dharmacon.com, Qiagen, http://www1.Qiagen.com, EURIT (European Union RNA Interference Technology) http://www.eurit-network.org).

4.11.1
Production of Bar Code-Supported shRNA Libraries

The groups of Greg Hannon and Rene Bernards simultaneously reported the construction of large-scale shRNA libraries comprising between 23 742 and 28 659 shRNAs targeting 7914 to 9610 human genes, and 9119 shRNAs targeting 5563 mouse genes. The silencing was assured by three to nine different shRNAs per gene. To construct the libraries, only coding sequences were designed bearing at least three mismatches with any other target. The shRNA encoding DNA sequence is then chemically synthesized and cloned into the appropriate Pol III vectors (pSHAG-MAGIC (Paddison et al. 2004); pRetroSUPER (Berns et al. 2004)) (Figure 4.60).

The pSHAG-MAGIC vector contains a variety of novel features (for more information, see Paddison et al. 2004), one of which is the introduction of the so-called "Mating-Assisted Genetically Integrated Cloning" (MAGIC). The pSHAG-MAGIC is used as a donor vector, in which the shRNA-expression cassette is flanked by two different 50-bp homology-regions, H1 and H2. Those sites are further flanked with linked I-*SceI* sites, which can be used for the excision of the flanked sequence by I-*SceI* protein present in the bacterial strain hosting the recipient vector also containing H1, H2, and I-*SceI* sites. Transfer of pSHAG-MAGIC into the recipient host by bacterial mating induces the cleavage and the homologous recombination at the H1 and H2 sites, allowing an easy transfer from one vector into the other, which is usually a viral vector.

Both libraries (Berns et al. 2004; Paddison et al. 2004) cover shRNAs silencing-known kinases and phosphatases, components of the cell cycle, transcription regula-

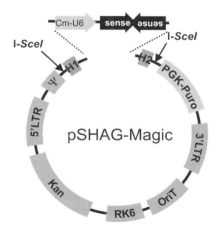

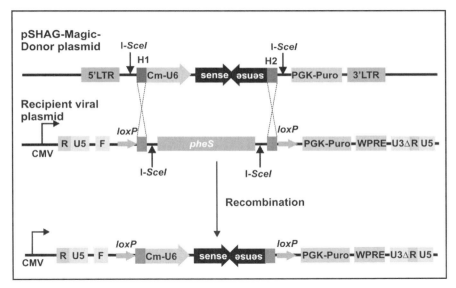

Fig. 4.60 Map of the pSHAG-MAGIC vector modified from Paddison et al. (2004). The shuttle vector can be used to transfer the shRNA cassette into other vectors and into the viral genome by recombination via the H1 sites, when the recipient contains the appropriate recombination sites.

tion, signaling cascades, etc. with an average efficiency of approximately 70% for about 70% of the targets.

A key feature of those libraries is the bar code-based evaluation procedure of the cell screens. Both libraries are designed to function either for genetic selection, which is based on shRNA vector/virus pools, or large screens. Both laboratories have developed a bar code-based strategy to facilitate the screening procedure of such ge-

netic screens, making use of microarrays with the corresponding bar code oligonucleotides to identify each shRNA. While Bernards and colleagues use the shRNA sequence as a unique sequence, Hannon and coworkers used an additional unique bar
code sequence for each vector (Figure 4.61).

For a genetic screen, cells were transfected or infected with individual shRNA-con-
structs in a 96-well or 348-well format. Analysis of the screen can be performed by
harvesting the DNA from cells that show interesting phenotypes, and by amplification of the shRNA-expressing cassette including the bar code sequence. The PCR
products are eventually labeled with the fluorescent dyes Cy3 or Cy5 and hybridized
to the microarray either in single experiments or in pools to identify their sequence.
Using this technology, both libraries were successfully tested in biological assays to
screen either for proteins involved in the p53 pathway or in proteasome function. It
becomes clear that this technology will become a new and powerful tool for RNAi in
mammalian systems facilitating the detection of new members of well-known pathways, even if the libraries have to be completed for genome-wide screens.

4.11.2
Enzymatically Retrieved Genome-Wide shRNA Libraries

Almost simultaneous to the large-scale approach described above, the construction
of genome-wide shRNA expression libraries has been reported that were retrieved
by enzymatic processing of cDNA (Luo et al. 2004; Shirane et al. 2004). Both approaches take advantage of the properties of the restriction enzyme *MmeI* that can
cleave DNA 19/21 nt from its recognition motif, thus creating fragments with 3'-
2 nt protruding ends such as RNase III. After partial digestion of a cDNA library
with restriction enzymes (Luo et al. 2004) or DNase (Shirane et al. 2004), fragments
are ligated to a hairpin-shaped linker containing an *MmeI* restriction site and an additional restriction site for subsequent cloning (Figure 4.62). As described above,
MmeI is used to cleave the DNA 19/21 nt upstream of the restriction site, thus
creating a siRNA-related size with 3'-2 nt protruding ends. After ligation of a second
adapter (no hairpin linker) to the opposite site of the fragment, a primer extension
with DNA polymerase is performed to convert the single-stranded hairpin structure
into a double-stranded inverted repeat DNA. The inverted repeat is eventually digested and cloned into the respective Pol III vector to generate recombinant viral
DNA. The recently published systems are called SPEED (Luo et al. 2004) or EPRIL
(Shirane et al. 2004), and were used to generate libraries of 10^5–10^6 independent
shRNA constructs, which would allow screening of the whole genome. Compared
to the techniques described above, both systems have their advantages and limitations. Enzymatically retrieved libraries can easily cover the whole genome, but off-
target effects cannot be excluded by evaluation of the sequence. The systems are
much cheaper than DNA oligonucleotide-based libraries, which are also limited in
their target number. However, the cloning of unique shRNA-expressing DNA oligos
allows the identification in microarrays using the shRNA sequence as a bar code.
Nevertheless, both approaches will facilitate the genome wide analysis of cellular
pathway in more detail.

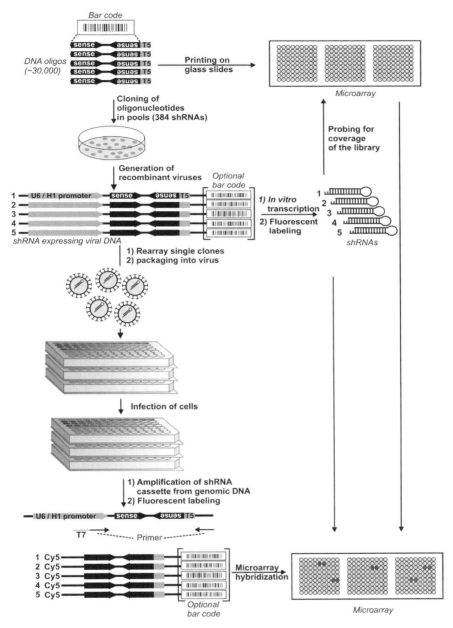

Fig. 4.61 Schematic view of the generation and screening of bar code shRNA libraries for large-scale RNAi screens in mammals. Figure modified from Berns et al. (2004) and Paddison et al. (2004). The sequence of the shRNA is usually used as a unique recognition sequence of the viral DNA (depicted here as a bar code). Optio- nally, Hannon and coworkers have added a unique bar code sequence to each construct (optional bar code). The resulting DNA clones can be transcribed in vitro using the T7 promoter on the vector, and the resulting shRNA can either be spotted on the slide or used to evaluate the coverage of the library.

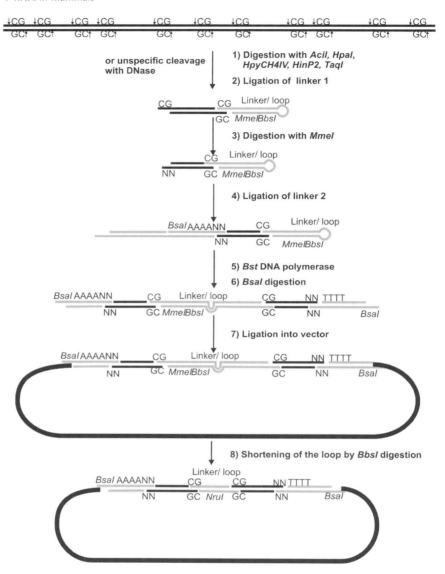

Fig. 4.62 Outline of the two cloning strategies SPEED and EPRIL. EPRIL differs from the SPEED system in that the starting cDNA material is randomly digested with DNase rather than directly digested with restriction enzymes. The flowchart illustrates the different cloning steps. Modified from Luo et al. (2004) and Shirane et al. (2004).

4.11.3
siRNA Validation for Genome-Wide Screens

There are several initiatives that are validating thousands of synthetic siRNAs for either single knock-down experiments or genome-wide approaches. Many of those validated siRNAs are combined in siRNA libraries that are commercially available at Dharmacon (si*ARRAY*™) (http://www.dharmacon.com) and Qiagen (http://www1.qiagen.com). They provide libraries for the insulin pathway, tyrosine kinases, cell cycle, SARS (Dharmacon), cancer genes, and the druggable human genome (Qiagen). The size of those libraries is rather small (up to 140 targets) except for the Qiagen "Human Druggable Genome siRNA Set" with 10 000 siRNAs for about 5000 target genes. This library comprises kinases, proteases, G-protein-coupled receptors, oncogenes and tumor suppressors, nuclear receptors, structural proteins, cell-surface receptors, ion channels, transcription factors, cytokines, cell-cycle control genes, genes involved in apoptosis and many hypothetical targets.

Those libraries are designed using the novel algorithms to ensure high efficiency and quality of the library. Compared to the genome-wide approaches using shRNA-expressing vectors, siRNAs can be perfectly designed to minimize interferon response and off-target effects, they are easy to use, and do not require laborious cloning procedures. However, they are almost unaffordable for academic scientists. The "Tyrosine Kinase Library" from Dharmacon that comprises 85 different siRNAs can be purchased as an array of a 1 nmol scale for $4760.00 (2003 price).

European Union for RNA Interference Technology

National siRNA validation initiatives and resource centers may clearly become an alternative to purchasing a library from a company. One of those initiatives is the EURIT Network (European Union for RNA Interference Technology) (http://www.eurit-network.org). EURIT aims to contribute to a comprehensive understanding of gene function and of the regulatory network systems in cells. This goal is achieved through an efficient coordination and integration of European research activities, and through enhancement of communication and the spreading of know-how and research materials between scientists. EURIT is highly engaged to reducing the cost and increasing knowledge on RNAi in mammals by providing siRNA to EURIT members at lower costs, and also by defining functional RNAi inhibitors using central high-throughput validation platform and bioinformatic tools. By sharing information about methodological progress on methods and new developments with other research groups, collaborations and exchanges are supported.

EURIT is a European network of scientific institutions, active research groups and key industrial partners. It acts as a non-profit scientific consortium of registered members, and is based on an initiative of the Max Planck Institute for Infection Biology, Berlin (MPIIB). Additional sponsorship is currently obtained from Qiagen GmbH, and the German Resource Center, RZPD GmbH.

EURIT has initiated the siRNAs collection and exchange platform, a collection of published siRNA sequences, a siRNA design service, a siRNA validation service and other projects such as high-throughput screenings as collaborations between re-

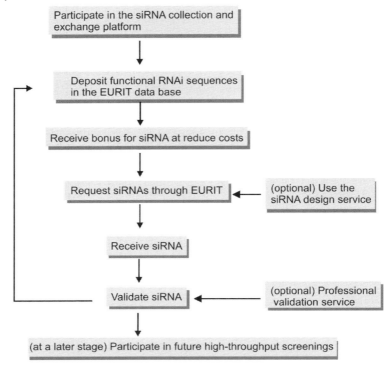

Fig. 4.63   Flowchart of the bonus program and data acquisition of the EURIT network.

search members and the EURIT platform. A library for vector-mediated RNAi (psi RNAi) and the development of alternative techniques for the generation of siRNAs is planned.

Many siRNA that have already been validated are listed on the EURIT web site. As a member, one can collect bonus points by submitting a complete validation report of novel siRNA that are not listed so far. Bonus points can later be exchanged for information on other siRNAs, or for rebates on synthetic siRNAs – thus making many experiments more affordable.

To participate in the EURIT program, go to the respective website (http:// www.eurit-network.org) and request an authorization for website entry (*eurit@mpiib-berlin.mpg.de*). Alternatively, contact the EURIT Headquarters, Elke Müller at the Max Planck Institute for Infection Biology, Berlin.

4.11.4
RNAi Microarrays

Another method that facilitates large-scale phenotypic screening in mammals is the use of siRNA microarrays. The large-scale methods described above require either multiple transfection or infection with shRNA-expressing viruses, which makes them very expensive and laborious. Thus, microarray technology offers a great advantage to speed up those screening procedures. It has already been shown that microarray technology allows analysis even in whole cells by seeding a cell monolayer onto a glass slide spotted with an array of DNA-cationic lipid complexes (Ziauddin and Sabatini 2001). These microarrays require a broader spotting, so that the cells can be transfected in situ with a single DNA construct. With regard to this technology, different microarray assays have been generated for the phenotypic screening of either multiple siRNAs or shRNAs expressing vectors (Kumar et al. 2003; Mousses et al. 2003; Silva et al. 2004). By printing nine spots of cationic lipid complexed with shRNA and reporter gene expressing vectors onto a glass slide fusing a 3 × 3 spot square to a single spot of 400–500 µm, sufficient surface is generated to allow for cell transfection (Figure 4.64).

The glass slide is placed into a 10-cm diameter culture dish and covered with 10^7 cells. After a 60-h incubation time, the microarray is phenotypically analyzed using various reporter assays, depending on the desired readout (i.e. cell cycle control, proteasome function) (Figure 4.64). This method has proved to be very cost-effective as it allows about 100–500 transfections to be performed with the same amount of material that is usually required for the transfection of a single well of a 96-well plate. Furthermore, it facilitates the simultaneous screening of several thousand shRNAs for the desired phenotype, which can be supported by the feasibility of cotransfection of reporter genes in the same assay (Kumar et al. 2003; Mousses et al. 2003). In summary, this RNAi microarray platform, together with the generation of large-scale human or mammalian siRNA and shRNA libraries, will clearly facilitate genomic-scale, cell-based analyses of gene function.

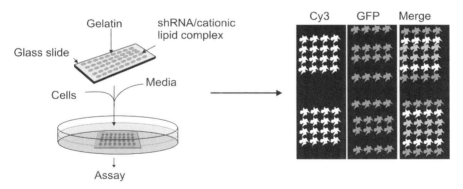

Fig. 4.64 Schematic view of cellular microarrays (modified from Ziauddin and Sabatini 2001). siRNA or shRNA-reporter gene transfection mixtures are printed on glass slides such that each spot has a diameter of 400–500 µm; this allows reasonable cell growth on the spot. The glass slide is placed into a 10-cm diameter culture dish and covered with cells and media. After incubation with the transfection mixture, the microarray is removed from the dish and analyzed for the phenotypes, using a microarray reader.

4.12
Useful Web Pages and Links

4.12.1
Academic Resources

- Advanced homology search
  http://www.paralign.org
  http://www.ncbi.nih.gov
  http://fasta.bioch.virginia.edu/fasta/home.html

- Novel algorithms for siRNA design and siRNA design programs
  http://jura.wi.mit.edu/pubint/http://iona.wi.mit.edu/siRNAext/
  http://rnaidesigner.invitrogen.com/sirna/

- Protocols
  Tuschl lab: http://www.rockefeller.edu/labheads/tuschl/sirna.html
  Bartel lab cloning protocol:
  http://web.wi.mit.edu/bartel/pub/protocols/miRNAcloning.pdf
  Hannon lab: http://www.cshl.edu/public/SCIENCE/hannon.html.
  van Parijs lab: http://web.mit.edu/ccrhq/vanparijs/
  http://screeninc.nki.nl/.
  http://www.protocol-online.org/

- Other links
  http://www.eurit-network.org

4.12.2
Company Resources

- siRNAs and large-scale libraries
  http://www.dharmacon.com
  http://www1.qiagen.com
  http://www.proligo.com
  http://www.ambion.com
  http://www.invitrogen.com
  http://screeninc.nki.nl
  http://www.oligoengine.com
  http://www.promega.com
  http://www.cenix-bioscience.com/

- Protocols
  http://www.dharmacon.com
  http://www.ambion.com

http://Screeninc.nki.nl
http://www.protocol-online.org/
siRNA tracker: http://www.mobitec.de/download/companies/siRNA_Tracker.pdf

4.12.3
Other Useful Links

- Biosafety requirements
  CDC: http://www.cdc.gov/od/ohs/biosfty/bmbl4/bmbl4toc.htm)
  Robert Koch Institut: http://www.rki.de/GENTEC/GENTEC.HTM

- Literature
  Hannon Gregorey (ed.), *RNAi: A guide to gene silencing*. Cold Spring Harbor Press, Cold Spring Harbor, USA, 2003.
  http://www.orbigen.com/RNAi_Orbigen.html: a daily updated list of interesting RNAi publications.

- Conferences
  The most exciting conferences on siRNAs, miRNAs, and RNAi are organized as Keystone meetings. Refer to the Keystone conference web page: http://www.keystonesymposia.org.

4.13
References

Aihara H, Miyazaki J (**1998**) Gene transfer into muscle by electroporation in vivo. *Nature Biotechnol* 16: 867–870.

Akusjarvi G, Svensson C, Nygard O (**1987**) A mechanism by which adenovirus virus-associated RNAI controls translation in a transient expression assay. *Mol Cell Biol* 7: 549–551.

Allikian MJ, Deckert-Cruz D, Rose MR, Landis GN, Tower J (**2002**) Doxycycline-induced expression of sense and inverted-repeat constructs modulates phosphogluconate mutase (Pgm) gene expression in adult *Drosophila melanogaster*. *Genome Biol* 3: research 0021.

Altschul SF, Gish W, Miller W, Myers EW, Lipman DJ (**1990**) Basic local alignment search tool. *J Mol Biol* 215: 403–410.

Altschul SF, Madden TL, Schaffer AA, Zhang J, Zhang Z, Miller W, Lipman DJ (**1997**) Gapped BLAST and PSI-BLAST: a new generation of protein database search programs. *Nucleic Acids Res* 25: 3389–3402.

Amarzguioui M, Holen T, Babaie E, Prydz H (**2003**) Tolerance for mutations and chemical modifications in a siRNA. *Nucleic Acids Res* 31: 589–595.

Arendt CW, Tang G, Zilberstein A (**2003**) Vector systems for the delivery of small interfering RNAs: Managing the RISC. *Chembiochem* 4: 1129–1136.

Arenz C, Schepers U (**2003**) RNA interference: from an ancient mechanism to a state of the art therapeutic application? *Naturwissenschaften* 90: 345–359.

Ashfield R, Patel AJ, Bossone SA, Brown H, Campbell RD, Marcu KB, Proudfoot NJ (**1994**) MAZ-dependent termination between closely spaced human complement genes. *EMBO J* 13: 5656–5667.

Bass BL (**2002**) RNA editing by adenosine deaminases that act on RNA. *Annu Rev Biochem* 71: 817–846.

BASS BL, WEINTRAUB H (**1988**) An unwinding activity that covalently modifies its double-stranded RNA substrate. *Cell* 55: 1089–1098

BEAUCAGE SL, CARUTHERS MH (**1996**) Bio-organic Chemistry: Nucleic Acids. Oxford University Press, Oxford.

BERNATOWICZ MS, MATSUEDA R, MATSUEDA GR (**1986**) Preparation of Boc-[S-(3-nitro-2-pyridi-nesulfenyl)]-cysteine and its use for unsymmetrical disulfide bond formation. *Int J Pept Protein Res* 28: 107–112.

BERNS K, HIJMANS EM, MULLENDERS J, BRUMMELKAMP TR, VELDS A, HEIMERIKX M, KERKHOVEN RM, MADIREDJO M, NIJKAMP W, WEIGELT B, AGAMI R, GE W, CAVET G, LINSLEY PS, BEIJERSBERGEN RL, BERNARDS R (**2004**) A large-scale RNAi screen in human cells identifies new components of the p53 pathway. *Nature* 428: 431–437.

BILLY E, BRONDANI V, ZHANG H, MULLER U, FILIPOWICZ W (**2001**) Specific interference with gene expression induced by long, double-stranded RNA in mouse embryonal teratocarcinoma cell lines. *Proc Natl Acad Sci USA* 98: 14428–14433.

BIRNBOIM HC, DOLY J (**1979**) A rapid alkaline extraction procedure for screening recombinant plasmid DNA. *Nucleic Acids Res* 7: 1513–1523.

BOCKAMP E, MARINGER M, SPANGENBERG C, FEES S, FRASER S, ESHKIND L, OESCH F, ZABEL B (**2002**) Of mice and models: improved animal models for biomedical research. *Physiol Genomics* 11: 115–132.

BODEN D, PUSCH O, LEE F, TUCKER L, SHANK PR, RAMRATNAM B (**2003**) Promoter choice affects the potency of HIV-1 specific RNA interference. *Nucleic Acids Res* 31: 5033–5038.

BOUTLA A, DELIDAKIS C, LIVADARAS I, TSAGRIS M, TABLER M (**2001**) Short 5′-phosphorylated double-stranded RNAs induce RNA interference in *Drosophila*. *Curr Biol* 11: 1776–1780.

BRADFORD MM (**1976**) A rapid and sensitive method for the quantitation of microgram quantities of protein utilizing the principle of protein-dye binding. *Anal Biochem* 72: 248–524.

BRENNER SE, CHOTHIA C, HUBBARD TJ (**1998**) Assessing sequence comparison methods with reliable structurally identified distant evolutionary relationships. *Proc Natl Acad Sci USA* 95: 6073–6078.

BRIDGE AJ, PEBERNARD S, DUCRAUX A, NICOULAZ AL, IGGO R (**2003**) Induction of an interferon response by RNAi vectors in mammalian cells. *Nature Genet* 34: 263–264.

BRUGIDOU J, LEGRAND C, MERY J, RABIE A (**1995**) The retro-inverso form of a homeobox-derived short peptide is rapidly internalised by cultured neurones: a new basis for an efficient intracellular delivery system. *Biochem Biophys Res Commun* 214: 685–693.

BRUMMELKAMP TR, BERNARDS R, AGAMI R (**2002**) A system for stable expression of short interfering RNAs in mammalian cells. *Science* 296: 550–553.

BUCHSCHACHER GL, JR., WONG-STAAL F (**2000**) Development of lentiviral vectors for gene therapy for human diseases. *Blood* 95: 2499–2504.

BZYMEK M, LOVETT ST (**2001**) Evidence for two mechanisms of palindrome-stimulated deletion in *Escherichia coli*: single-strand annealing and replication slipped mispairing. *Genetics* 158: 527–540.

CALEGARI F, HAUBENSAK W, YANG D, HUTTNER WB, BUCHHOLZ F (**2002**) Tissue-specific RNA interference in postimplantation mouse embryos with endoribonuclease-prepared short interfering RNA. *Proc Natl Acad Sci USA* 99: 14236–14240.

CAPLEN NJ, PARRISH S, IMANI F, FIRE A, MORGAN RA (**2001**) Specific inhibition of gene expression by small double-stranded RNAs in invertebrate and vertebrate systems. *Proc Natl Acad Sci USA* 98: 9742–9747.

CHIU YL, RANA TM (**2003**) siRNA function in RNAi: A chemical modification analysis. *Rn.a – A Publication of the Rn.a Society* 9: 1034–1048.

CHRISTIAENS B, SYMOENS S, VERHEYDEN S, ENGELBORGHS Y, JOLIOT A, PROCHIANTZ A, VANDEKERCKHOVE J, ROSSENEU M, VANLOO B, VANDERHEYDEN S (**2002**) Tryptophan fluorescence study of the interaction of penetratin peptides with model membranes. *Eur J Biochem* 269: 2918–2926.

CHUN HJ, ZHENG L, AHMAD M, WANG J, SPEIRS CK, SIEGEL RM, DALE JK, PUCK J, DAVIS J, HALL CG, SKODA-SMITH S, ATKINSON TP, STRAUS SE, LENARDO MJ (**2002**) Pleiotropic defects in lymphocyte activation caused by caspase-8 mutations lead to human immunodeficiency. *Nature* 419: 395–399.

CLARKE PA, MATHEWS MB (**1995**) Interactions between the double-stranded RNA binding motif and RNA: definition of the binding site for the interferon-induced protein kinase DAI (PKR) on adenovirus VA RNA. *Rn.a* 1: 7–20.

CLEMENS MJ (**1997**) Pkr – a protein kinase regulated by double-stranded RNA. *Int J Biochem Cell Biol* 29: 945–949.

CLEMENS MJ, ELIA A (**1997**) The double-stranded RNA-dependent protein kinase PKR – structure and function [Review]. *J Interferon Cytokine Res* 17: 503–524.

CORBEAU P, KRAUS G, WONG-STAAL F (**1996**) Efficient gene transfer by a human immuno-deficiency virus type 1 (HIV-1)-derived vector utilizing a stable HIV packaging cell line. *Proc Natl Acad Sci USA* 93: 14070–14075.

CZAUDERNA F, SANTEL A, HINZ M, FECHTNER M, DURIEUX B, FISCH G, LEENDERS F, ARNOLD W, GIESE K, KLIPPEL A, KAUFMANN J (**2003**) Inducible shRNA expression for application in a prostate cancer mouse model. *Nucleic Acids Res* 31: e127.

DAHL BJ, BJERGARDE K, HENRIKSEN L, DAHL O (**1990**) *Acta Chem Scand* 44: 639–641.

DAS AT, BRUMMELKAMP TR, WESTERHOUT EM, VINK M, MADIREDJO M, BERNARDS R, BERK-HOUT B (**2004**) Human immunodeficiency virus type 1 escapes from RNA interference-mediated inhibition. *J Virol* 78: 2601–2605.

DAVISON A, LEACH DR (**1994**) The effects of nucleotide sequence changes on DNA second-ary structure formation in Escherichia coli are consistent with cruciform extrusion in vivo. *Genetics* 137: 361–368.

DEROSSI D, CHASSAING G, PROCHIANTZ A (**1998**) Trojan peptides – the penetratin system for intracellular delivery. *Trends Cell Biol* 8: 84–87.

DEROSSI D, JOLIOT AH, CHASSAING G, PROCHIANTZ A (**1994**) The third helix of the Antennapedia homeodomain translocates through biological membranes. *J Biol Chem* 269: 10444–10450.

DIALLO M, ARENZ C, SCHMITZ K, SANDHOFF K, SCHEPERS U (**2003a**) Long endogenous dsRNAs can induce a complete gene silencing in mammalian cells and primary cultures. *Oligonucleotides* 13: 381–392.

DIALLO M, ARENZ C, SCHMITZ K, SANDHOFF K, SCHEPERS U (**2003b**) RNA Interference: A new method to analyze the function of glyco-proteins and glycosylating proteins in mam-malian cells. Knockout experiments with UDP-glucose/ceramide-glucosyltransferase. *Methods Enzymol* 363: 173–190.

DULL T, ZUFFEREY R, KELLY M, MANDEL RJ, NGUYEN M, TRONO D, NALDINI L (**1998**) A third-generation lentivirus vector with a conditional packaging system. *J Virol* 72: 8463–8471.

DUNNE J, DRESCHER B, RIEHLE H, HADWIGER P, YOUNG BD, KRAUTER J, HEIDENREICH O (**2003**) The apparent uptake of fluorescently labeled siRNAs by electroporated cells de-pends on the fluorochrome. *Oligonucleotides* 13: 375–380.

EASTMAN SJ, BASKIN KM, HODGES BL, CHU Q, GATES A, DREUSICKE R, ANDERSON S, SCHEULE RK (**2002**) Development of catheter-based procedures for transducing the isolated rabbit liver with plasmid DNA. *Hum Gene Ther* 13: 2065–2077.

ELBASHIR SM, HARBORTH J, LENDECKEL W, YALCIN A, WEBER K, TUSCHL T (**2001a**) Duplexes of 21-nucleotide RNAs mediate RNA interference in cultured mammalian cells. *Nature* 411: 494–498.

ELBASHIR SM, LENDECKEL W, TUSCHL T (**2001b**) RNA interference is mediated by 21- and 22-nucleotide RNAs. *Genes Dev* 15: 188–200.

ELBASHIR SM, MARTINEZ J, PATKANIOWSKA A, LENDECKEL W, TUSCHL T (**2001c**) Functional anatomy of siRNAs for mediating efficient RNAi in *Drosophila melanogaster* embryo lysate. *EMBO J* 20: 6877–6888.

ELLIOTT G, OHARE P (**1997**) Intercellular traf-ficking and protein delivery by a herpesvirus structural protein. *Cell* 88: 223–233.

FISCHER PM, KRAUSZ E, LANE DP (**2001**) Cellu-lar delivery of impermeable effector molecules in the form of conjugates with peptides cap-able of mediating membrane translocation. *Bioconjug Chem* 12: 825–841.

FRITSCH L, MARTINEZ LA, SEKHRI R, NAGUIBNEVA I, GERARD M, VANDROMME M, SCHAEFFER L, HAREL-BELLAN A (**2004**) Conditional gene knock-down by CRE-depen-dent short interfering RNAs. *EMBO Rep* 5: 178–182.

GAUR RK, PALIWAL S, SHARMA P, GUPTA KC (**1989**) A simple and sensitive spectrophoto-metric method for the quantitative determina-tion of solid supported amino groups. *J Bio-chem Biophys Methods* 18: 323–329.

GAUR RK, SHARMA P., GUPTA KC (**1989**) A simple method of thiol group at 5′-termini of oligo-deoxynucleotides. *Nucleic Acids Res* 17: 4404.

GE Q, MCMANUS MT, NGUYEN T, SHEN CH, SHARP PA, EISEN HN, CHEN J (**2003**) RNA interference of influenza virus production by directly targeting mRNA for degradation and indirectly inhibiting all viral RNA transcription. *Proc Natl Acad Sci USA* 100: 2718–2723.

GILADI H, KETZINEL-GILAD M, RIVKIN L, FELIG Y, NUSSBAUM O, GALUN E (**2003**) Small interfering RNA inhibits hepatitis B virus replication in mice. Mol Ther 8: 769–776.

GITLIN L, KARELSKY S, ANDINO R (**2002**) Short interfering RNA confers intracellular antiviral immunity in human cells. *Nature* 418: 430–434.

GOH KC, HAQUE SJ, WILLIAMS BR (**1999**) p38 MAP kinase is required for STAT1 serine phosphorylation and transcriptional activation induced by interferons. *EMBO J* 18: 5601–5608.

GOTO T, NISHI T, TAMURA T, DEV SB, TAKESHIMA H, KOCHI M, YOSHIZATO K, KURATSU J, SAKATA T, HOFMANN GA, USHIO Y (**2000**) Highly efficient electro-gene therapy of solid tumor by using an expression plasmid for the herpes simplex virus thymidine kinase gene. *Proc Natl Acad Sci USA* 97: 354–359.

GRATTON JP, YU J, GRIFFITH JW, BABBITT RW, SCOTLAND RS, HICKEY R, GIORDANO FJ, SESSA WC (**2003**) Cell-permeable peptides improve cellular uptake and therapeutic gene delivery of replication-deficient viruses in cells and in vivo. *Nature Med* 9: 357–362.

GRIFFITHS JC, HARRIS SJ, LAYTON GT, BERRIE EL, FRENCH TJ, BURNS NR, ADAMS SE, KINGSMAN AJ (**1993**) Hybrid human immunodeficiency virus Gag particles as an antigen carrier system: induction of cytotoxic T-cell and humoral responses by a Gag:V3 fusion. *J Virol* 67: 3191–3198.

GUNNERY S, MA Y, MATHEWS MB (**1999**) Termination sequence requirements vary among genes transcribed by RNA polymerase III. *J Mol Biol* 286: 745–757.

HAMASAKI K, NAKAO K, MATSUMOTO K, ICHIKAWA T, ISHIKAWA H, EGUCHI K (**2003**) Short interfering RNA-directed inhibition of hepatitis B virus replication. *FEBS Lett* 543: 51–54.

HAMILTON DL, ABREMSKI K (**1984**) Site-specific recombination by the bacteriophage P1 lox-Cre system. Cre-mediated synapsis of two lox sites. *J Mol Biol* 178: 481–486.

HANAHAN D (**1983**) Studies on transformation of *Escherichia coli* with plasmids. *J Mol Biol* 166: 557–580.

HAQUE SJ, WILLIAMS BR (**1998**) Signal transduction in the interferon system. *Semin Oncol* 25: 14–22.

HARBORTH J, ELBASHIR SM, BECHERT K, TUSCHL T, WEBER K (**2001**) Identification of essential genes in cultured mammalian cells using small interfering RNAs. *J Cell Sci* 114: 4557–4565.

HEIDENREICH O, KRAUTER J, RIEHLE H, HADWIGER P, JOHN M, HEIL G, VORNLOCHER HP, NORDHEIM A (**2003**) AML1/MTG8 oncogene suppression by small interfering RNAs supports myeloid differentiation of t(8;21)-positive leukemic cells. *Blood* 101: 3157–3163.

HELLER L, JAROSZESKI MJ, COPPOLA D, POTTINGER C, GILBERT R, HELLER R (**2000**) Electrically mediated plasmid DNA delivery to hepatocellular carcinomas in vivo. *Gene Ther* 7: 826–829.

HEMANN MT, FRIDMAN JS, ZILFOU JT, HERNANDO E, PADDISON PJ, CORDON-CARDO C, HANNON GJ, LOWE SW (**2003**) An epi-allelic series of p53 hypomorphs created by stable RNAi produces distinct tumor phenotypes in vivo. *Nature Genet* 33: 396–400.

HILLEMAN MR (**2003**) Critical overview and outlook: pathogenesis, prevention, and treatment of hepatitis and hepatocarcinoma caused by hepatitis B virus. *Vaccine* 21: 4626–4649.

HILLEN W, BERENS C (**1994**) Mechanisms underlying expression of Tn10 encoded tetracycline resistance. *Annu Rev Microbiol* 48: 345–369.

HOESS RH, ZIESE M, STERNBERG N (**1982**) P1 site-specific recombination: nucleotide sequence of the recombining sites. *Proc Natl Acad Sci USA* 79: 3398–3402.

HOGAN B, BEDDIGTON R, COSTANTINI F, LACY E (**1994**) *Manipulating the mouse embryo*, 2nd edn. Cold Spring Harbor Press, Cold Spring Harbor.

HOLEN T, AMARZGUIOUI M, WIIGER MT, BABAIE E, PRYDZ H (**2002**) Positional effects of short interfering RNAs targeting the human coagulation trigger Tissue Factor. *Nucleic Acids Res* 30: 1757–1766.

HOUK J, SINGH R, WHITESIDES GM (**1987**) Measurement of thiol-disulfide interchange reactions and thiol pKa values. *Methods Enzymol* 143: 129–140.

HUANG Y, CARMICHAEL GC (**1996**) Role of polyadenylation in nucleocytoplasmic transport of mRNA. *Mol Cell Biol* 16: 1534–1542.

HUBBARD TJ, AILEY B, BRENNER SE, MURZIN AG, CHOTHIA C (**1998**) SCOP, Structural Classification of Proteins database: applications to evaluation of the effectiveness of sequence alignment methods and statistics of protein structural data. *Acta Crystallogr D Biol Crystallogr* 54: 1147–1154.

ILVES H, BARSKE C, JUNKER U, BOHNLEIN E, VERES G (**1996**) Retroviral vectors designed for targeted expression of RNA polymerase III-driven transcripts: a comparative study. *Gene* 171: 203–208.

JACKSON AL, BARTZ SR, SCHELTER J, KOBAYASHI SV, BURCHARD J, MAO M, LI B, CAVET G, LINSLEY PS (**2003**) Expression profiling reveals off-target gene regulation by RNAi. *Nature Biotechnol* 21: 635–637.

JACQUE JM, TRIQUES K, STEVENSON M (**2002**) Modulation of HIV-1 replication by RNA interference. *Nature* 418: 435–438.

JANAKI C, JOSHI RR (**2003**) Accelerating comparative genomics using parallel computing. *In Silico Biol* 3: 429–440.

JENNINGS PA, MOLLOY PL (**1987**) Inhibition of SV40 replicon function by engineered antisense RNA transcribed by RNA polymerase III. *EMBO J* 6: 3043–3047.

JI J, WERNLI M, KLIMKAIT T, ERB P (**2003**) Enhanced gene silencing by the application of multiple specific small interfering RNAs. *FEBS Lett* 552: 247–252.

JIANG M, MILNER J (**2002**) Selective silencing of viral gene expression in HPV-positive human cervical carcinoma cells treated with siRNA, a primer of RNA interference. *Oncogene* 21: 6041–6048.

KASIM V, MIYAGISHI M, TAIRA K (**2003**) Control of siRNA expression utilizing Cre-loxP recombination system. *Nucleic Acids Res Suppl*: 255–256.

KATO Y, TAIRA K (**2003**) Expression of siRNA from a single transcript that includes multiple ribozymes in mammalian cells. *Oligonucleotides* 13: 335–343.

KAWAGISHI-KOBAYASHI M, CAO C, LU J, OZATO K, DEVER TE (**2000**) Pseudosubstrate inhibition of protein kinase PKR by swine pox virus C8L gene product. *Virology* 276: 424–434.

KAWASAKI H, SUYAMA E, IYO M, TAIRA K (**2003**) siRNAs generated by recombinant human Dicer induce specific and significant but target site-independent gene silencing in human cells. *Nucleic Acids Res* 31: 981–987.

KAWASAKI H, TAIRA K (**2003**) Short hairpin type of dsRNAs that are controlled by tRNA(Val) promoter significantly induce RNAi-mediated gene silencing in the cytoplasm of human cells. *Nucleic Acids Res* 31: 700–707.

KHVOROVA A, REYNOLDS A, JAYASENA SD (**2003**) Functional siRNAs and miRNAs exhibit strand bias. *Cell* 115: 209–216.

KIM DH, LONGO M, HAN Y, LUNDBERG P, CANTIN E, ROSSI JJ (**2004**) Interferon induction by siRNAs and ssRNAs synthesized by phage polymerase. *Nature Biotechnol* 22: 321–325.

KIMURA T, OHYAMA A (**1994**) Interaction with the Rev response element along an extended stem I duplex structure is required to complete human immunodeficiency virus type 1 rev-mediated trans-activation in vivo. *J Biochem (Tokyo)* 115: 945–952.

KISHIDA T, ASADA H, GOJO S, OHASHI S, SHIN-YA M, YASUTOMI K, TERAUCHI R, TAKAHASHI KA, KUBO T, IMANISHI J, MAZDA O (**2004**) Sequence-specific gene silencing in murine muscle induced by electroporation-mediated transfer of short interfering RNA. *J Gene Med* 6: 105–110.

KJEMS J, FRANKEL AD, SHARP PA (**1991**) Specific regulation of mRNA splicing in vitro by a peptide from HIV-1 Rev. *Cell* 67: 169–178.

KNIGHT SW, BASS BL (**2002**) The role of RNA editing by ADARs in RNAi. *Mol Cell* 10: 809–817.

KRAUSS M, KINUTA M, WENK MR, DE CAMILLI P, TAKEI K, HAUCKE V (**2003**) ARF6 stimulates clathrin/AP-2 recruitment to synaptic membranes by activating phosphatidylinositol phosphate kinase type I{gamma}. *J Cell Biol* 162: 113–124.

KRICHEVSKY AM, KOSIK KS (**2002**) RNAi functions in cultured mammalian neurons. *Proc Natl Acad Sci USA* 99: 11926–11929.

KUMAR M, CARMICHAEL GG (**1997**) Nuclear antisense RNA induces extensive adenosine modifications and nuclear retention of target transcripts. *Proc Natl Acad Sci USA* 94: 3542–3547.

KUMAR R, CONKLIN DS, MITTAL V (**2003**) High-throughput selection of effective RNAi probes for gene silencing. *Genome Res* 13: 2333–2340.

LAKSO M, SAUER B, MOSINGER B, JR, LEE EJ, MANNING RW, YU SH, MULDER KL, WESTPHAL H (**1992**) Targeted oncogene activation by site-specific recombination in transgenic mice. *Proc Natl Acad Sci USA* 89: 6232–6236.

LAWRENCE D (**2002**) RNAi could hold promise in the treatment of HIV. *Lancet* 359: 2007.

LEACH DR (**1994**) Long DNA palindromes, cruciform structures, genetic instability and secondary structure repair. *BioEssays* 16: 893–900.

LEE NS, DOHJIMA T, BAUER G, LI H, LI MJ, EHSANI A, SALVATERRA P, ROSSI J (**2002**) Expression of small interfering RNAs targeted against HIV-1 rev transcripts in human cells. *Nature Biotechnol* 20: 500–505.

LEWIS DL, HAGSTROM JE, LOOMIS AG, WOLFF JA, HERWEIJER H (**2002**) Efficient delivery of siRNA for inhibition of gene expression in postnatal mice. *Nature Genet* 32: 107–108.

LEWIS J, IZAURRALDE E (**1997**) The role of the cap structure in RNA processing and nuclear export. *Eur J Biochem* 247: 461–469.

LEWIS JD, TOLLERVEY D (**2000**) Like attracts like: getting RNA processing together in the nucleus. *Science* 288: 1385–1389.

LI HL, CHELLADURAI BS, ZHANG K, NICHOLSON AW (**1993**) Ribonuclease III cleavage of a bacteriophage T7 processing signal. Divalent cation specificity, and specific anion effects. *Nucleic Acids Res* 21: 1919–1925.

LINDENBACH BD, RICE CM (**2002**) RNAi targeting an animal virus: news from the front. *Mol Cell* 9: 925–927.

LIU F, SONG Y, LIU D (**1999**) Hydrodynamics-based transfection in animals by systemic administration of plasmid DNA. *Gene Ther* 6: 1258–1266.

LOIS C, HONG EJ, PEASE S, BROWN EJ, BALTIMORE D (**2002**) Germline transmission and tissue-specific expression of transgenes delivered by lentiviral vectors. *Science* 295: 868–872.

LUO B, HEARD AD, LODISH HF (**2004**) Small interfering RNA production by enzymatic engineering of DNA (SPEED). *Proc Natl Acad Sci USA* 101: 5494–5499.

LUO MJ, REED R (**1999**) Splicing is required for rapid and efficient mRNA export in metazoans. *Proc Natl Acad Sci USA* 96: 14937–14942.

MALIM MH, BOHNLEIN S, FENRICK R, LE SY, MAIZEL JV, CULLEN BR (**1989**a) Functional comparison of the Rev trans-activators encoded by different primate immunodeficiency virus species. *Proc Natl Acad Sci USA* 86: 8222–8226.

MALIM MH, FENRICK R, BALLARD DW, HAUBER J, BOHNLEIN E, CULLEN BR (**1989**b) Functional characterization of a complex protein-DNA-binding domain located within the human immunodeficiency virus type 1 long terminal repeat leader region. *J Virol* 63: 3213–3219.

MALIM MH, HAUBER J, LE SY, MAIZEL JV, CULLEN BR (**1989**c) The HIV-1 rev trans-activator acts through a structured target sequence to activate nuclear export of unspliced viral mRNA. *Nature* 338: 254–257.

MATSUKURA S, JONES PA, TAKAI D (**2003**) Establishment of conditional vectors for hairpin siRNA knockdowns. *Nucleic Acids Res* 31: e77.

MATTEUCCI MD, CARUTHERS MH (**1981**) Synthesis of desoxyoligonucleotides on a polymeric support. *J Am Chem Soc* 103: 3185–3191.

MATTEUCCI MD, CARUTHERS MH (**1992**) Synthesis of deoxyoligonucleotides on a polymer support. 1981. *Biotechnology* 24: 92–98.

McCAFFREY AP, MEUSE L, PHAM TTT, CONKLIN DS, HANNON GJ, KAY MA (**2002**) Gene expression – RNA interference in adult mice. *Nature* 418: 38–39.

McCAFFREY AP, NAKAI H, PANDEY K, HUANG Z, SALAZAR FH, XU H, WIELAND SF, MARION PL, KAY MA (**2003**) Inhibition of hepatitis B virus in mice by RNA interference. *Nature Biotechnol* 21: 639–644.

McKENDRICK L, THOMPSON E, FERREIRA J, MORLEY SJ, LEWIS JD (**2001**) Interaction of eukaryotic translation initiation factor 4G with the nuclear cap-binding complex provides a link between nuclear and cytoplasmic functions of the m(7) guanosine cap. *Mol Cell Biol* 21: 3632–3641.

MILLIGAN JF, GROEBE DR, WITHERELL GW, UHLENBECK OC (**1987**) Oligoribonucleotide synthesis using T7 RNA polymerase and synthetic DNA templates. *Nucleic Acids Res* 15: 8783–8798.

MINKS MA, BENVIN S, MARONEY PA, BAGLIONI C (**1979**) Synthesis of 2'5'-oligo(A) in extracts of interferon-treated HeLa cells. *J Biol Chem* 254: 5058–5064.

MIYAGISHI M, TAIRA K (**2002**a) Development and application of siRNA expression vector. *Nucleic Acids Res Suppl*: 113–114.

MIYAGISHI M, TAIRA K (**2002**b) U6 promoter driven siRNAs with four uridine 3' overhangs efficiently suppress targeted gene expression in mammalian cells. *Nature Biotechnol* 20: 497–500.

MORSE DP, ARUSCAVAGE PJ, BASS BL (**2002**) RNA hairpins in noncoding regions of human brain and *Caenorhabditis elegans* mRNA are edited by adenosine deaminases that act on RNA. *Proc Natl Acad Sci USA* 99: 7906–7911.

MOUSSES S, CAPLEN NJ, CORNELISON R, WEAVER D, BASIK M, HAUTANIEMI S, ELKAHLOUN AG, LOTUFO RA, CHOUDARY A, DOUGHERTY ER, SUH E, KALLIONIEMI O (**2003**) RNAi microarray analysis in cultured mammalian cells. *Genome Res* 13: 2341–2347.

MYERS JW, JONES JT, MEYER T, FERRELL JE, JR (**2003**) Recombinant Dicer efficiently converts large dsRNAs into siRNAs suitable for gene silencing. *Nature Biotechnol* 21: 324–328.

MYSLINSKI E, AME JC, KROL A, CARBON P (**2001**) An unusually compact external promoter for RNA polymerase III transcription of the human H1RNA gene. *Nucleic Acids Res* 29: 2502–2509.

NAGAHARA H, VOCERO-AKBANI AM, SNYDER EL, HO A, LATHAM DG, LISSY NA, BECKER-HAPAK M, EZHEVSKY SA, DOWDY SF (**1998**) Transduction of full-length TAT fusion proteins into mammalian cells: TAT-p27Kip1 induces cell migration. *Nature Med* 4: 1449–1452.

NALDINI L (**1998**) Lentiviruses as gene transfer agents for delivery to non-dividing cells. *Curr Opin Biotechnol* 9: 457–463.

NALDINI L (**1999**) In vivo gene delivery by lentiviral vectors. *Thromb Haemost* 82: 552–554.

NALDINI L, BLOMER U, GALLAY P, ORY D, MULLIGAN R, GAGE FH, VERMA IM, TRONO D (**1996**) In vivo gene delivery and stable transduction of nondividing cells by a lentiviral vector. *Science* 272: 263–267.

NALDINI L, VERMA IM (**2000**) Lentiviral vectors. *Adv Virus Res* 55: 599–609.

NGO TT (**1986**) A simple spectrophotometric determination of solid supported amino groups. *J Biochem Biophys Methods* 12: 349–354.

NICHOLSON RH, NICHOLSON AW (**2002**) Molecular characterization of a mouse cDNA encoding Dicer, a ribonuclease III ortholog involved in RNA interference. *Mamm Genome* 13: 67–73.

NISHIKURA K, YOO C, KIM U, MURRAY JM, ESTES PA, CASH FE, LIEBHABER SA (**1991**) Substrate specificity of the dsRNA unwinding/modifying activity. *EMBO J* 10: 3523–3532.

OGILVIE KK, SADANA KL, THOMPSON EA, QUILLIAM JB, WESTMORE JB (**1974**) *Tetrahedron Lett* 15: 2861.

OHASHI S, KUBO T, KISHIDA T, IKEDA T, TAKAHASHI K, ARAI Y, TERAUCHI R, ASADA H, IMANISHI J, MAZDA O (**2002**) Successful genetic transduction in vivo into synovium by means of electroporation. *Biochem Biophys Res Commun* 293: 1530–1535.

OHKAWA J, TAIRA K (**2000**) Control of the functional activity of an antisense RNA by a tetracycline-responsive derivative of the human U6 snRNA promoter. *Hum Gene Ther* 11: 577–585.

OHTSUKA E, TANAKA S, IKEHARA M (**1974**) Studies on transfer ribonucleic acids and related compounds. IX. Ribooligonucleotide synthesis using a photosensitive o-nitrobenzyl protection at the 2′-hydroxyl group. *Nucleic Acids Res* 1: 1351–1357.

OJWANG JO, HAMPEL A, LOONEY DJ, WONG-STAAL F, RAPPAPORT J (**1992**) Inhibition of human immunodeficiency virus type 1 expression by a hairpin ribozyme. *Proc Natl Acad Sci USA* 89: 10802–10806.

OLIVEIRA DM, GOODELL MA (**2003**) Transient RNA interference in hematopoietic progenitors with functional consequences. *Genesis* 36: 203–208.

OSHIMA Y, SAKAMOTO T, YAMANAKA I, NISHI T, ISHIBASHI T, INOMATA H (**1998**) Targeted gene transfer to corneal endothelium in vivo by electric pulse. *Gene Ther* 5: 1347–1354.

PADDISON PJ, CAUDY AA, BERNSTEIN E, HANNON GJ, CONKLIN DS (**2002**a) Short hairpin RNAs (shRNAs) induce sequence-specific silencing in mammalian cells. *Genes Dev* 16: 948–958.

PADDISON PJ, CAUDY AA, HANNON GJ (**2002**b) Stable suppression of gene expression by RNAi in mammalian cells. *Proc Natl Acad Sci USA* 99: 1443–1448.

PADDISON PJ, HANNON GJ (**2003**) siRNAs and shRNAs: skeleton keys to the human genome. *Curr Opin Mol Ther* 5: 217–224.

PADDISON PJ, SILVA JM, CONKLIN DS, SCHLABACH M, LI M, ARULEBA S, BALIJA V, O'SHAUGHNESSY A, GNOJ L, SCOBIE K, CHANG K, WESTBROOK T, CLEARY M, SACHIDANANDAM R, MCCOMBIE WR, ELLEDGE SJ, HANNON GJ (**2004**) A resource for large-scale RNA-interference-based screens in mammals. *Nature* 428: 427–431.

PADDISON PJC, A.; HANNON, G. J. (**2002**) Stable suppression of gene expression by RNAi in mammalian cells. *Proc Natl Acad Sci USA* 99: 1443–1448.

PARRISH S, FLEENOR J, XU SQ, MELLO C, FIRE A (**2000**) Functional anatomy of a dsRNA trigger: Differential requirement for the two trigger strands in RNA interference. *Mol Cell* 6: 1077–1087.

PATTERSON JB, SAMUEL CE (**1995**) Expression and regulation by interferon of a double-stranded-RNA-specific adenosine deaminase from human cells: evidence for two forms of the deaminase. *Mol Cell Biol* 15: 5376–5388.

PAUL CP, GOOD PD, WINER I, ENGELKE DR (**2002**) Effective expression of small interfering RNA in human cells. *Nature Biotechnol* 20: 505–508.

PAULE MR, WHITE RJ (**2000**) Survey and summary: transcription by RNA polymerases I and III. *Nucleic Acids Res* 28: 1283–1298.

PEARSON WR (**1991**) Searching protein sequence libraries: comparison of the sensitivity and selectivity of the Smith-Waterman and FASTA algorithms. *Genomics* 11: 635–650.

PEARSON WR, LIPMAN DJ (**1988**) Improved tools for biological sequence comparison. *Proc Natl Acad Sci USA* 85: 2444–2448.

PEITZ M, PFANNKUCHE K, RAJEWSKY K, EDENHOFER F (**2002**) Ability of the hydrophobic FGF and basic TAT peptides to promote cellular uptake of recombinant Cre recombinase: a tool for efficient genetic engineering of mammalian genomes. *Proc Natl Acad Sci USA* 99: 4489–4494.

PFEIFER A, IKAWA M, DAYN Y, VERMA IM (**2002**) Transgenesis by lentiviral vectors: lack of gene silencing in mammalian embryonic stem cells and preimplantation embryos. *Proc Natl Acad Sci USA* 99: 2140–2145.

PITSCH S, WEISS PA, JENNY L, STUTZ A, WU XL (**2001**) Reliable chemical synthesis of oligoribonucleotides (RNA) with 2′-O-[(triisopropylsilyl)oxy]methyl(2 ,-O-tom)-protected phosphoramidites. *Helv Chim Acta* 84: 3773–3795.

PITSCH S, WEISS PA, XU XL, ACKERMANN D, HONEGGER T (**1999**) Fast and reliable automated synthesis of RNA and partially 2′-O-protected precursors (‚caged RNA') based on two novel, orthogonal 2′-O-protecting groups. *Helv Chim Acta* 82: 1753–1761.

POLSON AG, BASS BL (**1994**) Preferential selection of adenosines for modification by double-stranded RNA adenosine deaminase. *EMBO J* 13: 5701–5711.

POOGA M, KUT C, KIHLMARK M, HALLBRINK M, FERNAEUS S, RAID R, LAND T, HALLBERG E, BARTFAI T, LANGEL U (**2001**) Cellular translocation of proteins by transportan. *FASEB J* 15: 1451–1453.

POOGGIN M, SHIVAPRASAD PV, VELUTHAMBI K, HOHN T (**2003**) RNAi targeting of DNA virus in plants. *Nature Biotechnol* 21: 131–132.

PROCHIANTZ A (**1996**) Getting hydrophilic compounds into cells – lessons from homeopeptides – Commentary. *Curr Opin Neurobiol* 6: 629–634.

PROVOST P, DISHART D, DOUCET J, FRENDEWEY D, SAMUELSSON B, RADMARK O (**2002**) Ribonuclease activity and RNA binding of recombinant human Dicer. *EMBO J* 21: 5864–5874.

PUSCH O, BODEN D, SILBERMANN R, LEE F, TUCKER L, RAMRATNAM B (**2003**) Nucleotide sequence homology requirements of HIV-1-specific short hairpin RNA. *Nucleic Acids Res* 31: 6444–6449.

RAJEWSKY K, GU H, KUHN R, BETZ UA, MULLER W, ROES J, SCHWENK F (**1996**) Conditional gene targeting. *J Clin Invest* 98: 600–603.

RANDALL G, GRAKOUI A, RICE CM (**2003**) Clearance of replicating hepatitis C virus replicon RNAs in cell culture by small interfering RNAs. *Proc Natl Acad Sci USA* 100: 235–240.

REDDY MP, HANNA NB, FAROOQUI F (**1994**) Fast cleavage and deprotection of oligonucleotides. *Tetrahedron Lett* 35: 4311–4314.

REYNOLDS A, LEAKE D, BOESE Q, SCARINGE S, MARSHALL WS, KHVOROVA A (**2004**) Rational siRNA design for RNA interference. *Nature Biotechnol* 00: 00–00.

RICHARD JP, MELIKOV K, VIVES E, RAMOS C, VERBEURE B, GAIT MJ, CHERNOMORDIK LV, LEBLEU B (**2003**) Cell-penetrating peptides – A reevaluation of the mechanism of cellular uptake. *J Biol Chem* 278: 585–590.

ROGNES T, SEEBERG E (**1998**) SALSA: improved protein database searching by a new algorithm for assembly of sequence fragments into gapped alignments. *Bioinformatics* 14: 839–845.

ROGNES T, SEEBERG E (**2000**) Six-fold speed-up of Smith-Waterman sequence database searches using parallel processing on common microprocessors. *Bioinformatics* 16: 699–706.

ROLS MP, DELTEIL C, GOLZIO M, DUMOND P, CROS S, TEISSIE J (**1998**) In vivo electrically mediated protein and gene transfer in murine melanoma. *Nature Biotechnol* 16: 168–171.

RUBINSON DA, DILLON CP, KWIATKOWSKI AV, SIEVERS C, YANG L, KOPINJA J, ROONEY DL, IHRIG MM, MCMANUS MT, GERTLER FB,

Scott ML, Van Parijs L (2003) A lentivirus-based system to functionally silence genes in primary mammalian cells, stem cells and transgenic mice by RNA interference. *Nature Genet* 33: 401–406.

Sadowski PD (1993) Site-specific genetic recombination: hops, flips, and flops. *FASEB J* 7: 760–767.

Sambrook J, Russell DW (2001) *Molecular Cloning: A Laboratory Manual,* 3rd edn. Cold Spring Harbor Press, New York.

Scadden AD, Smith CW (2001a) RNAi is antagonized by A_I hyper-editing. *EMBO Rep* 2: 1107–1111.

Scadden ADJ, Smith CWJ (2001b) Specific cleavage of hyper-edited dsRNAs. *EMBO J* 20: 4243–4252.

Scaringe SA (2001) RNA oligonucleotide synthesis via 5′-silyl-2′-orthoester chemistry. *Methods* 23: 206–217.

Scaringe SA, Wincott FE, Caruthers MH (1998) Novel RNA synthesis method using 5′-O-silyl-2′-O-orthoester protecting groups. *J Am Chem Soc* 120: 11820–11821.

Scherr M, Battmer K, Eder M, Schule S, Hohenberg H, Ganser A, Grez M, Blomer U (2002) Efficient gene transfer into the CNS by lentiviral vectors purified by anion exchange chromatography. *Gene Ther* 9: 1708–1714.

Scherr M, Eder M (2002) Gene transfer into hematopoietic stem cells using lentiviral vectors. *Curr Gene Ther* 2: 45–55.

Scherr M, Morgan MA, Eder M (2003) Gene silencing mediated by small interfering RNAs in mammalian cells. *Curr Med Chem* 10: 245–256.

Schmitz K, Diallo M, Mundegar R, Schepers U (submitted) pepsiRNAs for RNAi in mammals.

Schmitz K, Schepers U (2004) Cell penetrating peptides in siRNA delivery. *Expert Opin Biol Ther,* in press.

Schwachtgen JL, Ferreira V, Meyer D, Kerbiriou-Nabias D (1994) Optimization of the transfection of human endothelial cells by electroporation. *Biotechniques* 17: 882–887.

Schwartz A, Rahmouni AR, Boudvillain M (2003) The functional anatomy of an intrinsic transcription terminator. *EMBO J* 22: 3385–3394.

Schwartz ME, Breaker RR, Asteriadis GT, Gough GR (1995) a universal adapter for chemical synthesis of DNA or RNA on any single type of solid support. *Tetrahedron Lett* 36: 27–30.

Schwarz DS, Hutvagner G, Du T, Xu Z, Aronin N, Zamore PD (2003) Asymmetry in the assembly of the RNAi enzyme complex. *Cell* 115: 199–208.

Schwarze SR, Dowdy SF (2000) In vivo protein transduction: intracellular delivery of biologically active proteins, compounds and DNA. *Trends Pharmacol Sci* 21: 45–48.

Schwarze SR, Ho A, Vocero-Akbani A, Dowdy SF (1999) In vivo protein transduction: Delivery of a biologically active protein into the mouse. *Science* 285: 1569–1572.

Semizarov D, Frost L, Sarthy A, Kroeger P, Halbert DN, Fesik SW (2003) Specificity of short interfering RNA determined through gene expression signatures. *Proc Natl Acad Sci USA* 100: 6347–6352.

Shinagawa T, Ishii S (2003) Generation of Ski-knockdown mice by expressing a long double-strand RNA from an RNA polymerase II promoter. *Genes Dev* 17: 1340–1345.

Shirane D, Sugao K, Namiki S, Tanabe M, Iino M, Hirose K (2004) Enzymatic production of RNAi libraries from cDNAs. *Nature Genet* 36: 190–196.

Silhol M, Tyagi M, Giacca M, Lebleu B, Vives E (2002) Different mechanisms for cellular internalization of the HIV-1 Tat-derived cell penetrating peptide and recombinant proteins fused to Tat. *Eur J Biochem* 269: 494–501.

Silva JM, Mizuno H, Brady A, Lucito R, Hannon GJ (2004) RNA interference microarrays: High-throughput loss-of-function genetics in mammalian cells. *Proc Natl Acad Sci USA* 101: 6548–6452.

Sinden RR, Zheng GX, Brankamp RG, Allen KN (1991) On the deletion of inverted repeated DNA in *Escherichia coli*: effects of length, thermal stability, and cruciform formation in vivo. *Genetics* 129: 991–1005.

Sledz CA, Holko M, de Veer MJ, Silverman RH, Williams BR (2003) Activation of the interferon system by short-interfering RNAs. *Nature Cell Biol* 5: 834–839.

Smith NA, Singh SP, Wang MB, Stoutjedijk PA, Green AG, Waterhouse PM (2000) Total silencing by intron spliced hairpin RNAs. *Nature* 407: 319–320.

Smith TF, Waterman MS (1981) Identification of common molecular subsequences. *J Mol Biol* 147: 195–197.

Song E, Lee SK, Wang J, Ince N, Ouyang N, Min J, Chen J, Shankar P, Lieberman J (**2003**) RNA interference targeting Fas protects mice from fulminant hepatitis. *Nature Med* 9: 347–351.

Soukup GA, Breaker RR (**2000**) Allosteric nucleic acid catalysts. *Curr Opin Struct Biol* 10: 318–325.

Spencer DM, Wandless TJ, Schreiber SL, Crabtree GR (**1993**) Controlling signal transduction with synthetic ligands. *Science* 262: 1019–1024.

Srivastava SP, Davies MV, Kaufman RJ (**1995**) Calcium depletion from the endoplasmic reticulum activates the double-stranded RNA-dependent protein kinase (PKR) to inhibit protein synthesis. *J Biol Chem* 270: 16619–16624.

Srivastava SP, Kumar KU, Kaufman RJ (**1998**) Phosphorylation of eukaryotic translation initiation factor 2 mediates apoptosis in response to activation of the double-stranded RNA-dependent protein kinase. *J Biol Chem* 273: 2416–2423.

Stark GR, Kerr IM, Williams BR, Silverman RH, Schreiber RD (**1998**) How cells respond to interferons. *Annu Rev Biochem* 67: 227–264.

Sui G, Soohoo C, Affar el B, Gay F, Shi Y, Forrester WC (**2002**) A DNA vector-based RNAi technology to suppress gene expression in mammalian cells. *Proc Natl Acad Sci USA* 99: 5515–5520.

Sullenger BA, Gallardo HF, Ungers GE, Gilboa E (**1990**) Overexpression of TAR sequences renders cells resistant to human immunodeficiency virus replication. *Cell* 63: 601–608.

Suzuki T, Shin BC, Fujikura K, Matsuzaki T, Takata K (**1998**) Direct gene transfer into rat liver cells by in vivo electroporation. *FEBS Lett* 425: 436–440.

Svoboda P, Stein P, Hayashi H, Schultz RM (**2000**) Selective reduction of dormant maternal mRNAs in mouse oocytes by RNA interference. *Development* 127: 4147–4156.

Svoboda P, Stein P, Schultz RM (**2001**) RNAi in mouse oocytes and preimplantation embryos: effectiveness of hairpin dsRNA. *Biochem Biophys Res Commun* 287: 1099–1104.

Symons RH (**1992**) Small catalytic RNAs. *Annu Rev Biochem* 61: 641–671.

Takeuchi S, Hirayama K, Ueda K, Sakai H, Yonehara H (**1958**) Blasticidin S, a new antibiotic. *J Antibiot (Tokyo)* 11: 1–5.

Tiscornia G, Singer O, Ikawa M, Verma IM (**2003**) A general method for gene knockdown in mice by using lentiviral vectors expressing small interfering RNA. *Proc Natl Acad Sci USA* 100: 1844–1848.

Tuschl T, Zamore PD, Lehmann R, Bartel DP, Sharp PA (**1999**) Targeted mRNA degradation by double-stranded RNA in vitro. *Genes Dev* 13: 3191–3197.

Uhlenbeck OC (**1987**) A small catalytic oligoribonucleotide. *Nature* 328: 596–600.

van de Wetering M, Oving I, Muncan V, Pon Fong MT, Brantjes H, van Leenen D, Holstege FC, Brummelkamp TR, Agami R, Clevers H (**2003**) Specific inhibition of gene expression using a stably integrated, inducible small-interfering-RNA vector. *EMBO Rep* 4: 609–615.

Vanbever R, Leroy MA, Preat V (**1998**) Transdermal permeation of neutral molecules by skin electroporation. *J Control Release* 54: 243–250.

Verrecchia F, Tacheau C, Wagner EF, Mauviel A (**2003**) A central role for the JNK pathway in mediating the antagonistic activity of pro-inflammatory cytokines against transforming growth factor-beta-driven SMAD3/4-specific gene expression. *J Biol Chem* 278: 1585–1593.

Vickers TA, Koo S, Bennett CF, Crooke ST, Dean NM, Baker BF (**2003**) Efficient reduction of target RNAs by small interfering RNA and RNase H-dependent antisense agents. A comparative analysis. *J Biol Chem* 278: 7108–7118.

Vigna E, Cavalieri S, Ailles L, Geuna M, Loew R, Bujard H, Naldini L (**2002**) Robust and efficient regulation of transgene expression in vivo by improved tetracycline-dependent lentiviral vectors. *Mol Ther* 5: 252–261.

Villa R, Folini M, Lualdi S, Veronese S, Daidone MG, Zaffaroni N (**2000**) Inhibition of telomerase activity by a cell-penetrating peptide nucleic acid construct in human melanoma cells. *FEBS Lett* 473: 241–248.

Vives E, Brodin P, Lebleu B (**1997**) A truncated HIV-1 Tat protein basic domain rapidly translocates through the plasma membrane and accumulates in the cell nucleus. *J Biol Chem* 272: 16010–16017.

Vives E, Lebleu B (**1997**) Selective coupling of a highly basic peptide to an oligonucleotide. *Tetrahedron Lett* 38: 1183–1186.

VIVES E, RICHARD JP, RISPAL C, LEBLEU B (**2003**) TAT peptide internalization: seeking the mechanism of entry. *Curr Protein Peptide Sci* 4: 125–132.

WALTERS DK, JELINEK DF (**2002**) The effectiveness of double-stranded short inhibitory RNAs (siRNAs) may depend on the method of transfection. *Antisense Nucleic Acid Drug Dev* 12: 411–418.

WENDER PA, JESSOP TC, PATTABIRAMAN K, PELKEY ET, VANDEUSEN CL (**2001**) An efficient, scalable synthesis of the molecular transporter octaarginine via a segment doubling strategy. *Organic Lett* 3: 3229–3232.

WIANNY F, ZERNICKA-GOETZ M (**2000**) Specific interference with gene function by double-stranded RNA in early mouse development. *Nature Cell Biol* 2: 70–75.

WILLIAMS EJ, DUNICAN DJ, GREEN PJ, HOWELL FV, DEROSSI D, WALSH FS, DOHERTY P (**1997**) Selective inhibition of growth factor-stimulated mitogenesis by a cell-permeable Grb2-binding peptide. *J Biol Chem* 272: 22349–22354.

WINCOTT F, DIRENZO A, SHAFFER C, GRIMM S, TRACZ D, WORKMAN C, SWEEDLER D, GONZA-LEZ C, SCARINGE S, USMAN N (**1995**) Synthesis, deprotection, analysis and purification of RNA and ribozymes. *Nucleic Acids Res* 23: 2677–2684.

WIZNEROWICZ M, TRONO D (**2003**) Conditional suppression of cellular genes: lentivirus vector-mediated drug-inducible RNA interference. *J Virol* 77: 8957–8961.

WUNDERBALDINGER P, JOSEPHSON L, WEISS-LEDER R (**2002**) Tat peptide directs enhanced clearance and hepatic permeability of magnetic nanoparticles. *Bioconjug Chem* 13: 264–268.

WYMAN AR, WOLFE LB, BOTSTEIN D (**1985**) Propagation of some human DNA sequences in bacteriophage lambda vectors requires mutant *Escherichia coli* hosts. *Proc Natl Acad Sci USA* 82: 2880–2884.

XIA H, MAO Q, PAULSON HL, DAVIDSON BL (**2002**) siRNA-mediated gene silencing in vitro and in vivo. *Nature Biotechnol* 20: 1006–1010.

XIA XG, ZHOU H, DING H, AFFAR EL B, SHI Y, XU Z (**2003**) An enhanced U6 promoter for synthesis of short hairpin RNA. *Nucleic Acids Res* 31: e100.

YAMAGUCHI H, YAMAMOTO C, TANAKA N (1965) Inhibition of protein synthesis by blasticidin S. I. Studies with cell-free systems from bacterial and mammalian cells. *J Biochem (Tokyo)* 57: 667–677.

YAMAMOTO T, OMOTO S, MIZUGUCHI M, MIZU-KAMI H, OKUYAMA H, OKADA N, SAKSENA NK, BRISIBE EA, OTAKE K, FUJII YR (**2002**) Double-stranded nef RNA interferes with human immunodeficiency virus type 1 replication. *Microbiol Immunol* 46: 809–817.

YANG D, BUCHHOLZ F, HUANG Z, GOGA A, CHEN CY, BRODSKY FM, BISHOP JM (**2002**a) Short RNA duplexes produced by hydrolysis with *Escherichia coli* RNase III mediate effective RNA interference in mammalian cells. *Proc Natl Acad Sci USA* 99: 9942–9947.

YANG D, LU H, ERICKSON JW (**2000**) Evidence that processed small dsRNAs may mediate sequence-specific mRNA degradation during RNAi in *Drosophila* embryos. *Curr Biol* 10: 1191–1200.

YANG PL, ALTHAGE A, CHUNG J, CHISARI FV (**2002**b) Hydrodynamic injection of viral DNA: a mouse model of acute hepatitis B virus infection. *Proc Natl Acad Sci USA* 99: 13825–13830.

YANISCH-PERRON C, VIEIRA J, MESSING J (**1985**) Improved M13 phage cloning vectors and host strains: nucleotide sequences of the M13mp18 and pUC19 vectors. *Gene* 33: 103–119.

YEE JK, FRIEDMANN T, BURNS JC (**1994**a) Generation of high-titer pseudotyped retroviral vectors with very broad host range. *Methods Cell Biol* 43 Pt A: 99–112.

YEE JK, MIYANOHARA A, LAPORTE P, BOUIC K, BURNS JC, FRIEDMANN T (**1994**b) A general method for the generation of high-titer, pantropic retroviral vectors: highly efficient infection of primary hepatocytes. *Proc Natl Acad Sci USA* 91: 9564–9568.

YONAHA M, PROUDFOOT NJ (**2000**) Transcriptional termination and coupled polyadenylation in vitro. *EMBO J* 19: 3770–3777.

YOSHIZATO K, NISHI T, GOTO T, DEV SB, TAKESHIMA H, KINO T, TADA K, KIMURA T, SHIRAISHI S, KOCHI M, KURATSU JI, HOFMANN GA, USHIO Y (**2000**) Gene delivery with optimized electroporation parameters shows potential for treatment of gliomas. *Int J Oncol* 16: 899–905.

YU JY, DE RUITER SL, TURNER DL (**2002**) RNA interference by expression of short-interfering RNAs and hairpin RNAs in mammalian cells. *Proc Natl Acad Sci USA* 99: 6047–6052.

Yu Y, Bradley A (**2001**) Engineering chromosomal rearrangements in mice. *Nature Rev Genet* 2: 780–790.

Yuan B, Lewitter F (**2003**) *siRNA Selection Program*. Whitehead Institute for Biomedical Research, Cambridge.

Zamore PD (**2001**) RNA interference: listening to the sound of silence. *Nature Struct Biol* 8: 746–750.

Zeng Y, Cullen BR (**2003**) Sequence requirements for micro RNA processing and function in human cells. *Rn.a – A Publication of the Rn.a Society* 9: 112–123.

Zhang BL, Cui ZY, Sun LL (**2001**) Synthesis of 5'-deoxy-5'-thioguanosine-5'-monophosphorothioate and its incorporation into RNA 5'-termini. *Organic Lett* 3: 275–278.

Zhang G, Budker V, Wolff JA (**1999**) High levels of foreign gene expression in hepatocytes after tail vein injections of naked plasmid DNA. *Hum Gene Ther* 10: 1735–1737.

Zhang H, Kolb FA, Brondani V, Billy E, Filipowicz W (**2002**) Human Dicer preferentially cleaves dsRNAs at their termini without a requirement for ATP. *EMBO J* 21: 5875–5885.

Zheng GX, Kochel T, Hoepfner RW, Timmons SE, Sinden RR (**1991**) Torsionally tuned cruciform and Z-DNA probes for measuring unrestrained supercoiling at specific sites in DNA of living cells. *J Mol Biol* 221: 107–122.

Zhou DM, Taira K (**1998**) The hydrolysis of RNA: from theoretical calculations to the hammerhead ribozyme-mediated cleavage of RNA. *Chem Rev* 98: 991–1026.

Ziauddin J, Sabatini DM (**2001**) Microarrays of cells expressing defined cDNAs. *Nature* 411: 107–110

Zufferey R, Dull T, Mandel RJ, Bukovsky A, Quiroz D, Naldini L, Trono D (**1998**) Self-inactivating lentivirus vector for safe and efficient in vivo gene delivery. *J Virol* 72: 9873–9880.

Appendix 1: Abbreviations

A	adenine/adenosine
Ac$_2$O	acetic anhydride
ACD	active cell death
ACE	[bis(2-acetoxyethoxy)methyl orthoester]
ADAR	adenosine deaminase that acts on dsRNA
ADP	adenosine diphosphate
Ago-2	Argonaute-like protein 2
Amp	ampicillin
Amp^r	ampicillin resistance
AntP	peptide derived from the Antennapedia protein
APS	ammonium persulfate
asRNA	antisense RNA
ATCC	American Type Culture Collection
ATP	adenosine triphosphate
attB	recombination site
BCA	bicinchonic acid
BDGP	Berkeley Drosophila Genome Project
BGH	bovine growth hormone (gene)
BLAST	basic local alignment search tool
C	cytosine/cytidine
C. elegans	*Caenorhabditis elegans*
ccdB	blasticidin resitance gene
CDC	Centers for Disease Control
cDNA	copy DNA
CIP	calf intestinal phosphatase
CmR	chloramphenicol resistance (gene)
CMV	cytomegalovirus
CPG	controlled pore glass
CPP	cell penetrating peptide
Cppt	central polypurine tract
Cre	site that causes recombination
CTD	C-terminal domain
Dcr-1	Dicer-1 phenotype (*C. elegans*)

RNA Interference in Practice: Principles, Basics, and Methods for Gene Silencing in C. elegans, Drosophila, and Mammals. Ute Schepers
Copyright © 2005 WILEY-VCH Verlag GmbH & Co. KGaA, Weinheim
ISBN: 3-527-31020-7

DCR-1/2	Dicer 1/2 (*Drosophila*)
DDM1	chromatin remodeling complex in *Arabidopsis*
DEAD-box	Asp-Glu-Ala-Asp motif
DEC	1-(3-dimethylaminopropyl)-3-ethylcarbodiimide hydrochloride
DEPC	diethylpyrocarbonate
DexH/DEAH	RNA helicase domain
DMEM	Dulbecco's modified medium
DMF	dimethylformamide
DMSO	dimethylsulfoxide
DMT	dimethoxytrityl
DNase	deoxyribonuclease
DOD	[bis(trimethylsiloxy)cyclododecyloxysilyl ether]
DRSC	*Drosophila* RNAi Screening Center
d-siRNA	Dicer generated siRNA
dsOligo	double stranded oligonucleotide
dsRDB	double stranded RNA binding domain
DSRM	double stranded RNA binding motif
dsRNA	double stranded RNA
dT	deoxythymidine
DTT	dithiothreitol
dUTP	deoxyuridine triphosphate
E. coli	*Escherichia coli*
E3L	double stranded RNA binding protein of vaccinia
EB	elution buffer from *Qiagen*
ECL	enhanced chemiluminescence
EDTA	ethylenediamine tetraacetate
EGFP	enhanced green fluorescent protein
EGTA	ethyleneglycol tetraacetic acid
eIF2C1/2	translation initiation factor 2 C1/2
eIF2α	translation initiation factor 2 alpha
EMBL	European Molecular Biology Laboratory
endA$^-$	endonuclease deficient strain
env	HIV gene encoding an envelope protein
EPRIL	enzymatic production of RNAi library
EURIT	European Union for RNA Interference Technology
ERS	end-restriction site
ESC	embryonic stem cell
esiRNA	endoribonuclease-prepared siRNA
EST	expressed sequence tag
FACS	fluorescence assisted cell sorting
FCS	fetal calf serum
fed	RNAi defective mutant in *C. elegans* when fed with dsRNA
FITC	fluoresceine isothiocyanate
FLAP	human 5-lipoxygenase activating protein
FPMP	[1-(2-fluorophenyl)-4-methoxypiperidin-4-yl]

G	guanine/guanosine
GAL4	yeast transactivator
GAPDH	glyceraldehyde-3-phosphate dehydrogenase
GFP	green fluorescent protein
Glac	glacial
globin IVS	the second intron (beta IVS-II) of the human beta-globin gene
GSMP	5'-desoxy-5'-thioguanoside-monophosphorothioate
GST	glutathione-S-transferase
H1	human promoter region
H1/2	homology regions
HBS	HEPES buffered saline
HDV	hepatitis delta virus
HEPES	2-[4-(2-hydroxyethyl)-1piperazino-ethane sulfonic acid
HGMP	MRC bioinformatics resource center
Hi5	insect cells from *Trichoplusia ni*
HIV	human immunodeficiency virus
HRE	hormone response element
HRP	horseradish peroxidase
HSC	hematopoietic stem cell
HSV	human stomatitis virus
hsp	heat shock protein
Hz	Hertz
ID	identity
IFN-α	interferon α
int	bacteriophage lambda gene with nicking and closing activity
IPRS	inversion point restriction site
IPTG	isopropyl-1-thio-β-D-galactoside
I-SceI	mitochondrial endonuclease that recognizes and cuts an 18-bp restriction site
ISG	interferon stimulated gene
JAK	Janus kinase
K3L	double-stranded RNA binding protein of vaccinia
kan	kanamycin
L1−4	larvae stages of *C. elegans*
lacI	lactose operon repressor
lacUV5	constitutive promoter region in the lac operon
lacZ	gene encoding β-galactosidase
LB	Luria Bertani (culture medium)
let-7	lethal-7 phenotype in *C. elegans*
let-858	ubiquitously expressed *C. elegans* gene
LexGAD	LexA protein transactivator
lhRNA	long hairpin RNA
lin-4	lineage abnormal 4 phenotype in *C. elegans*
LL	LexGAD response element
loxP	locus of crossover P1

LTR	long terminal repeat
m^7G	7-methylguanosine
M9	minimal culture medium
MAGIC	mating assisted genetically integrated cloning
MAZ	RNA polymerase pausing site
MCS	multiple cloning site
MDCK cells	Madin-Darby canine kidney cells
F	Farad (unit of electrical capacitance)
MEM	minimal essential medium
MET1	methyltransferase 1
mir-23	micro RNA 23
miRNA	micro RNA
MOI	multiplicity of infection
MRC	Medical Research Council (UK)
mRNA	messenger RNA
MTD	modification of common transfer RNA-derived
mut-7	mutant phenotype in *C. elegans*
myo-2	*C. elegans* gene expressed in pharyngeal muscle
myo-3	*C. elegans* gene expressed in body wall muscle
N	any nucleotide
N. crassa	*Neurospora crassa*
N3	buffer from *Qiagen*
NCBI	National Center for Bioinformatics (USA)
NEB	New England Biolabs
Neo^r	neomycin resistance (gene)
NIH	National Institutes of Health (USA)
Ni-NTA	nickel-nitrilotriacetic acid
NLS	nuclear localization signal
NMD	nonsense mediated RNA decay
N-Me-Imid	*N*-methylimidazole
NMR	nuclear magnetic resonance
nt	nucleotide
OAS	2'-5'-oligoadenylate synthase
OD	optical density
ORF	open reading frame
P1,2,3	buffers from *Qiagen*
PAGE	polyacrylamide gel electrophoresis
PAZ domain	domain conserved in PIWI, Argonaute, Zwille (*Drosophila*)
PB	buffer from *Qiagen*
PBS	phosphate buffered saline
PCD	programmed cell death
pCMV	CMV promoter
PCR	polymerase chain reaction
pDNA	plasmid DNA
PE	buffer from *Qiagen*

PEI	polyethyleneimine
P-element	*Drosophila* transposon
pepsiRNA	peptide coupled siRNA
pfu	plaque forming units
PGK-Puro	phosphoglycerate kinase promoter-puromycin resistance
PI	propidium iodide
Pi	pyrophosphate
PKR	protein kinase R
Pol II	RNA polymerase II
Pol III	RNA polymerase III
pol	HIV gene encoding a polymerase gene
pre-miRNA	precursor miRNA
pri-miRNA	primary transcript of miRNA
psiRNA	plasmid based siRNA
PTD	protein transduction domain
PTGS	post transcriptional gene silencing
puro	puromycin
PVDF	polyvinylidene fluoride
Ψ	Psi element (HIV packaging sequence)
QDE-1	quelling deficient phenotype in *Neurospora crassa*
R2D2	tandem dsRNA binding domain (R2)-Dicer-2 (D2) complex
RDE-1	RNAi deficient-1 phenotype in *C. elegans*
RDE-4	RNAi deficient-4 phenotype in *C. elegans*
RdRp	RNA dependent (directed) RNA polymerase
recA-	recombinase deficiency in *E. coli*
re-hDicer	recombinant-human Dicer
REV	HIV gene
RISC	RNA induced silencing complex
RISC*	activated RNA induced silencing complex
RNAi	RNA interference
RNase III	ribonuclease type III
RNP	ribonucleotide protein complex
rNTP	ribonucleotide triphosphate
r.p.m.	revolutions per minute
RPMI	media formulation
rps-5/28	*C. elegans* gene expressed in multiple tissues
RRE	Rev response element
RRF-1/2/3	RdRp deficient phenotype in *C. elegans*
rrf-3	putative RNA-directed RNA polymerase in *C. elegans*
rRNA	ribosomal RNA
RS	restriction site
RSV	Rous sarcoma virus
RT	room temperature
RT-PCR	reverse transcription-PCR
RZPD	Deutsches Resourcenzentrum für Genomforschung

S2	Schneider cells, *Drosophila* cells
S$_2$Na$_2$	disodium-2-carbamoyl-2-cyanoethylene-1,1dithiolate trihydrate
SARS	severe acute respiratory syndrome
sbCD	*E. coli* exonuclease
SDS	sodium dodecyl sulphate
Sf9	insect cells from *Spodoptera frugiperda*
sgs2/sde2	PTGS deficient mutant of *C. elegans*
SH2	Src homology domain
shRNA	short hairpin RNA
sid	systemic RNAi defective mutants of *C. elegans*
Sil	silyl
SIN-LTR	self inactivating -LTR
siRNA	small (short) interfering RNA
snb-1	*C. elegans* maternal gene
SOS	repair system in *E. coli*
SPEED	small interfering RNA production by enzymatic engineering of DNA
sRNA	sense RNA
SSearch	Smith-Waterman algorithm
ssRNA	single stranded RNA
STAT	signal transducer and activator of transcription
stRNA	small temporal RNA
SURE II	*E. coli* strain
T	thymine/thymidine
T/A cloning	cloning with single 3'-T overhangs into a vector with 5'-A overhangs
T5	Pol III termination signal
T7 promoter	recognition sequence for T7 RNA polymerase
T7/T3/Sp6	RNA polymerases (from bacteriophages)
TAE	tris-acetate-EDTA buffer
TAT	HIV protein
TATA	promoter element
TAT-NLS	fusion peptide from TAT peptide and nuclear localization sequence
TB	terrific broth culture medium
TBDMS	tert-butyldimethylsilyl
TBE	tris-borate-EDTA buffer
TBS-T	tris buffered saline plus Tween 20
TCA	trichloroethane
TE	tris-EDTA buffer
TEAHF	triethylammonium fluoride
TEMED	N,N,N',N'-tetramethylene diamine
tet	tetracycline
tetO	tetracycline operon
tetR	tetracycline repressor

TGS	transcriptional gene silencing
TOM	[(triisopropylsilyl)oxy]methyl group
TRE	tTA response element
Tris	trishydroxymethylaminomethane
tRNA	transfer RNA
TRz	trimming ribozyme
tTA	tetracycline controlled transactivator
TU	transducing units
Tudor SN	Tudor staphylococcal nuclease
Tyk2	Janus family tyrosine kinase
u	unit
U	uracil/uridine
U6	gene/promoter element in mice
UAS	upstream activation sequence
unc-119	*C. elegans* gene expressed in neurons
unc-22	*C. elegans* gene expressed in adults
UTR	untranslated region
uvrC	UV repair system
VA RNA	two virus-associated RNA
Vif	viral infectivity factor of HIV
vit-2	*C. elegans* gene expressed in intestine
VSV	vesicular stomatitis virus
VSVG	vesicular stomatitis virus glycoprotein
WRE	Woodchuck hepatitis B virus RNA regulatory element
WPRE	Woodchuck hepatitis virus posttranscriptional regulatory element
WT	wild type
x-gal	5-chloro-4-bromo-3-indolyl-β-D-galactoside
XYT	culture medium formulation
zeo[r]	Zeocin resistance

Appendix 2: List of Protocols

Protocol 1: *In vitro* dsRNA Transcription for RNAi in *C. elegans*

Protocol 2: Delivery of dsRNA in *C. elegans*: Microinjection Protocol

Protocol 3a: Delivery of dsRNA in *C. elegans*: Soaking Plain dsRNA

Protocol 3b: Soaking Liposome-embedded dsRNA or Inverted Repeat DNA Constructs

Protocol 4: Generation of Inverted Repeat Constructs for RNAi in *C. elegans*

Protocol 5: Delivery of dsRNA in *C. elegans*: RNAi Feeding Protocol

Protocol 6: Mounting Animals for Microscopy

Protocol 7: *In vitro* dsRNA Transcription for RNAi in *Drosophila*

Protocol 8: Generation of Inverted Repeat DNA for RNAi in *Drosophila*

Protocol 9: DsRNA or Inverted Repeat DNA Preparation for Injection of *Drosophila* Embryos

Protocol 10: *Drosophila* Embryo Collection and Preparation for Injection

Protocol 11: Thawing and Maintenance of *Drosophila* S2 Cells

Protocol 12: *Drosophila* S2 Cell Freezing Protocol

Protocol 13: Delivery of dsRNA in *Drosophila*: dsRNA Soaking of S2 Cells

Protocol 14: RNAi in Mammals: siRNA Design

Protocol 15: *In vitro* siRNA Transcription

Protocol 16: Expression of Dicer in Hi5 Insect Cells

Protocol 17: DsRNA Digestion by Recombinant Dicer

Protocol 18: Production of Recombinant RNaseIII from *E. coli*

Protocol 19: Delivery of siRNAs: Transfection of siRNAs

Protocol 20: Delivery of siRNAs: Electroporation of siRNAs

Protocol 21: Enzymatic Synthesis of 5'-Thiol-Modified siRNAs

Protocol 22: Coupling of Cys-modified CPPs to siRNAs

Protocol 23: Delivery of siRNAs: Treatment of Cells with pepsiRNAs

Protocol 24: Analysis of the siRNAs: PAGE of siRNAs

Protocol 25: Determination of dsRNA and siRNAs by Agarose Gel Electrophoresis

Protocol 26: Simultaneous Detection of siRNA and mRNA: Non-denaturing Gels

Protocol 27: RNAi with Short Hairpin RNAs (shRNA): Cloning of pSUPER-shRNA Expression Vectors

Protocol 28: Lentiviral Approach: Cloning of pLentiLox3.7

Protocol 29: Lentiviral Approach: Cloning of the pLenti6/GW/U6 Vector

RNA Interference in Practice: Principles, Basics, and Methods for Gene Silencing in C. elegans, Drosophila, and Mammals. Ute Schepers
Copyright © 2005 WILEY-VCH Verlag GmbH & Co. KGaA, Weinheim
ISBN: 3-527-31020-7

Protocol 30: Lentivirus Production by Calcium Phosphate Transfection
Protocol 31: Lentivirus Production by Lipofectamine 2000™ Transfection
Protocol 32: Titration of the Virus: Preparation of Cells
Protocol 33: Titration of the Virus
Protocol 34: RNAi with Long Hairpin RNAs: Cloning of Inverted Repeat DNA
Protocol 35: High-throughput Inverted Repeat Isolation
Protocol 36: Generation of lhRNA from Direct Repeat DNA
Protocol 37: Generation of Inverted Repeat DNA Missing 5'- Cap and Poly(A) Tail
Protocol 38: Test for Interferon Response: Semi-Quantitative RT-PCR Analysis of 2'- 5'- OAS Activation
Protocol 39: Test for Interferon Response: DNA Fragmentation Analysis
Protocol 40: Test for Interferon Response: eIF2α Phosphorylation by Western Blot Analysis
Protocol 41: Test for Interferon Response: Autophosphorylation Activity of PKR
Protocol 42: Application of RNAi in Mice: *In vivo* Electroporation
Protocol 43: Generation of Transgenic Mice by Pronucleus Injection

Appendix 3: Suppliers of RNAi-related Chemicals and Probes

3M
http://www.3m.com/index.jhtml

Amaxa GmbH
http://amaxa.com

Ambion
http://www.ambion.com

Amersham Biosciences
http://www.amershambiosciences.com

Applied Biosystems
http://europe.appliedbiosystems.com

ATCC, American Type Culture Collection
http://www.atcc.org

BDGP, Berkeley Drosophila Genome Project
http://www.fruitfly.org

BioRad
http://www.bio-rad.com

Bloomington Drosophila Stock Center
http://flystocks.bio.indiana.edu/

Caenorhabditis Genetics Center (CGC)
http://biosci.umn.edu/CGC/CGChomepage.htm

Carnation (Nestlé)
http://www.nestle.com

Celera
http://www.celera.com/

Cell Signaling Technologies
http://www.cellsignal.com/

Cenix Bioscience
http://www.cenix-bioscience.com

RNA Interference in Practice: Principles, Basics, and Methods for Gene Silencing in C. elegans, Drosophila, and Mammals. Ute Schepers
Copyright © 2005 WILEY-VCH Verlag GmbH & Co. KGaA, Weinheim
ISBN: 3-527-31020-7

Cyclacel Ltd.
http://www.cyclacel.com/

Dharmacon
http://www.dharmacon.com

Duke University Non-Mammalian Model Systems Flyshop

http://www.biology.duke.edu/model-system/services.htm

DuPont NEN
http://www.nenlifesci.com/

EMBL, European Molecular Biology Laboratory
http://www.embl-heidelberg.de

Eppendorf
http://www.eppendorf.com/

EURIT, European Union for RNA Interference Technology
http://www.eurit-network.org

Eurogentec
http://www.eurogentec.com

FASTA
http://fasta.bioch.virginia.edu/fasta/home.html,
http://www2.igh.cnrs.fr/bin/fasta-guess.cgi

Fisher Scientific
http://www.fisherscientific.com

Genetic Services Inc.
http://www.geneticservices.com

German Resource Center, RZPD GmbH
http://www.rzpd.de/

Roche Diagnostics
http://www.roche-applied-science.com

Hyclone
http://www.hyclone.com/

Invitrogen
http://www.invitrogen.com

Max-Planck Institute for Infection Biology, MPIIB
http://mpiib-berlin.mpg.de

MRC gene service
http://www.hgmp.mrc.ac.uk/geneservice/index.shtml

MWG Biotech
http://www.mwgdna.com

Nalgene
http://nalgenelab.nalgenunc.com

NCBI, National Center for Bioinformatics
http://www.ncbi.nih.gov

NEB, New England Biolabs
http://www.neb.com/

Novagen
http://www.emdbiosciences.com

OligoEngine
http://www.oligoengine.com

Open Biosystems
http://www.openbiosystems.com/

ParAlign
http://www.paralign.org/

PerkinElmer
http://www.perkinelmer.com/

Pharmingen (BD Bioscience)
http://www.bdbiosciences.com/pharmingen/

Proligo
http://www.proligo.com

Promega
http://www.promega.com

Q-biogene
http://www.qbiogene.com

Qiagen
http://www1.qiagen.com

Roche Diagnostics
http://www.roche-applied-science.com/LabFAQs/index.htm

Roth
http://www.carl-roth.de/

Santa Cruz Biotechnology
http://www.scbt.com/

Sigma
http://www.sigmaaldrich.com

Staples
http://www.staples.com/

Stratagene
http://www.stratagene.com

Takara
http://www.takara-bio.co.jp/english/

Wellcome CRC Institute
http://www.hgmp.mrc.ac.uk/geneservice/reagents/products/

Appendix 4: Glossary of Terms

A-form RNA
In contrast to DNA, double-stranded RNA is forming an A-form helix due to the interaction with more water molecules. It has tilted base pairs and more base pairs per turn than the B-form helix of DNA.

Abdomen
Third, hind major division of an insect body, for reproduction, digestion and excretion.

Accession number
This refers to the unique GenBank identifier a sequence has been assigned. This number can be used to search GenBank records at SGD for a specific sequence.

Active transport
Movement of a molecule across a membrane or other barrier driven by energy other than that stored in the concentration gradient or electrochemical gradient of the transported molecule.

ADAR
Adenosine deaminase acting on RNA. Deaminates primarily RNA hairpins in primary transcripts.

Adenovirus
A group of DNA-containing viruses which cause respiratory disease, including one form of the common cold. Adenoviruses can also be genetically modified and used in gene therapy to treat cystic fibrosis, cancer, and (potentially) other diseases.

Aldehyde
Organic compound that contains a –CH=O group. An example is glyceraldehyde. Can be oxidized to an acid, or reduced to an alcohol.

Alkyl group
General term for a group of covalently linked carbon and hydrogen atoms such as methyl ($-CH_3$) or ethyl ($-CH_2CH_3$) groups; these groups can be formed by removing a hydrogen atom from an alkane.

RNA Interference in Practice: Principles, Basics, and Methods for Gene Silencing in C. elegans, Drosophila, and Mammals. Ute Schepers
Copyright © 2005 WILEY-VCH Verlag GmbH & Co. KGaA, Weinheim
ISBN: 3-527-31020-7

Allosteric ribozymes (allozymes)
Allozymes are a form of catalytic RNA and/or DNA-based molecules, the ability of which to catalyze a reaction is modulated by their interaction with an effector molecule. Regulators of allozymes include a diverse range of compounds, including proteins, nucleic acids, peptides, small molecules, and ions. Allozyme reactivity can be monitored by a variety of different methods, making these molecules useful sensors for both in-vitro and in-vivo applications

Amide
Molecule containing a carbonyl group linked to an amine. Adjacent amino acids in a protein molecule are linked by amide groups.

Amino group
Weakly basic functional group derived from ammonia (NH_3) in which one or more hydrogen atoms are replaced by another atom. In aqueous solution it can accept a proton and carry a positive charge.

Amphipathic
Having both hydrophobic and hydrophilic regions, as in a phospholipid or a detergent molecule.

Amphipathic helix
A protein structure that serves in part as an interface between polar and nonpolar phases; an A-form helix that displays nonpolar residues on one side and polar residues on the other (e.g., in many globular proteins).

Annealing
Generally synonymous with hybridization. The spontaneous pairing of complementary DNA or RNA sequences by hydrogen bonding to form a double-stranded polynucleotide.

Antennapedia (= antp)
Homeobox containing gene of *Drosophila*, controlling thoracic/head fate determination. In addition to the homeobox, it contains a so-called *protein transduction domain* (PTD) or *cell-penetrating peptide* (CPP), which comprises positively charged amino acids, and is responsible for the membrane penetration of the protein and the translocation.

Anterior
Situated toward the head end of the body.

Antisense technology
The method is based on the $1:1$ hybridization of antisense DNA oligos, RNAs, or PNAs to the mRNA, thus preventing its translation by blocking the translocation of the mRNA through the ribosome.

Antisense
Antisense RNA (also: ASO, antisense oligonucleotides) is complementary to the mRNA transcribed from a gene. These anti-mRNAs, or shorter oligonucleotides de-

rived from the sequence of the mRNA, can form complexes with the primary RNA transcript of a gene. It further describes the noncoding strand in double-stranded DNA or dsRNA. The antisense strand serves as the template for mRNA synthesis.

Apoptosis
Apoptosis (from a Greek word meaning the dropping of leaves from a tree) is a term referring to cellular self-destruction observed in all eukaryotes. This process of cell death has been termed as **p**rogrammed **c**ell **d**eath (PCD) or **a**ctive **c**ell **d**eath (ACD) because it requires controlled gene expression, which is activated in response to a variety of external or internal stimuli.

Aptamer
An aptamer (from Latin *aptus*, meaning 'fitting') is a short strand of synthetic nucleic acids (usually RNA but also DNA) selected from randomized combinatorial nucleic acid libraries by virtue of its ability to bind to a predetermined specific target molecule with high affinity and specificity. The procedure generating such aptamers has been termed SELEX (**s**ystematic **e**volution of **l**igands by **ex**ponential enrichment). Aptamers have been generated against a large variety of molecules ranging from amino acids, disaccharides, and antibiotics to complex proteins. Selection of a target-specific aptamer requires numerous liquid-handling steps, including repeated PCR amplification, and has been automated.

ATCC
American Type Culture Collection; maintains collections of yeast strains and clones.

Baculovirus
A group of DNA viruses that are known to multiply only in invertebrates and are now classified in the family Baculoviridae. Their genome consists of double-stranded circular DNA. Because of their host range they have potential as pest-control agents. Baculovirus vectors are valuable as a means of expressing certain animal proteins.

BLAST (Basic Local Alignment Search Tool)
Software program from NCBI for searching public databases for homologous sequences or proteins. Designed to explore all available sequence databases regardless of whether query is protein or DNA.

Blastocyst
A stage in the development of a mammalian embryo just after the blastula stage, where the hollow ball of cells becomes two layers of cells: one layer, called the trophoblast, is at one end of the embryo and attaches it to the wall of the uterus so that it can receive nutrition during its development.

Blastoderm
In an insect embryo, the layer of cells that completely surrounds an internal mass of yolk.

Blastula
Early stage of an animal embryo, usually consisting of a hollow ball of cells, before gastrulation begins.

Blood–brain barrier
The semipermeable membranous barrier that regulates the passage of dissolved materials from the blood into the cerebrospinal fluid that bathes the brain.

Breed
Group of animals or plants presumably related by descent from common ancestors and phenotypically similar in most traits.

Caenorhabditis elegans (C. elegans)
A nematode; a small worm favored by developmental and molecular biologists because of its ability to grow under laboratory conditions, its short generation time, and its transparency. Because of its simplicity, it is possible to trace the development of the zygote to each of the approximately 1000 cells of the adult.

Cap
A 7-methylguanosine in 5′-5′ triphosphate linkage added in reverse polarity (i.e., 3′pMeG5′ppp5′NpNp3′) to the 5′-end of eukaryotic mRNA during transcription initiation. It is added post-transcriptionally, and is not encoded in the DNA. The cap binds a cap binding protein and acts as an initial binding site for ribosomes during translation.

Capsid
The protein shell of a virus.

Carpel factory gene
CAF (CARPEL FACTORY) gene; important for normal flower morphogenesis of *Arabidopsis* and to the SUS1 (SUSPENSOR1) gene essential for embryogenesis. SIN1/SUS1/CAF has sequence similarity to the *Drosophila melanogaster* gene Dicer, which encodes a multidomain ribonuclease specific for double-stranded RNA, first identified by its role in RNA silencing.

Catalytic RNA
RNA which contains an intron sequence that has an enzyme-like catalytic activity. This intron sequence has been shown to fold up to form a complex surface that can function like an enzyme in reactions with other RNA molecules and thus synthesize new molecules, even in the absence of protein.

cDNA library
A collection of DNA sequences generated from mRNA sequences. This type of library contains only protein-coding DNA (genes), and does not include any noncoding DNA.

Centromere
Constricted region of a mitotic chromosome that holds sister chromatids together; also the site on the DNA where the kinetochore forms and then captures microtubules from the mitotic spindle.

Chorion
A hard shell external to the vitelline membrane of the *Drosophila* egg.

Chromatin
The substance of chromosomes; now known to include DNA, chromosomal proteins, and chromosomal RNA. The nucleoprotein material of the eukaryotic chromosome.

cis
Refers to an effect on a gene directed by the sequence of that gene in contrast to *trans* effects, which are produced by other factors such as transcription factors encoded by other genes. The terms are commonly used to describe factors that influence gene expression.

cis-acting element
An arrangement of sequences on a contiguous piece of DNA.

Cistron
The DNA segment in a genome determining a single gene product.

Cleft
The space between domains of a protein, often the binding or catalytic site of an enzyme.

Complementary RNA
Synthetic RNA produced by reverse transcription from a specific DNA single-stranded template.

Conditional expression
Technique of constructing cell lines, or transgenic animals, in which the expression of a gene may be controlled by the level of an exogenous compound. One such system, tet-on, involves a plasmid with the *Escherichia coli* tetracycline-dependent operator upstream from the coding sequence of the targeted gene: in the presence of a tetracycline, the gene is expressed. In the tet-off system, a regulator plasmid is also present, and encodes the tetracycline-controlled transactivator protein, which brings the tetracycline operator into proximity with the promoter sequence of the gene and allows gene expression to proceed. If, however, tetracycline is present, expression is interrupted.

Conformational change
The adjustment of a protein's tertiary structure in response to external factors (e. g., pH, temperature, solute concentration) or to binding of a ligand.

Consensus sequence
A minimum nucleotide sequence found to be common (although not necessarily identical) in different genes and in genes from different organisms that is associated with a specific function. Examples include binding sites for transcription factors and splicing machinery.

Conserved sequence
A base sequence in a DNA molecule (or an amino acid sequence in a protein) that has remained essentially unchanged throughout evolution. A conserved gene should express for basic vital functions so that mutations could not persist and were eliminated.

Constitutive
Produced in constant amount; opposite of regulated. Constitutive secretion, for example, occurs continuously without requiring an external stimulus.

Contig
Group of clones representing overlapping regions of a genome.

Co-suppression
Refers to the specific case of gene silencing, in which RNA from a transgene and a homologous endogenous gene are suppressed at the same time.

Covalent bond
Stable chemical link between two atoms produced by sharing one or more pairs of electrons.

Covalent modification
As applied to enzymes, the regulation of activity by modifications that may be reversible (e.g., phosphorylation or adenylation) or irreversible (e.g., limited proteolysis). As applied to RNA/DNA modification of either the 5′- or 3′-end or the base.

CPP
Cell-penetrating peptide. Peptides with mainly positively charged amino acid sequence (such as many Arg residues).

Cre-loxP system
Method for the introduction of genetic modifications into specific genes by homologous recombination using Cre site-specific, bacteriophage P1-derived recombinase. The Cre recombinase can either excise, recombine, or invert loxP-site flanked genes or DNA sequences. Cre recombinase can be expressed under the control of a tissue-specific promoter. Thus, the enzyme is expressed only in the desired tissue, and it deletes or modifies the gene of interest via the loxP target sites.

Cre recombinase
Causes recombination in bacteriophage P1.

Crossing
Fertilization of an organism from two organism with a different genetic constitution.

Cross-link
In protein chemistry, a natural or synthetic covalent bond between protein side chains. In carbohydrate and nucleic acid chemistry, a synthetic bridging group.

Cruciform DNA
A region of DNA having a sequence at one end repeated but inverted at the other end, so that each strand may pair with itself to form a helix extending sideways from the main helix.

Cuticle
The noncellular (nonliving) skin or integument of insects consisting of chitin, structural proteins and pigments; in larvae it is shed at intervals to allow growth.

DEAD box protein
One of a group of proteins that have in common an Asp-Glu-Ala-Asp (DEAD in the one-letter code).

DEAD-box helicases
Family of ATP-dependent DNA or RNA helicases with a 4 amino acid consensus sequence – D-E-A-D – or a related sequence that resembles an ATP binding site and functions in the ATP-dependent processing of RNA.

Deamination
The abstraction of the elements of ammonia from a compound, for example, from histidine by the histidine lyase reaction, or from AMP in the adenylate deaminase reaction, or by ADARs.

Dendrimer
A dendrimer (from Greek *dendra* for tree) is an artificially manufactured or synthesized molecule built up from branched units called monomers. Such processes involve working on the scale of nanometers. Technically, a dendrimer is a polymer, which is a large molecule comprised of many smaller ones linked together. Dendrimers have some proven applications, and numerous potential applications

Dicer
RNase III type nuclease of the RNAi-pathway witch specifically cuts the target-mRNA species into pieces of 21–23 bp.

Disulfide bond (–S–S–)
Covalent linkage formed between two sulfhydryl groups on cysteines. Common way to join two proteins or to link together different parts of the same protein in the extracellular space.

DNA methylation
A phenomenon that represses expression of regions of the genome. Transcription is prevented when the DNA is methylated and folded into nucleosomes. Eukaryotic DNA is methylated almost exclusively as 5-methyl cytosine; prokaryotic DNA is methylated also as 6-methyl adenosine.

Domain
A discrete portion of a protein (and corresponding segment of gene) with its own function. The combination of domains in a single protein determines its overall function. A protein may have several different domains, and the same domain may be found in different proteins.

Dorsal
Relating to the back of an animal; also the upper surface of a leaf, wing, etc.

Dorsal appendage
The respiratory structures on the anterior dorsal side of the *Drosophila* egg.

Double-stranded RNA (dsRNA)
Generally refers to long or full-length RNA duplexes. These large dsRNA initiate a general host cell shutdown in most mammalian cell types; the cells subsequently begin to decrease their expression of nontargeted genes and ultimately undergo apoptosis. Further dsRNA occurs in the nucleus when complementary sequences are forming stem-loop structures.

Downstream
The region extending in the 3′ direction from a gene.

Drosophila melanogaster
Species of small fly, commonly called a fruit fly, much used in genetic studies of development.

Druggable genome
Mainly receptors and extracellular proteins, hormone precursors, transcription factors.

dsRBD
dsRNA binding domain.

DSRM
dsRNBA binding motif.

dsRNA
See double-stranded RNA.

E. coli
Abbreviation of the bacterium *Escherichia coli*, a common Gram-negative bacterium useful for cloning experiments. Present in the human intestinal tract. Hundreds of strains of *E. coli* exist. One strain, K-12, has been completely sequenced. Laboratory strains of *E. coli* are derived from the strain *E. coli* K12, a strain that does not survive outside the laboratory conditions.

Ecdysis
Shedding of the larval or pupal cuticle.

Ectoderm
One of the three primordial germ layers formed during early embryogenesis; a precursor of the central nervous system, sensory organs, adrenal medulla, etc.

Ectopic
Occurring in an abnormal position or in an unusual manner or form, e.g., ectopic recombination.

Ectopic expression
The occurrence of gene expression in a tissue in which it is normally not expressed. Such ectopic expression can be caused by the juxtaposition of novel enhancer elements to a gene during genetic manipulation of transgenic organisms.

Electroelution

A technique to remove a sample previously purified by electrophoresis on a solid support, by electrophoresing it into a buffer.

Electroporation

A technique for transfecting cells by the application of a high-voltage electric pulse. High-voltage pulses of electricity are used to open pores in cell membranes, through which foreign DNA can pass.

Elongation factors

Proteins necessary for the proper elongation and translocation processes during translation at the ribosome in prokaryotes. Replaced by eEF1 and eEF2 in eukaryotes.

Endoderm

One of the three primordial germ layers formed during early embryogenesis; a precursor of the gastrointestinal system, digestive glands, liver, lungs, etc.

Endonuclease

An enzyme which digests nucleic acids starting in the middle of the strand (as opposed to an exonuclease, which must start at an end). Examples include the restriction enzymes, DNase I and RNase A.

Enhancer

A *cis*-regulatory sequence that can elevate levels of transcription from an adjacent promoter. Many tissue-specific enhancers can determine spatial patterns of gene expression in higher eukaryotes. Enhancers can act on promoters as far as kilobases of DNA away and can be 5' or 3' to the promoter they regulate. A eukaryotic DNA sequence that increases transcription of a region even if the enhancer is distant from the region being transcribed.

Enhancer trap

A transgenic construction inserted in a chromosome, which is used to identify tissue-specific enhancers in the genome. In such a construct, a promoter sensitive to enhancer regulation is fused to a reporter gene, such that expression patterns of the reporter gene identify the spatial regulation conferred by nearby enhancers.

ENV

(env.) The env gene gives rise to the two major viral glycoproteins (gp120 and gp41) that are associated with the membrane envelope surrounding each HIV-1 virion.

Envelope

The outer coat, or envelope, of HIV is composed of two layers of fat-like molecules called lipids taken from the membranes of human cells. Embedded in the envelope are numerous cellular proteins, as well as mushroom-shaped HIV proteins that protrude from the surface. Each mushroom is thought to consist of a cap made of four glycoprotein molecules (called gp120) and a stem consisting of four gp41 molecules embedded in the envelope. The virus uses these proteins to attach to and infect cells.

Epigenesis
Development of a plant or animal from an egg or spore through a series of processes in which unorganized cell masses differentiate into organs and organ systems.

Epigenetic
Relating to, or produced by the chain of developmental processes in epigenesis that lead from genotype to phenotype after the initial action of the genes.

Episome
Term (from the Greek: *epi*, upon; *soma*, body) used for genetic elements that can either exist independently in a cell or become integrated into the host chromosome.

EST
See expressed sequence tag.

Ester
Molecule formed by the condensation reaction of an alcohol group with an acidic group. Most phosphate groups are esters.

Euchromatin
Parts of chromosomes showing the normal cycle of condensation and normal staining properties at nuclear divisions (from the Greek: *eu*, true). It is thought to contain active or potentially active genes.

Expressed sequence tag (EST)
Expressed sequence tag (EST). A unique DNA sequence derived from a cDNA library (therefore from a sequence which has been transcribed in some tissue or at some stage of development). The EST can be mapped, by a combination of genetic mapping procedures, to a unique locus in the genome and serves to identify that gene locus.

F1
Denotes the different generations involved in breeding experiments. F_1 is the first filial or filial-one generation – that is, the progeny after mating or genetically crossing two types of parents with different genotypes or phenotypes (the parents are known as the P generation).

F2
F_2 is the second filial or filial-two generation – that is, the progeny of self-fertile or intercrossing F_1 individuals, and so on. Members of this generation are two generations remote from the original parent generation.

FACS
Fluorescence-activated cell sorting; a technique for separating cells according to their fluorescence after attachment of a fluorophore to specific cells; for example with a fluorescent antibody. Droplets that contain no more than one cell are passed by a device that imposes an electrical charge on fluorescent droplets and deflects them into their own receptacle.

FASTA

An algorithm used for database similarity searching. The program looks for optimal local alignments by scanning the sequence for small matches called 'words'. Initially, the scores of segments in which there are multiple word hits are calculated ('init1'). Later, the scores of several segments may be summed to generate an 'initn' score. An optimized alignment that includes gaps is shown in the output as 'opt'. The sensitivity and speed of the search are inversely related and controlled by the 'k-tup' variable which specifies the size of a 'word'. (Pearson and Lipman) [NCBI Bioinformatics]. More rigorous and slower than BLAST.

Fertilization

Fusion of a male and a female gamete (both haploid) to form a diploid zygote, which develops into a new individual.

Flow cytometry (FACS)

Analysis of biological material by detection of the light-absorbing or fluorescing properties of cells or subcellular fractions (i.e., chromosomes) passing in a narrow stream through a laser beam. An absorbance or fluorescence profile of the sample is produced. Automated sorting devices are used to fractionate samples and sort successive droplets of the analyzed stream into different fractions depending on the fluorescence emitted by each droplet.

Fluorescent dye

Molecule that absorbs light at one wavelength and responds by emitting light at another wavelength; the emitted light is of longer wavelength (and hence of lower energy) than the light absorbed.

Foregut

An anterior part of the alimentary canal derived from the ectoderm.

Functional Genomics

Functional Genomics deals with the development and application of experimental approaches to assess gene function. It makes use of the information provided by genome sequencing and mapping.

gag

The gene encoding the polyprotein translation product that contains the capsid proteins of a retrovirus; the group-specific antigens of a retrovirus.

Gal4

A yeast transcription factor, which in attendance of galactose switches on the genes of galactose metabolism. It can be recognized by Gal4 binding sites (see UAS) and is used in the Gal4/UAS conditional expression system.

Gamete

Specialized haploid cell, either a sperm or an egg, serving for sexual reproduction.

Gastrula

An embryonic stage at which complex morphogenetic movements result in the formation of the three germ layers; ectoderm, mesoderm, and endoderm.

Gastrulation

During early embryogenesis, the invagination and reshaping of the cells of the blastoderm that results in differentiation into ectoderm, endoderm and mesoderm.

Gel filtration

A type of chromatography in which the components of a mixture are separated according to molecular size.

Gene targeting

The introduction of a homologous DNA sequence into a specific site in the genome of a cell, either by replacement of the former sequence (i. e., *sequence replacement*), or by insertion into the former sequence (i. e., *sequence insertion*). A vector that is homologous and co-linear with a partial sequence of the targeted gene is introduced into an appropriate cell, for example, a stem cell of some sort; it is incorporated into the genome of some of the cell's progeny and replaces the former sequence. If, however, homologous regions of the vector hybridize to the gene, and it is then incorporated into the gene by homologous recombination, the result is a disruption of the gene by insertion of the vector sequence.

Gene therapy

Gene therapy in its original meaning is the substitution of a defective gene by a functional one in an organism suffering from a genetic disease. Recombinant DNA techniques are used to isolate the functioning gene and insert it into the target cells.

Gene transfer

The transfer of genes into a cell by any of a number of different methods available.

Genotype

Term for the hereditary constitution of an individual, or of particular nuclei within its cells.

Germ cells

Precursor cells that give rise to gametes.

Germline

Mature male or female reproductive cell (sperm or ovum) with a haploid set of chromosomes (23 for humans). The lineage of germ cells (which contribute to the formation of a new generation of organisms), as distinct from somatic cells (which form the body and leave no descendants).

Germline mosaicims

Presence of two or more cell lines that differ in genetic make-up among gametes (germ cells); implies risk of transmission of mutations present in the gonads to offspring.

Gonad
The organs of the reproductive system in which the germ cells reside.

H3K9
Histone H3 at lysine K9; major methylation site.

Hairpin
A section of single-stranded DNA that curls back onto itself, creating a partial double helix that resembles a hairpin.

Hairpin ribozymes
Named for their conserved domains that form a hairpin-like shape, they are *trans*-cleaving molecules that tend to be 70 nucleotides in length. Like hammerhead ribozymes, they catalyze simple self-cleaving reactions of phosphodiester bonds and can be produced naturally or synthetically.

Hammerhead ribozymes
Small *trans*-cleaving ribozymes characterized by a hammerhead motif – three short helices flanking a central catalytic core of 15 conserved nucleotides. The best understood of the ribozymes.

Haploid
A single set of chromosomes (half the full set of genetic material), present in the egg and sperm cells of animals and in the egg and pollen cells of plants (from the Greek: *haploos*, single). Human beings have 23 chromosomes in their reproductive cells.

HeLa cell
Line of human epithelial cells that grows vigorously in culture. Derived from a human cervical carcinoma.

Hermaphrodite
An individual with both male and female genitalia. (1) A plant species in which male and female organs occur in the same flower of a single individual (compare monoecious plant). (2) An animal with both male and female sex organs.

Heterochromatin
Densely condensed chromosomal regions, believed to be for the most part genetically inert. Chromatin that remains tightly coiled (and darkly staining) throughout the cell cycle. Composed of repetitive DNA, stains dark bands in G-banding.

Heterodimer
A dimer in which the two subunits are different.

High throughput
Although the adjective 'high throughput' was originally coined in a drug-screening context, high-throughput strategies to accelerate and automate earlier steps in the drug discovery pipeline have already been introduced. With the introduction of genomics-based drug discovery strategies, the concept of high throughput has extended to areas such as gene expression analysis, where microarrays allow the simultaneous expression profiling of thousands of genes in diseased versus normal samples. In

the early stages of disease-gene research, when one wishes to identify alterations in gene expression that are associated with a disease state with significant societal impact and potential market value, a microarray-based approach provides significant acceleration over traditional methods to evaluate candidate genes one at a time.

Histone
Protein associated with DNA in chromosomes in the nucleus of the cell.

Histone deacetylase (HDAC)
These enzymes remove an acetyl group from histones, which allows histones to bind DNA and inhibit gene transcription.

Homeo-box
A consensus sequence of about 180 base pairs discovered in homeotic genes in *Drosophila*. Also found in other developmentally important genes from yeast to human beings. A family of quite similar 180 base pair DNA sequences that encode a polypeptide sequence called a homeodomain, a sequence-specific DNA binding sequence. While the homeo-box was first discovered in all homeotic genes, it is now known to encode a much more widespread DNA-binding motif.

Homeo-domain
An approximately 60-amino acid protein domain translated from the homeo-box. A highly conserved family of protein domain sequences 60 amino acids in length found within a large number of transcription factors that can form a helix-turn-helix structure and bind DNA in a sequence-specific manner.

Homeotic genes
These genes have been identified originally by mutations (homeotic mutations) causing aberrant segment development or replacement of one body structure by a different one in the fruit fly *Drosophila melanogaster*. Examples of such genes are anatennapedia, bithorax, engrailed. Homeotic genes comprise a large gene family. The genes are organized in clusters expressed in the developing embryo in complex temporal and spatial patterns and with a positional hierarchy. The proteins encoded by homeotic genes have been found to contain an evolutionarily highly conserved region, the so-called homeobox or homeodomain. Proteins containing a homeobox sequence domain are sequence-specific DNA-binding proteins that bind to gene regulatory sequences in many different genes. The proteins encoded by these genes control gene expression and modify expression patterns of genes both in developing, as well as in adult, tissues. They play an important role in mammalian embryonic pattern formation and are involved in oncogenic processes. Some of these genes play a crucial role in local pattern formation, while others are tissue-specific or ubiquitous transcription regulators.

Homologous recombination
Substitution of a segment of DNA by another that is identical homologous, or nearly so. Occurs naturally during meiotic recombination; also used in the laboratory for gene targeting to modify the sequence of a gene.

Homology
Similarities in DNA, RNA, or protein sequences between individuals of the same species or among different species. These two types of homology are called parology and orthology.

Homozygote
Term for a diploid individual derived from two diploid individuals, which carries only one member of the alleles of a gene, that is, a zygote derived from the union of gametes identical (from the Greek: *homos*, alike) in respect of a particular gene.

Housekeeping gene
Gene serving a function required in all the cell types of an organism, regardless of their specialized role.

Hybrid
A cross-bred, heterozygous organism or cell, an individual from any cross involving parents of differing genotypes. In molecular genetics a DNA, dsRNA, or DNA/RNA molecule with strands of different origin.

Hybridization
Process whereby two complementary nucleic acid strands form a double helix during an annealing period; a powerful technique for detecting specific nucleotide sequences.

Hydrodynamic transfection
Injection of a relatively large volume of buffer and DNA (2.6 ml/20 g mouse) into the tail vein of a mouse within a short period of time (5–7 s).

Hydrophilic
Polar molecule or part of a molecule that forms enough hydrogen bonds to water to dissolve readily in water (literally, 'water-loving').

Hydrophobic (lipophilic)
Nonpolar molecule or part of a molecule that cannot form favorable bonding interactions with water molecules and therefore does not dissolve in water (literally, 'water-hating').

Hydroxyl (–OH)
Chemical group consisting of a hydrogen atom linked to an oxygen, as in an alcohol.

Hypertonic
Describes any medium with a sufficiently high concentration of solutes to cause water to move out of a cell due to osmosis (from the Greek: *huper*, over).

Hypotonic
Describes any medium with a sufficiently low concentration of solutes to cause water to move into a cell due to osmosis (from the Greek: *hupo*, under).

Immortalization
Production of a cell line capable of an unlimited number of cell divisions. Can be the result of a chemical or viral transformation or of fusion with cells of a tumor line.

Immune response
Response made by the immune system of a vertebrate when a foreign substance or microorganism enters its body.

Immunoprecipitation
A purification technique that separates antigenic material from a soluble mixture by precipitation with an appropriate antibody; an essential step in radioimmunoassays.

in silico
Method to test biological models, drugs and medical interventions using sophisticated computer models rather than expensive laboratory (in vitro) and animal experiments (in vivo). Integrated methods of genomics and life science informatics.

in vitro
Outside a living organism. In a glass (test-tube).

in vivo
Within a living organism.

Inbreeding
The mating of genetically related individuals. Mating between relatives. Breeding through a succession of parents belonging to the same stock.

Inducible operon
A gene system, often encoding a coordinated group of enzymes involved in a catabolic pathway, is inducible if an early metabolite in the pathway causes activation, usually by interaction with and inactivation of a repressor, of transcription of the genes encoding the enzymes.

Infection
Process of cellular entry of a protozoan, fungal, bacterial, or viral pathogen or parasite.

Initiation factor
An accessory protein that is necessary for assembly of the ribosome-mRNA complex and the start of protein synthesis.

Inoculation
Process of introducing a substance into an organism.

Insert
In molecular genetics refers to a DNA sequence of interest that has been inserted into a cloning vector such as a plasmid or bacteriophage.

Instar
Period between the hatching of the egg and the first larval molts (ecdysis), and the period between two successive ecdyses; synonym for stadium.

Integumentary system
The organ system that forms the covering layer of the animal.

Interferon
A signaling molecule that triggers a signal cascade to induce cells to resists viral replication and induces programmed cell death.

Intergenic DNA
Intergenic DNA is the DNA found between two genes. The term is often used to mean nonfunctional DNA (or at least DNA with no known importance to the two genes flanking it). Alternatively, one might speak of the 'intergenic distance' between two genes as the number of base pairs from the poly(A) site of the first gene to the cap site of the second. This usage might therefore include the promoter region of the second gene.

Internal ribosome entry site (IRES)
A feature in the secondary structure near the 5'-end of a picornaviral RNA genome that allows eukaryotic ribosomes to bind and begin translation without binding to a 5'-capped end. Or: In the 5'-cap-independent initiation of translation, the place on the mRNA that first attaches a ribosome subunit.

Intestine
Synonym for gut.

Inverted repeat
A repeat of a nucleotide sequence in the opposite orientation of the base sequence and mostly below 50 base pairs.

IRES
see Internal ribosome entry site.

JAK
Janus kinase; Family of intracellular tyrosine kinases (120–140 kDa) that associate with cytokine receptors (particularly, but not exclusively, interferon receptors) and are involved in the signaling cascade. JAK is so-called either from Janus kinase (Janus was the gatekeeper of heaven) or 'just another kinase'.

Kinase
A kinase is in general an enzyme that catalyzes the transfer of a phosphate group from ATP to something else.

Knock-down
A strategy for down-regulation of expression of a gene by incorporation into the genome an antisense oligodeoxynucleotide or ribozyme sequence that is directed against the targeted gene.

Knockout
Inactivation of specific genes. Knockouts are often created in laboratory organisms such as yeast or mice so that scientists can study the knockout organism as a model for a particular disease.

Knockout experiment
A technique for deleting, mutating or otherwise inactivating a gene in a mouse. This laborious method involves transfecting a crippled gene into cultured embryonic stem cells, searching through the resulting clones for one in which the crippled gene exactly replaced the normal one (by homologous recombination), and inserting that cell back into a mouse blastocyst. The resulting mouse will be chimeric, but its germ cells will carry the deleted gene. After a few rounds of breeding, progeny are produced in which both copies of the gene are inactivated.

Lamins
Intermediate filament proteins that form the fibrous matrix (nuclear lamina) on the inner surface of the nuclear envelope.

Larva (pl. larvae)
The feeding, wingless, sexually immature, developmental stage of an insect after emerging from the egg.

Lentivirus
'Slow' virus characterized by a long interval between infection and the onset of symptoms. HIV is a lentivirus, as is the simian immunodeficiency virus (SIV) that infects nonhuman primates.

let-7
The 21-nucleotide let-7 micro RNA in *C. elegans* that regulates developmental timing.

lhRNA
long hairpin RNA. A dsRNA stem-loop structure of more than 50 nucleotides.

lin-4
lin-4 is a micro RNA that is essential for the normal temporal control of diverse postembryonic developmental events in *C. elegans*. lin-4 acts by negatively regulating the level of LIN-14 protein, creating a temporal decrease in LIN-14 protein starting in the first larval stage (L1).

Lineage
Linear evolutionary sequence from an ancestral species through all intermediates to a descendant species.

Linker
A small piece of synthetic double-stranded DNA which contains something useful, such as a restriction site. A linker might be ligated onto the end of another piece of DNA to provide a desired restriction site.

Liposomes
A spherical particle in an aqueous (watery) medium (e.g., inside a cell) formed by a lipid bilayer enclosing an aqueous compartment. Microscopic globules of lipids are manufactured to enclose medications. The fatty layer of the liposome is supposed to protect and confine the enclosed drug until the liposome binds with the outer mem-

brane of target cells. By delivering treatments directly to the cells needing them, drug efficacy may be increased while overall toxicity is reduced.

Locus
The position of a gene on a chromosome. The use of the term locus is sometimes restricted to main regions of DNA that are expressed.

Long hairpin RNA
See lhRNA.

Long terminal repeat (LTR)
The genetic material at each end of the HIV genome. When the HIV genome is integrated into a cell's own genome, the LTR interacts with cellular and viral factors to trigger the transcription of the HIV-integrated HIV DNA genes into an RNA form that is packaged in new virus particles. Activation of LTR is a major step in triggering HIV replication.

LoxP site
A 34-bp sequence consisting of two 13-bp inverted repeats and an 8-bp asymmetrical core spacer region. Recognition site for the Cre recombinase. Locus of cross over in P1.

LR Clonase™
An enzyme mix that facilitates recombination between *att*L and *att*R sites in an LxR Gateway® recombination reaction, to allow transfer of DNA sequences from an entry vector to a destination vector.

MALDI-MS
Matrix-assisted laser desorption/ionization mass spectrometry.

Mating
The act of pairing a male and female organism for reproductive purposes.

MAZ
Zinc finger protein that is involved in transcriptional pausing and poly adenylation of mRNA.

Meiosis
Special type of cell division by which eggs and sperm cells are produced, involving a diminution in the amount of genetic material. Comprises two successive nuclear divisions with only one round of DNA replication, which produces four haploid daughter cells from an initial diploid cell (from the Greek: *meiosis*, diminution).

Melanocyte
Cell that produces the dark pigment melanin; responsible for the pigmentation of skin and hair.

Meristem
An organized group of dividing cells whose derivatives give rise to the tissues and organs of a flowering plant. Key examples are the root apical meristem and shoot apical meristem.

Mesoderm
One of the three primordial germ layers formed during early embryogenesis; a precursor of muscle, adipose tissue, blood vessels, the gastrointestinal tract, etc.

Methyl (–CH_3)
Hydrophobic chemical group derived from methane (CH_4).

Methylation
Attachment of methyl groups (–CH_3) to DNA most commonly at cytosine residues. May be involved in the regulation of gene expression. Also may prevent some restriction endonucleases from cutting DNA at their recognition sites.

Microarray technology
A new way of studying how large numbers of genes interact with each other and how a cell's regulatory networks control vast batteries of genes simultaneously. The method uses a robot to precisely apply tiny droplets containing functional DNA to glass slides. The attachment of fluorescent labels to DNA from the cell facilitates the study. The labeled probes are allowed to bind to complementary DNA strands on the slides. The slides are then placed into a scanning microscope that can measure the brightness of each fluorescent dot; brightness reveals how much of a specific DNA fragment is present, an indicator of how active it is.

Microarrays
An ordered series of small (200 µm) spots of material (nucleic acid or protein) immobilized on a solid surface (see also Microchip) such that their interaction with a target molecule in solution can be observed. Microarrays usually contain thousands of such sample spots.

Microinjection
The insertion of a substance into a cell through a microelectrode. Typical applications include the injection of drugs, histochemical markers (such as horseradish peroxidase or lucifer yellow) and RNA or DNA in molecular biological studies. To extrude the substances through the very fine electrode tips, either hydrostatic pressure (pressure injection) or an electric current (ionophoresis) is employed.

Micro-RNA (miRNA)
A large class of 21- to 24-nucleotide noncoding RNAs that are related to siRNA. The miRNAs have diverse expression patterns during development. The abundance of these tiny RNAs, their expression patterns, and their evolutionary conservation imply that, as a class, miRNAs have broad regulatory functions in animals. They are processed from long stem-loop structures in the nucleus and are transported into cytosol where they are processed by Dicer into 21- to 23-mers and act as translational repressors.

Mispairing

The presence of a nucleotide in one nucleotide chain of a DNA molecule, which is not the complement of that at the corresponding position in the other chain.

Mobile genetic element

A transposon or an insertion sequence; a polynucleotide sequence that can move from a chromosome or plasmid to another chromosome or plasmid.

MOI

See multiplicity of infection.

Molt

In insects, the formation of a new cuticle followed by shedding of the old cuticle (ecdysis).

Monoclonal

Derived from a single clone.

Morpholinos

Morpholino oligos are assembled from four different Morpholino subunits, each of which contains one of the four genetic bases (adenine, cytosine, guanine, and thymine) linked to a 6-membered morpholine ring. 18 to 25 subunits of these four subunit types are joined in a specific order by nonionic phosphorodiamidate intersubunit linkages to give a Morpholino oligo. Commercially available antisense types have suffered from such limitations as poor specificity, instability, unpredictable targeting and undesirable non-antisense effects. Morpholino antisense oligos overcome these limitations, and can be readily delivered into cultured cells – making them the best tools for most genetic studies and for drug target validation programs.

Mosaicism

Condition in which an individual harbors two or more genetically distinct cell lines; results from a genetic change after formation of a zygote.

Mouse model

A laboratory mouse useful for medical research because it has specific characteristics that resemble a human disease or disorder. Strains of mice having natural mutations similar to human ones may serve as models of such conditions. Scientists can also create mouse models by transferring new genes into mice or by inactivating certain existing genes in them.

Multiplicity of infection (MOI)

Average ratio of infectious virus particles to target cells in a given infection.

NEF

(nef) One of the regulatory genes of HIV. Three HIV regulatory genes – tat, rev, and nef – and three so-called auxiliary genes – vif, vpr, and vpu – contain information necessary for the production of proteins that control the virus; ability to infect a cell, produce new copies of itself, or cause disease. See rev; tat.

Nematodes

Group of organisms also known as the Roundworms. Nematodes have what can only be described as a typical 'worm' shape – long, tapered at the ends, and round in cross-section (think of the shape of an earthworm, but earthworms are *not* nematodes). They have an internal body cavity, with recognizable digestive and reproductive tracts. They reproduce by laying eggs, or larvae which hatch from their eggs inside the body of the female worm. They are among the most common multicellular parasite of humans in the world, although the majority of nematodes are not parasitic, and live in the soil. Examples of parasitic roundworms include Human Roundworm (*Ascaris*), Pinworm/Threadworm, Whipworm, Hookworm and Filarial Worms.

Nick translation

Method used to introduce ^{32}P into a DNA probe so that the probe can be detected.

NLS

Nuclear localization sequence; positively charged nuclear transport sequence or nuclear targeting sequence, which direct the protein into the cell nucleus after receptor-mediated endocytosis. No strict consensus sequences have been found for nuclear targeting sequences. Passage through the nuclear pore complex is an energy-dependent process that requires recognition of the NLS by import receptors and interaction with import factors (importin-alpha, karypherin-alpha (importin-beta, karypherin-beta).

Nomarski interference optics

A microscope optical system for visualization of differences in refractive indices so as to observe unstained specimens.

Noncoding strand

Anti-sense strand.

Noncovalent bond

Chemical bond in which, in contrast to a covalent bond, no electrons are shared. Noncovalent bonds are relatively weak, but they can sum together to produce strong, highly specific interactions between molecules.

Nonsense mutation

Mutation that changes a codon or an amino acid to a termination or stop codon and leads to premature termination of translation.

Nuclease

An enzyme which degrades nucleic acids. A nuclease can be DNA-specific (a DNase), RNA-specific (RNase), or nonspecific. It may act only on single-stranded nucleic acids, or only on double-stranded nucleic acids, or it may be nonspecific with respect to the strand. A nuclease may degrade only from an end (an exonuclease), or may be able to start in the middle of a strand (an endonuclease).

Nuclease protection assay

See 'RNase protection assay'.

Nucleofection™
Nucleofection™ is a method for gene transfer. Based on electroporation, the Nucleofector™ concept uses a combination of electrical parameters and cell type-specific Nucleofector™ buffer solutions. The Nucleofector™ technology is unique in its ability to transfer DNA directly into the nucleus of a cell. Thus, cells with limited ability to divide, such as primary cells and hard-to-transfect cell lines, are made accessible for efficient gene transfer.

Offspring
See Progeny.

OAS (2′-5′-oligoadenylate synthase)
An enzyme of the interferon response cascade that is activated by double-stranded RNA (dsRNA) and catalyzes the conversion of ATP to 2'-5′ A oligonucleotides which further activates RNase L, leading to nonspecific RNA degradation.

Oligonucleotide
A short fragment of single-stranded DNA, typically 5 to 50 nucleotides long.

Oncogene
One of a large number of genes that can help make a cell cancerous. Typically, a mutant form of a normal gene (proto-oncogene) involved in the control of cell growth or division.

Oocyte
Developing egg; usually a large and immobile cell. The gamete in females.

Operator
The region of a regulated bacterial gene to which the product of a regulatory gene binds to modulate transcription.

Operon
In a bacterial chromosome, a group of contiguous genes that are transcribed into a single mRNA molecule.

Origin
The point of specific sequence at which DNA replication is initiated.

Orthology
Describes the evolutionary origin of a locus. Loci in two species are said to be orthologous when they arise from the same locus of a common ancestor. For example, gene A in species 1 and 2 are orthologous. In contrast, gene B1, which has arisen by gene duplication in species 2, is paralogous to gene B in species 1.

Osmotic pressure
Pressure that must be exerted on the high solute concentration side of a semipermeable membrane to prevent the flow of water across the membrane due to osmosis.

Oxidation

Loss of electron density from an atom, as occurs during the addition of oxygen to a molecule or when a hydrogen is removed. Opposite of reduction.

P element

A *Drosophila* transposable element that has been used as a tool for insertion mutagenesis and for germline transformation.

P1

Parental generation.

P53

A gene which normally regulates the cell cycle and protects the cell from damage to its genome. A 53-kDa protein, the product of a tumor suppressor gene. The loss of this protein due to mutation is a primary event in the formation of many types of cancer (breast, colon, lung, leukemia, liver). P53 is involved in regulating the activity of some other genes. It also prevents cells entering cell division. P53 levels are increased after DNA is damaged by UV and ionizing radiation, and cells are arrested in cell division until either the damage is repaired or they die by apoptosis (programmed cell death).

Packaging signal

Specific sequence of bases within the genome of a virus that functions in the association and insertion of the genome into the procapsid.

Palindromic sequence

Nucleotide sequence that is identical to its complementary strand when each is read in the same chemical direction; for example, GATC.

Partial digest

A restriction digest that has not been allowed to go to completion and thus contains pieces of DNA with some restriction endonuclease sites that have not yet been cleaved.

Pathogen

Disease-causing parasite, often microorganisms.

PAZ

A protein domain that is found in a variety of proteins especially in the *Drosophila* proteins **PIWI**, argonaute, and zwille.

Peptide bond

Chemical bond between the carbonyl group of one amino acid and the amino group of a second amino acid – a special form of amide linkage.

Permutation

A permanent and heritable change in a gene that does not have phenotypic consequences (does not cause disease), but predisposes to a future mutation.

Persistence
Property of some plasmids that can be replicated in a host cell for a longer period of time, when they contain antibiotic resistance genes that allow for positive selection and persistence of the genotype.

Phagemid
A type of plasmid which carries within its sequence a bacteriophage replication origin. When the host bacterium is infected with 'helper' phage, the phagemid is replicated along with the phage DNA and packaged into phage capsids.

Pharynx
An anterior region of the fore gut. Connection between oral cavity and esophagus.

Phenotype
Term for the appearance (from the Greek: *phainein*, to appear) of an organism with respect to a particular character or group of characters (physical, biochemical, and physiologic), as a result of the interaction of its genotype and its environment. Often used to define the consequences of a particular mutation.

Phosphorothioate
A phosphate analogue, especially a nucleotide analogue, in which a sulfur atom replaces an oxygen in one of the phosphate groups.

Phosphorylation
Reaction in which a phosphate group becomes covalently coupled to another molecule.

PIWI
Protein domain that is conserved in proteins and encodes a nucleoplasmic factor present in both somatic and germline cells and has so far unknown function in *Drosophila*. This family of proteins is mainly implicated in translational control and gene silencing.

PKR
Protein kinase R; the kinase, activated by double-stranded RNA (dsRNA), phosphorylates and inactivated the translation initiation factor eIF2α, resulting in a global inhibition RNA translation.

Plaque assay
A technique for quantification of infectious phage particles by counting plaque-forming units.

Plaque-forming unit (PFU)
A unit of infectious virus determined by the ability of the virus to form a plaque or area of lysed cells on a 'lawn' of susceptible cells.

Plasmid
Autonomously replicating, extrachromosomal circular DNA molecules, distinct from the normal bacterial genome and nonessential for cell survival under nonselective conditions. Some plasmids are capable of integrating into the host genome and

are used as a cloning vector for small pieces of DNA (typically 50 to 5000 base pairs) by insertion into the plasmid. A number of artificially constructed plasmids are used as cloning vectors.

Pol II
RNA polymerase II. An RNA polymerase that transcribes messenger RNA (mRNA) and most of the small nuclear RNAs of eukaryotes in conjunction with various transcription factors.

Pol III
RNA polymerase III. An RNA polymerase that transcribes stable RNAs such as transfer RNAs, 5S ribosomal RNA, 7SL RNA and U6 small nuclear RNAs that are usually not translated into protein.

Pol III promoter
A pol III promoter, U6 or H1, is typically used to drive the production of shRNA. Pol III promoters have all the elements required for the initiation of transcription upstream of a defined start site.

Pol III terminator
Four or more Ts for Pol III-dependent transcription termination.

Pole cell
The cells at the posterior of the early embryo which give rise to the germ cells.

Poly(A) tail
After an mRNA is transcribed from a gene, the cell adds a stretch of A residues (typically 50–200) to its 3′-end. It is thought that the presence of this 'poly(A) tail' increases the stability of the mRNA (possibly by protecting it from nucleases). Note that not all mRNAs have a poly(A) tail; the histone mRNAs in particular do not.

Polyamine
Compounds with many amino groups that are associated with nucleic acids.

Polycistronic mRNA
An mRNA molecule that contains multiple open reading frames, usually found in prokaryotic mRNAs.

Polyribosome (polysome)
mRNA molecule to which are attached a number of ribosomes engaged in protein synthesis.

Polysome
The configuration of several ribosomes simultaneously translating the same mRNA. Shortened form of the term polyribosome.

Posterior
Situated toward the tail end of the body.

Post-transcriptional modification

The processing of RNA subsequent to its synthesis, which may include cleavage of phosphodiester bonds and modification of bases. For eukaryotic mRNA it includes capping of the 5'-end, polyadenylation of the 3'-end and splicing of introns.

Post-translational modification

Enzyme-catalyzed change to a protein made after it is synthesized. Examples are cleavage, glycosylation, phosphorylation, methylation, and prenylation.

Potter-Elvehjem homogenizer

A device used to disrupt tissues. A cylindrical glass or hard polymer pestle rotates in a close-fitting tube and a suspension of the tissue particles is subjected to shearing forces as the pestle moves up and down and presses the suspension through the space between the rotating pestle and the tube.

pre-miRNA

Pri-miRNAs that are trimmed into 70-nucleotide (nt) pre-miRNA forms, mainly in the nucleus.

pri-miRNA

Primary-precursor micro RNA.

Probe

A radioactively, fluorescent or immunologically labeled oligonucleotide (RNA or DNA) used to detect complementary sequences in a hybridization experiment, for example, to identify bacterial colonies that contain cloned genes or to detect specific nucleic acids following separation by gel electrophoresis.

Processing

The reactions occurring in the nucleus which convert the primary RNA transcript to a mature mRNA. Processing reactions include capping, splicing and polyadenylation. The term can also refer to the processing of the protein product, including proteolytic cleavages, glycosylation, etc.

Progeny

The subsequent generation following a mating or crossing of parents; offspring.

Promoter

The first few hundred nucleotides of DNA 'upstream' (on the 5' side) of a gene, which control the transcription of that gene. The promoter is part of the 5' flanking DNA – that is not transcribed into RNA, but without the promoter, the gene is not functional. The promoter sequence is located immediately upstream from the cap site, and includes binding sites for one or more transcription factors which cannot work if moved farther away from the gene.

Pronuclear fusion

The merging of two pronuclei in a fertilized egg to fuse and produce a single genome.

Pronucleus

Haploid nucleus of a sperm or an egg prior to fertilization. Sperm and egg cells carry half the number of chromosomes of other nonreproductive cells. When the pronucleus of a sperm fuses with the pronucleus of an egg, their chromosomes combine and become part of a single nucleus in the resulting embryo, which contains a full set of chromosomes.

Proteomics

Systematic analysis of protein expression of normal and diseased tissues that involves the separation, identification and characterization of all of the proteins in an organism.

Proteosome/proteasome

A complex structure within eukaryotic cells that is the site of protein degradation. Proteins destined for turnover at the proteosome have been tagged by the addition of ubiquitin.

PTD

Protein transduction domain; see CPP.

PTGS

Post transcriptional gene silencing; a phenomenon first identified in plants that has also been shown to occur in animals. Although PTGS was initially described as an endogenous method for viral defense and transposon silencing, it has now emerged as an exciting new research tool, RNA interference.

Pupa

The 'dormant' stage in the life-cycle of some insects where the larva changes into the adult.

Quelling

Cosuppression as described in *Neurospora crassa* – this term has only been used to describe silencing of a gene in fungi.

RdRps

RNA-directed (dependent) RNA polymerases necessary for the amplification of an RNA fragment to produce dsRNA.

Recombinant DNA

A novel DNA sequence formed by the joining, usually *in vitro*, of two nonhomologous DNA molecules.

Recombinase

Isomerase; this enzyme catalyzes the processes of DNA recombination, and is used in recombinant DNA technology. The process of site-specific DNA recombination is involved in viral integration, excision and chromosomal segregation. These processes are catalyzed by recombinase enzymes that recognize specific DNA sequences and catalyze the reciprocal exchange of DNA strands between these sites.

Recombination

The process by which progeny derive a combination of linked genes different from that of either parent. In higher organisms, this can occur by crossing over between their loci during meiosis. Recombination may come about through random orientation of nonhomologous chromosome pairs on the meiotic spindles, from crossing over between homologous chromosomes, from gene conversion, or by other means. See homologous recombination.

Reduction

Addition of electron density to an atom, as occurs during the addition of hydrogen to a molecule or the removal of oxygen from it. Opposite of oxidation.

Repressor

Protein that binds to a specific location (operator) on DNA and prevents RNA transcription from a specific gene or operon.

Response element

By definition, a 'response element' is a portion of a gene, which must be present in order for that gene to respond to some hormone or other stimulus. Response elements are binding sites for transcription factors. Certain transcription factors are activated by stimuli such as hormones or heat shock. A gene may respond to the presence of that hormone because the gene has in its promoter region a binding site for hormone-activated transcription factor. Examples: the glucocorticoid response element (GRE) or the tetracycline response element (TRE).

Retrotransposon

Special transposon whose sequence is transcribed to RNA in the cell. After generation of the RNA strand, a reverse transcriptase produced by the retrotransposon reconverts the RNA to DNA. This sequence is integrated into the original DNA strand at any position.

Retrovirus

Short for Reverse Transcriptase Onko Virus. A type of virus that contains RNA as its genetic material. The RNA of the virus is translated into DNA, which inserts itself into an infected cell's own DNA. Retroviruses can cause many diseases, including some cancers and AIDS and are used as vectors to introduce genes (or portions thereof) of interest into eukaryotic cells.

REV

(rev) One of the regulatory genes of HIV. Three HIV regulatory genes – tat, rev, and nef – and three so-called auxiliary genes – vif, vpr, and vpu – contain information necessary for the production of proteins that control the virus' ability to infect a cell, produce new copies of the virus, or cause disease. See nef; tat.

Reverse genetics

The experimental procedure that begins with a cloned segment of DNA, or a protein sequence, and uses this knowledge to introduce programmed mutations (through directed mutagenesis) back into the genome in order to investigate gene and protein function.

Ribozyme

Catalytic or autocatalytic RNA. RNA with enzymatic activity, for instance, self-splicing RNA molecules.

Ribozyme technology

The method is based on an RNA molecule that binds the target messenger RNA in a sequence-specific manner and catalyzes the cleavage of the latter by RNA catalytic activity.

RISC

RNA-induced silencing complex; this is the proposed complex made of multiple proteins that acts to bring the siRNA and the cellular mRNA together and activates a cleavage mechanism (likely endonucleolytic) so that the mRNA is released and degraded.

RNA

Nuclei acid species containing ribonucleotides. The structure of RNA is similar to that of DNA. There are several classes of RNA molecules, including messenger RNA, transfer RNA, ribosomal RNA, and other small RNAs, each serving a different purpose.

RNA decoys

RNAs used to competitively bind pathogenic protein molecules in order to modulate their activity.

RNA editing

A mechanism that is responsible for post-transcriptional alteration of genetic information. The term is used to identify any mechanism which will produce mRNA molecules with information not specifically encoded in DNA. Initially, the term referred to the insertion or deletion of particular bases (e.g., uridine), or some sort of base conversion (e.g., adenosine guanosine). Today, many more RNA editing mechanisms, have been observed.

RNA interference

'RNA interference' (also known as 'RNA silencing') is the mechanism by which small double-stranded RNAs can interfere with expression of any mRNA having a similar sequence. Those small RNAs are known as 'siRNA', for short interfering RNAs. The mode of action for siRNA appears to be via dissociation of its strands, hybridization to the target RNA, extension of those fragments by an RNA-dependent RNA polymerase, then fragmentation of the target. Importantly, the remnants of the target molecule appears to then act as an siRNA itself; thus, the effect of a small amount of starting siRNA is effectively amplified and can have long-lasting effects on the recipient cell. The RNAi effect has been exploited in numerous research programs to deplete the call of specific messages, thus examining the role of those messages by their lack.

RNA polymerase

The movement of RNA polymerase (RNAP) along DNA during transcription is a complex set of different activities, including initiation, elongation, pausing, backtracking, and arrest. A complete understanding of how this molecular machinery works requires characterization of the individual activities, when and why they occur, what structural components are required in each case, and what the biochemical parameters are.

RNA virus

A virus with an RNA genome that may be either an mRNA, (+)-RNA, or its complement, (–)-RNA. Class 1 contains (+)-RNA; class 2 (–)-RNA, which is the template for an RNA-dependent RNA polymerase; class 3, double-stranded RNA, in which (+)-RNA is synthesized by an RNA-dependent RNA polymerase; class 4, retrovirus, in which (+)-RNA is a template for an RNA-dependent DNA polymerase (a reverse transcriptase).

RNAi

Abbreviation of RNA interference.

RNAi knockdowns

We are using antisense and RNAi to knockdown target genes in cellular disease models. The methods are both fast and inexpensive, yet specific, and serve as useful validation tools.

RNase

Ribonuclease; an enzyme which degrades RNA. It is ubiquitous in living organisms and is exceptionally stable. The prevention of RNase activity is the primary problem in handling RNA.

RNase III

A family of ribonucleases that are specifically cleaving long, double-stranded RNA (dsRNA) leaving short fragments with characteristic 3′-2-nt overhangs and a recessed 5′-phosphate.

RNase H assay

A method that is used to silence gene function and to map endonuclease-sensitive sites. Single-stranded DNA oligonucleotides complementary to an mRNA are transfected into cells, leading to the formation of RNA-DNA hybrids, which are recognized by RNase H and degraded, thus preventing protein synthesis.

RNase L

A ribonuclease that is activated by 2′-5′-A oligonucleotides, leading to the nonspecific cleavage of several RNA species including ribosomal RNA and messenger RNA, thus preventing protein biosynthesis.

RNase protection assay

This is a sensitive method to determine: (1) the amount of a specific mRNA present in a complex mixture of mRNA; and/or (2) the sizes of exons which comprise the mRNA of interest. A radioactive DNA or RNA probe (in excess) is allowed to hybridize with a sample of mRNA (for example, total mRNA isolated from tissue), after which the mixture is digested with single-strand-specific nuclease. Only the probe which is hybridized to the specific mRNA will escape the nuclease treatment, and can be detected on a gel. The amount of radioactivity which was protected from nuclease is proportional to the amount of mRNA to which it hybridized. If the probe included both intron and exons, only the exons will be protected from nuclease and their sizes can be ascertained on the gel.

RT-PCR
Reverse transcription-polymerase chain reaction.

S2 cells
Schneider cells from embryonic lysates of *Drosophila* cells, which are used as a model system in genetic experiments.

SARS
Severe acute respiratory syndrome. A disease caused by a coronavirus (SARS-CoV) with significant pathology and mortality. The first SARS cases were identified in China and Singapore in early 2003.

SDS-PAGE
Short for Sodium Dodecyl Sulfate (SDS)-PolyAcrylamide Gel Electrophoresis (PAGE). Method to separate proteins by exposing them to the anionic detergent SDS and PAGE.

Sense strand
Equal to coding strand. A gene has two strands: the sense strand and the anti-sense strand. The sense strand is, by definition, the same 'sense' as the mRNA; that is, it can be translated exactly as the mRNA sequence can. Note however that when the RNA is transcribed from this sequence, the antisense strand is used as the template for RNA polymerization.

Short hairpin RNA (shRNA)
Short hairpin RNA; shRNA are endogenously expressed from a DNA inverted repeat of 21 bp each connected with a 4- to 9-bp loop sequence. In vivo, it can replace the siRNA and can decrease the expression of a gene with complementary sequences by RNAi. shRNAs are usually delivered through nonviral and viral plasmids. The concurrent development of similar but not identical shRNA designs and expression vectors in independent laboratories, using different model systems, suggests that this approach is robust, and likely to be applicable to a wide variety of biological questions.

siRNA (small interfering RNA)
Short (or small) interfering RNA. Small double-stranded RNAs (20–25 nt) that are generated from specific dsRNAs. siRNAs results from the cleavage of dsRNA by Dicer – a member of the RNase III family – and have the characteristic length of 21–25 nt and contain 3'-end 2-nt overhangs. It is involved in eliciting the RNAi response in mammalian cells.

Small molecules

Referred for drugs as they are orally available (unlike proteins which must be administered either by injection or topically). The size of small molecules is generally under 1000 Da, but many estimates seem to range between 300 to 700 Da.

Small temporal RNA (stRNA)

Two ~21-nt RNAs regulate *C. elegans* temporal development, and one of these RNAs is likely to regulate developmental timing in animals. Genome sequence comparisons and expression analyses among animals may reveal additional stRNAs that regulate other developmental transitions.

Smith-Waterman alignment

Amino acid sequence alignment that illustrates sequence similarity. The alignment is generated using the Smith-Waterman algorithm.

Snap-back sequence

Half the base-pairing sequence of a hairpin turn of a tRNA, which can easily rebind to its complementary sequence after denaturation.

Somatic cell

Any cell in the body except gametes and their precursors or stem cells.

Spermatheca

Synonym for male gametes.

Stable transfection

A form of transfection experiment designed to produce permanent lines of cultured cells with a new gene inserted into their genome. Usually this is done by linking the desired gene with a 'selectable' gene that is, a gene which confers resistance to an antibiotic (like G418, Geneticin). On adding the antibiotic to the culture medium, only those cells which incorporate the resistance gene will survive, and essentially all of those will also have incorporated the recombinant DNA.

STATs

(= signal transducers and activators of transcription). These contain SH2 domains that allow them to interact with phosphotyrosine residues in receptors, particularly cytokine-type receptors; they are then phosphorylated by JAKs (Janus kinase), dimerize and translocate to the nucleus where they act as transcription factors. Many STATs are known; some are relatively receptor-specific, others more promiscuous, so that a wide range of responses is possible with some STATs being activated by several different receptors, sometimes acting synergistically with other STATs.

Stem-loop structure

A lollipop-shaped structure formed when a single-stranded nucleic acid molecule loops back on itself to form a complementary double helix (stem) topped by a loop.

Steric hindrance

The constraint on a reaction or conformational change that is due to crowding of atoms within the van der Waals radii of other atoms.

Stoichiometry
A ratio of different molecules in a reaction.

Stringency
During nucleic acid hybridization or reassociation, the strictness with which the Watson–Crick base-pairing is required under specified conditions of temperature, pH, salt concentration, etc. Conditions of high stringency require all bases of one polynucleotide to be paired with complementary bases on the other; conditions of low stringency allow some bases to be unpaired.

Sulfhydryl (thiol, –SH)
Chemical group containing sulfur and hydrogen found in the amino acid cysteine and other molecules. Two sulfhydryls can join to produce a disulfide bond.

Syncytial blastoderm
The part of the blastoderm developmental stage at which time there are many nuclei not separated by cell walls. In insects, the syncytial stage of blastoderm preceding the formation of cell membranes around the individual nuclei of the early embryo. (An insect embryo in which all the cleavage nuclei are contained within a common cytoplasm.)

Syncytium
A single cell with many nuclei. Mass of cytoplasm containing many nuclei enclosed by a single plasma membrane. Typically the result either of cell fusion or of a series of incomplete division cycles in which the nuclei divide, but the cell does not.

Systemic
Concerning or affecting the body as a whole. A systemic therapy is one that the entire body is exposed to, rather than just the target tissues affected by a disease.

Systemic distribution
Referring to a substance or a signal or a phenotype incorporated by the body of a host (plant or animal) that is spreading via the vascular system to all parts.

Tandem repeats
Multiple copies of the same base sequence on a chromosome; used as a marker in physical mapping, when the number of repeats varies in the population.

Target
The material – DNA or RNA – that one exposes to the probes on a microarray so that hybridization can be measured subsequently. May also refer to molecules in the body that may be addressed by drugs to produce a therapeutic effect.

Targeted gene knockout
The introduction of a null mutation in a gene by a designed alteration in a cloned DNA sequence that is then introduced into the genome by homologous recombination and replacement of the normal allele.

TAT protein

Transactivator protein from lentiviruses, notably HIV; sequence-specific RNA binding protein that recognizes TAR RNA. Will induce endothelial cell migration and invasion *in vitro* and rapid angiogenesis *in vivo*. Peptides from this protein are potent neurotoxins, implying a possible route for HIV-mediated toxicity. It contains a membrane-penetrating peptide domain that is also called PTD or CPP, which is responsible for the transport of TAT into the nucleus of the cell. One of the regulatory genes of HIV. Three HIV regulatory genes – tat, rev, and nef – and three so-called auxiliary genes – vif, vpr, and vpu – contain information necessary for the production of proteins that control the virus' ability to infect a cell, produce new copies of the virus, or to cause disease.

Template

A molecule that serves as the pattern for synthesizing another molecule, for example, a single-stranded DNA molecule can be used as a template to synthesize the complementary nucleotide strand.

Termination

The site on a DNA sequence at which transcription and DNA replication stops.

Tet resistance

Resistance to tetracycline treatment.

Tetracycline

Broad-spectrum antibiotic that blocks binding of aminoacyl-tRNA to the ribosomes of both Gram-positive and Gram-negative organisms (and those of organelles). Produced by *Streptomyces aureofasciens*.

Tetracycline-inducible operon

An inducible gene expression system. The presence of tetracycline causes transcriptional activation, usually by interaction with and inactivation of, a repressor.

Thioester bond

High-energy bond formed by a condensation reaction between an acid (acyl) group and a thiol group (–SH); seen, for example, in acetyl CoA and in many enzyme–substrate complexes.

Thiol

See Sulfhydryl.

Tibial muscle

Muscle of the fourth segment of the leg, the shinbone muscle.

Tissue-specific expression

Gene function, which is restricted to a particular tissue or cell type. For example, the glycoprotein hormone alpha subunit is produced only in certain cell types of the anterior pituitary and placenta, not in lungs or skin; thus, expression of the glycoprotein hormone alpha-chain gene is said to be tissue-specific. Tissue-specific expression is usually the result of an enhancer which is activated only in the proper cell type.

Titer
A laboratory measurement of the amount – or concentration – of a given compound in solution.

trans
See *cis*.

trans-acting genetic elements
Genetic elements, transcripts, or proteins functioning throughout the cell in which it is expressed; the converse of a *cis*-acting element, which only functions on elements within the contiguous genome in which it occurs.

Transduction
Gene transfer from one cell to another brought about by a virus. Introduction of viral gene carriers into cells for the purpose of gene transfer.

Transfection
The process of introduction of nucleic acid into a cell by nonspecific chemical means.

Transformation
The alteration of a cell by insertion of one or more foreign or mutant genes.

Transformation (with respect to bacteria)
The process by which a bacterium acquires a plasmid and becomes antibiotic-resistant. This term most commonly refers to a bench procedure performed by the investigator, which introduces experimental plasmids into bacteria.

Transgenic
An experimentally produced organism in which DNA has been artificially introduced and incorporated into the organism's germline, usually by injecting the foreign DNA into the nucleus of a fertilized embryo.

Transgenic mouse
A mouse which carries experimentally introduced DNA. The procedure by which one makes a transgenic mouse involves the injection of DNA into a fertilized embryo at the pro-nuclear stage. The DNA is generally cloned, and may be experimentally altered. It will become incorporated into the genome of the embryo. That embryo is implanted into a foster mother, who gives birth to an animal carrying the new gene. Various experiments are then carried out to test the functionality of the inserted DNA.

Transient expression
The temporary expression of genetic information in a cell after the insertion of genome sequences into that cell by some artificial means (e.g., transfection). In this case, the new information is not stably incorporated into the genetic material of the cell, but is expressed temporarily during the lifetime of the cell.

Transient transfection
When DNA is transfected into cultured cells, it is able to stay in those cells for about 2–3 days, but then will be lost (unless steps are taken to ensure that it is retained; see Stable transfection). During those 2–3 days, the DNA is functional, and any functional genes it contains will be expressed. Investigators take advantage of this transient expression period to test gene function.

Transition state
Structure that forms transiently in the course of a chemical reaction and has the highest free energy of any reaction intermediate; a rate-limiting step in the reaction.

Translocation
1. The relocation of a chromosomal segment in a different position in the genome. A chromosomal configuration in which part of a chromosome becomes attached to a different chromosome. 2. Also a part of the translation process in which the mRNA is shifted one codon in relation to the ribosome.

Transposable element
Segment of DNA that can move from one position in a genome to another.

Transposon
DNA elements carrying genes for transposition and other genetic functions. The former genes enable those elements to move from one site on a chromosome to another. In many cases the latter genes enable bacteria to live in extreme environments. Transposons are much longer than IS elements. Some resemble, and may originate from, retroviruses. Also called transposable element or 'jumping gene'.

TRE
Tetracycline Response Element. A binding site in a promoter to which tetracycline can bind and transcription of the adjacent gene will be altered. See also Response element.

Tumor suppressors
A protective gene that normally limits the growth of tumors. When a tumor suppressor is mutated, it may fail to keep a cancer from growing. BRCA1 and p53 are well-known tumor suppressor genes.

Turnover
In-vivo degradation and resynthesis of a compound or macromolecule.

Tyrosine kinase
Enzyme that transfers the terminal phosphate of ATP to a specific tyrosine residue on its target protein.

U6 promoter
A pol III-type promoter that allows the production of shRNA with a defined end (see also Pol III promoter).

UAS
see Upstream activator sequence.

Upstream

The region extending in the 5′ direction from a gene.

Upstream activator sequence (UAS)

A binding site for transcription factors such as Gal4, generally part of a promoter region. A UAS may be found upstream of the TATA sequence (if there is one), and its function is (like an enhancer) to increase transcription. Unlike an enhancer, it can not be positioned just anywhere, or in any orientation.

3′-UTR

3′-untranslated region: a region of the DNA which is transcribed into mRNA and becomes the 3′-end or the message, but which does not contain protein-coding sequence. Everything between the stop codon and the poly(A) tail is considered to be 3′ untranslated. The 3′ untranslated region may affect the translation efficiency of the mRNA or the stability of the mRNA. It also has sequences which are required for the addition of the poly(A) tail to the message (including one known as the 'hexanucleotide', AAUAAA).

5′-UTR

5′-untranslated region: a region of a gene which is transcribed into mRNA, becoming the 5′-end of the message, but which does not contain protein-coding sequence. The 5′-untranslated region is the portion of the DNA starting from the cap site and extending to the base just before the ATG translation initiation codon. While not itself translated, this region may have sequences which alter the translation efficiency of the mRNA, or which affect the stability of the mRNA.

Vaccination

The process of using an inactivated or attenuated pathogen (or portion thereof) to induce an immune response in an individual prior to his or her exposure to the pathogen.

Ventral

Situated toward the belly surface of an animal; opposite of dorsal.

Viability

Capability of living.

Viable

Alive, even if in a dormant state

Virion

A virus particle.

Virus

A noncellular biological entity that can reproduce only within a host cell. Viruses consist of nucleic acid covered by protein; some animal viruses are also surrounded by a membrane. Inside the infected cell, the virus uses the synthetic capability of the host to produce progeny virus.

Vitelline membrane
Protective layer around the egg.

VSV-G
Glycoprotein of the vesicular stomatitis virus.

VSV-G tag
Epitope tag (YTDIEMNRLGK) derived from the vesicular stomatitis virus G protein.

Vulva
Female genitalia.

Wild-type
The normal condition, either with regard to a whole organism (wild-type strain), or with reference to a particular mutation (wild-type at that locus or site, denoted by a plus sign).

Wobble
The ability of certain bases at the third position of an anticodon in tRNA to form hydrogen bonds in various ways, causing alignment with several possible codons. Referring to the reduced constraint of the third base of an anticodon as compared with the other bases, thus allowing additional complementary base pairings.

Parts of the glossary are from the following sources:
Aids Info: Service of the US Department of Health and Human Services: Glossary of HIV/AIDS related Terms, 4th edition (http://aidsinfo.nih.gov/); the *Glossary of Biochemistry and Molecular Biology* by David M. Glick, Portland Press; and the Life Science Glossary http://www.Biology-Text.com.

Subject Index

RNA Interference in Practice: Principles, Basics, and Methods for Gene Silencing in C. elegans, Drosophila, and Mammals. Ute Schepers
Copyright © 2005 WILEY-VCH Verlag GmbH & Co. KGaA, Weinheim
ISBN: 3-527-31020-7